# Horticulture Research: Theory and Practice

# Horticulture Research: Theory and Practice

Edited by **Thelma Bosso**

New York

Published by Syrawood Publishing House,
750 Third Avenue, 9th Floor,
New York, NY 10017, USA
www.syrawoodpublishinghouse.com

**Horticulture Research: Theory and Practice**
Edited by Thelma Bosso



International Standard Book Number: 978-1-68286-160-8 (Hardback)



Printed in the United States of America.

# Contents

# Preface

This book has been an outcome of determined endeavour from a group of educationists in the field. The primary objective was to involve a broad spectrum of professionals from diverse cultural background involved in the field for developing new researches. The book not only targets students but also scholars pursuing higher research for further enhancement of the theoretical and practical applications of the subject.

Horticulture is an evolving field of agriculture that has seen tremendous shift in the last few decades. This book attempts to understand significant concepts in this field with the help of topics such as effects of external factors on plant growth, plant physiology, vegetable and fruit yield, etc. It includes advanced researches and explanatory case studies by internationally acclaimed experts which would help to provide a comprehensive insight to the readers.

It was an honour to edit such a profound book and also a challenging task to compile and examine all the relevant data for accuracy and originality. I wish to acknowledge the efforts of the contributors for submitting such brilliant and diverse chapters in the field and for endlessly working for the completion of the book. Last, but not the least; I thank my family for being a constant source of support in all my research endeavours.

**Editor**

# 1

# Boron affects pollen viability and seed set in sunflowers

**Arporn Krudnak, Sodchol Wonprasaid and Thitiporn Machikowa***

School of Crop Production Technology, Institute of Agricultural Technology, Suranaree University of Technology, Nakhon Ratchasima, 30000, Thailand.

**Boron (B) is an essential micronutrient for plant growth and development of sunflower (*Helianthus annuus* L.). Most sunflowers growing areas in Thailand are in coarse texture soils with low B availability. The objectives of this experiment were (1) to evaluate the response of sunflower to B applications and (2) to determine the optimum B levels for sunflower under field conditions in sandy loam soil. This experiment was conducted at Suranaree University of Technology, Nakhon Ratchasima, Thailand, in January 2011. Two varieties of sunflower, "S473" and "Pacific 77", were grown in sandy loam soil with available B of 0.14 ppm. Five levels of B (0, 3.13, 6.25, 9.38 and 12.50 kg B $ha^{-1}$) were applied as borax ($Na_2B_4O_7.10H_2O$). The data was recorded for leaf B concentration, B uptake, pollen viability and seed set. They were highly correlated between pollen viability and seed set. Pollen viability and seed set of both varieties responded to B application, but pollen viability was more responsive. In addition, variety "S473" responded to B application more than "Pacific 77". Regression analysis showed that the levels of B application produced maximum pollen viability ranged between 5.6 and 11.3 kg B $ha^{-1}$, but B application at higher rates tended to decrease pollen viability and seed set.**

**Key words:** *Helianthus annuus* L., borax, pollen viability, seed set, 2,3,5-triphenyltetrazolium chloride.

## INTRODUCTION

Boron (B) is one of the micronutrients required for normal growth and plant development of many crops. However, the optimum quantity range of proper B application is rather narrow, because high concentrations become toxic to plants (Goldberg and Glaubig, 1985). The roles of B in plant has been proposed including functions in cell wall structure, cell wall synthesis, sugar translocation, cell division, enzymatic reactions and plant growth regulation (Blevins and Lukaszewski, 1998). Boron has also been reported to be required for flowering, pollen germination, pollen tube growth and seed development (Cakmak and Römheld, 1997). However, many functions of B in plant growth are not fully understood.

Sunflower is one of the most sensitive crops to B deficiency (Rerkasem, 1986). Boron deficiency symptoms in sunflower become evident on leaves, stems, reproductive parts, dry matter, yield components and seed yield (Rerkasem, 1986; Blamey et al., 1997). Asad et al. (2002, 2003) reported that B requirement of sunflower during reproductive growth is higher than during vegetative growth.

At flowering, B deficiency can affect pollen viability and abortion of stamens and pistils which contribute to poor seed set due to malformed capitulums and consequently low seed yield (Dell and Longbin, 1997; Chatterjee and Nautiyal, 2000). In Thailand, sunflower is usually grown in well drained sandy soils which are most likely B deficient, because of leaching. Low yield with high percentage of unfilled grain are usually found in sunflower grown in this area which might be partly contributed by low B availability. Under B deficient conditions, foliar application of B increased vegetative and reproductive dry mass of sunflower (Asad et al., 2003). Rerkasem (1986) reported the responses of sunflower to low levels of B in sandy soil in the Northern part of Thailand. To overcome this problem, B was applied at planting date at the rate of 10 kg B $ha^{-1}$, which increased seed yield by 10.5%. However, little information exists on the B requirements of sunflower grown in sandy loam soil in other parts of

*Corresponding author. E-mail: machiko@sut.ac.th.

**Table 1.** Physical and chemical properties of soil samples (0 to 30 cm) from experimental field at NakhonRatchasima, Thailand.

| Soil property | Unit | Mean | Optimal level |
|---|---|---|---|
| pH | - | 7.25 | 6.0 - 7.0 |
| EC | dS/cm | 0.12 | 2.0 |
| Organic matter | % | 1.54 | 2.0 - 3.0 |
| Available P | mg kg$^{-1}$ | 50 | 35 - 60 |
| Exchangeable K | mg kg$^{-1}$ | 176 | 100 - 120 |
| Exchangeable Ca | mg kg$^{-1}$ | 1306 | 800 - 1500 |
| Exchangeable Na | mg kg$^{-1}$ | 42 | - |
| Available S | mg kg$^{-1}$ | 134 | - |
| Available B | mg kg$^{-1}$ | 0.14 | 0.26 |
| Textural class | Sandy loam soil | - | - |

Thailand. The objectives of this study were: (1) to evaluate the effect of different levels of B on pollen viability and seed set of two sunflower varieties and (2) to determine optimum B levels for sunflower under field condition in a sandy loam soil.

## MATERIALS AND METHODS

The experiment was conducted during the cropping season of 2011 at the experimental field of Suranaree University of Technology, Nakhon Ratchasima, in the Northeastern part of Thailand. Soil was collected from 0 to 30 cm layer for physical and chemical analysis. The soil type is classified as Typic Haplustulf (Soil Survey Staff, 2010). Physical and chemical properties of the soil were as follows: pH, 7.25; organic matter, 1.54%; exchangeable potassium (K), calcium (Ca), and sodium (Na) were 176, 1,306, and 134 ppm, respectively; available phosphorus (P), sulphur (S) and B concentrations were 50, 42 and 0.14 ppm, respectively (Table 1).

The soil was slightly basic, low in organic matter, total nitrogen (N) and B. These values revealed the low to medium fertility classes suggested for soils in the Northeastern Thailand. Factorial treatments (2 sunflower varieties and 5 levels of B application) were arranged in a randomized complete block design experiment with three replications. Sunflower varieties, "S473" (synthetic variety) and "Pacific 77" (hybrid variety), were used in this study. Five levels of B (0, 3.13, 6.25, 9.38 and 12.50 kg B ha$^{-1}$) were applied at planting as borax ($Na_2B_4O_7.10H_2O$).

Basal fertilizers were also applied at planting at the rate of 120 kg ha$^{-1}$ of N, 45 kg ha$^{-1}$ of $P_2O_5$ and 45 kg ha$^{-1}$ of $K_2O$.The plot size was 4 × 5 m and three seeds were sown in one hill with the plant spacing of 70 × 30 cm. Seedlings were thinned to a plant per hill at 15 days after planting.

Data was recorded for leaf B concentration, B uptake, pollen viability and seed set. The viability of pollen grains was examined by 2,3,5-triphenyltetrazolium chloride (TTC). At R5.1 stage (Schneiter and Miller, 1981), anthers of both varieties in each replication were randomly collected. They were placed in a solution of 0.5% TTC for 60 min. To determine viability, pollen grains were counted under a microscope. Pollen grains were considered viable if they turn red, whereas those that remained translucent were dead (Shivanna and Rangaswamy, 1992). At R5.1 stage, leaf B concentration was analyzed. In each treatment, the youngest fully expanded leaves from the top (3rd and 4$^{th}$ leaves) were collected and dried at 70°C for 48 h. The dried leaves were finely milled and digested in 0.01 M $CaCl_2$. Boron analysis was based on the Azomethine-H method (Bingham, 1982). Percent seed set was determined at maturity by randomly sampling, ten plants per plot. Average quantities of total B uptake were estimated based on the total of plant dry matter and B concentration of plant tissue (Yermiyahu et al., 2001).Data collected were subjected to analysis of variance (ANOVA) using R statistical software (Zuur et al., 2009). The Duncan's multiple range test was performed to determine significant differences among B application levels. Correlation and regression analysis were carried out to exhibit the relationship between the observed traits determined. Then, the optimal B levels for sunflower varieties were determined from the regression equations.

## RESULTS

No visual B deficiency symptoms were found on leaves of sunflower grown at various levels of B application. However, there were significant differences (Table 2) for pollen viability and seed set of sunflower varieties grown at various B levels.

### Pollen viability

Data on pollen viability are shown in Table 2 and Figure 1. There were significances in pollen viability in response to applied B compared with control. Sunflower varieties grown in the low B (without added B) soil exhibited low pollen viability, while B application caused an increase in pollen viability. When B supply was increased from 0 to 9.38 kg B ha$^{-1}$, pollen viability increased from 71.93 to 98.33%. These gave an increase of 34.76 and 19.83% relative to the control for S473 and Pacific 77, respectively. Pollen viability of S473 increased to 96.93% at 9.38 kg B ha$^{-1}$ after which there was a decline in the pollen viability with further increase in B levels. However, pollen viability of Pacific 77 increased up to 98.0% at 6.25 B ha$^{-1}$ and did not decrease with further increase in B levels. Addition of B was sufficient to correct the low level of B in the soil. However, negative effect of pollen viability was observed when the level of B in the soil was increase up to 12.50 kg ha$^{-1}$ in S473 variety.

**Table 2.** The effects of B applications on pollen viability, seed set, leaf B concentration and B uptake of sunflower varieties"S473" and "Pacific 77".

| Variety | Applied B level (kg B ha$^{-1}$) | Pollen viability (%) | Seed set (%) | B concentration (mg kg$^{-1}$) | B uptake (mg plant$^{-1}$) |
|---|---|---|---|---|---|
| | 0 | $71.93 \pm 2.4^{d}$ | $63.00 \pm 1.9^{bc}$ | $32.40 \pm 4.1^{e}$ | $4.67 \pm 0.8^{c}$ |
| | 3.13 | $82.60 \pm 1.9^{bc}$ | $63.47 \pm 3.1^{bc}$ | $44.20 \pm 3.6^{de}$ | $6.57 \pm 0.7^{bc}$ |
| S473 | 6.25 | $85.26 \pm 1.6^{bc}$ | $67.53 \pm 3.2^{ab}$ | $47.40 \pm 2.9^{cde}$ | $6.64 \pm 1.4^{bc}$ |
| | 9.38 | $96.93 \pm 1.1^{a}$ | $72.13 \pm 4.6^{a}$ | $50.04 \pm 3.0^{bcd}$ | $9.17 \pm 0.5^{a}$ |
| | 12.50 | $81.73 \pm 1.2^{c}$ | $71.00 \pm 3.6^{a}$ | $44.77 \pm 3.5^{de}$ | $7.94 \pm 1.6^{ab}$ |
| | 0 | $82.06 \pm 1.9^{bc}$ | $60.26 \pm 2.4^{c}$ | $33.87 \pm 3.4^{de}$ | $4.88 \pm 0.7^{c}$ |
| | 3.13 | $89.40 \pm 1.7^{abc}$ | $62.73 \pm 1.8^{bc}$ | $57.23 \pm 1.9^{ab}$ | $5.85 \pm 1.2^{bc}$ |
| Pacific 77 | 6.25 | $98.00 \pm 1.0^{a}$ | $65.00 \pm 2.8^{bc}$ | $57.98 \pm 2.1^{ab}$ | $6.18 \pm 0.6^{bc}$ |
| | 9.38 | $98.33 \pm 0.7^{a}$ | $63.89 \pm 2.8^{bc}$ | $58.46 \pm 2.2^{ab}$ | $6.65 \pm 1.3^{bc}$ |
| | 12.50 | $98.13 \pm 0.7^{a}$ | $62.67 \pm 2.9^{bc}$ | $61.06 \pm 2.9^{a}$ | $6.84 \pm 0.3^{bc}$ |
| F-test | | ** | * | ** | * |

*, **Significant differences at 0.05 and 0.01 levels, respectively. Means within a column followed by the different letters are significantly different at P < 0.05.

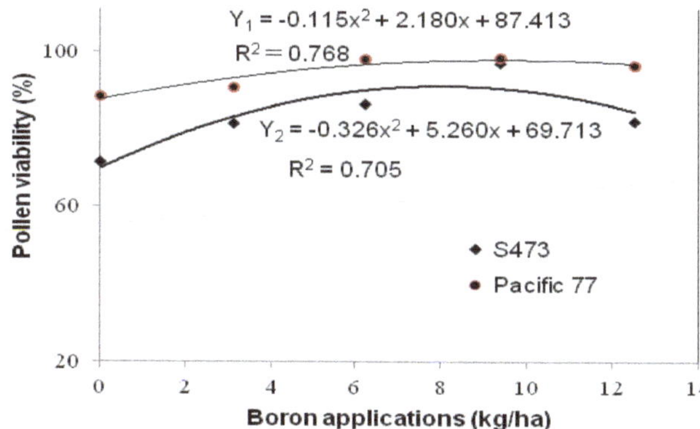


**Figure 1.** The relationship between pollen viability and B application levels.

## Seed set

Effects of B levels on the seed set of both varieties were significantly different. Seed set of both varieties increased with the increase in the application of B up to 9.38 kg B ha$^{-1}$. Pacific 77 applied with 6.25 kg B ha$^{-1}$ gave the highest seed set of 65.00%, whereas S473 variety recorded the highest seed set of 72.13% at 9.38 kg B ha$^{-1}$. Application of B up to 12.50 kg B ha$^{-1}$ resulted in no further benefit, but tended to decrease percentage of seed set in comparison with 9.38 kg B ha$^{-1}$.

## Leaf B concentration

There was low B concentration in leaves for both varieties grown under no B application. Increase in B application levels resulted to significant increase in leaf B concentration. Maximum B concentration of S473 (50.04 mg kg$^{-1}$) was obtained at B application level of 9.38 kg B ha$^{-1}$, while Pacific 77 (61.06 mg kg$^{-1}$) was obtained at 12.50 kg B ha$^{-1}$. It was observed that the B concentration in B added plants of Pacific 77 was greater than in S473, but the B concentrations in both varieties were similar in the control plants (no B application).

## B uptake

In this study, B uptake of both varieties significantly increased with increase in B application levels. Among B application levels, minimum B uptake in sunflower leaves was recorded in the control for both varieties (4.67 and 4.88 mg plant$^{-1}$). Maximum B uptake of 9.17 mg plant$^{-1}$ was obtained at 9.38 kg B ha$^{-1}$ for S473. Further increase in B applications tended to decrease B uptake in S473. For Pacific 77, B uptake gradually increased with increase in B applications and reached the maximum level (6.84 mg plant$^{-1}$) at the highest B application level.

## Correlation analysis

The correlation coefficients between traits are shown in Table 3. There was a strong positive association (r=0.876**) between the concentration of B in leaves at flowering stage and B uptake. There was a positive correlation between pollen viability and seed set (r=0.636*) and highly positive correlation between pollen viability and B concentration (r=0.844**) and B uptake (r=0.798**). Seed set had positive correlation with B concentration (r=0.534*) and had significant positive correlation with B uptake (r=0.863**). The results revealed

**Table 3.** Correlation coefficients among pollen viability, seed set, B concentration and B uptake of two sunflower varieties at different B levels.

| Trait | Pollen viability | Seed set | B concentration | B uptake |
|---|---|---|---|---|
| Pollen viability | 1.000 | 0.636* | 0.844** | 0.798** |
| Seed set | | 1.000 | 0.534* | 0.863** |
| B concentration | | | 1.000 | 0.876** |
| B uptake | | | | 1.000 |

*, ** Significant differences at 0.05 and 0.01 levels, respectively.

that seed set and pollen viability are also highly correlated with B uptake.

## Regression analysis

The regression analysis result (Figure 1) showed that pollen viability could be used for prediction of optimum B levels for sunflower grown in sandy loam soil ($R^2$>70%). The optimum B application levels determined from the range of B application that gave 98% of maximum pollen viability ranged between 5.6 to 11.3 kg B $ha^{-1}$ for Pacific 77 and 5.7 to 10.4 kg B $ha^{-1}$ for S473.

## DISCUSSION

There was a clear increase in B concentration, B uptake, pollen viability and seed set with the increasing levels of B application in both varieties.The results revealed high levels (32 to 60 mg $kg^{-1}$) of B concentrations in sunflower leaves, whereas critical concentration of 29 to 34 kg B $ha^{-1}$ in the top mature leaf blades of sunflower have been reported (Blamey et al., 1997). In wheat, B requirement in anthers for successful seed set is 10 mg B $kg^{-1}$ dry matter (Rerkasem et al., 1997).

In this study, increase in B applications caused an increase in B concentration and B uptake in both sunflower varieties. Boron concentration and B uptake in sunflower were closely related to pollen viability and seed set (Table 3). It can be concluded that increase in uptake of B enhanced the pollen viability and seed set. The efficacy of B on the pollen viability and grain set of wheat were studied by several researchers (Cheng and Rerkasem, 1993; Rerkasem et al., 1997). It is revealed that B tended to increase the pollen viability and grain set of wheat due to B role in plant growth and development. It has been reported that B plays essential roles in the structure and function of cell walls, cellular membranes, translocation of sugars, fruit and seed development (Cakmak and Römheld, 1997). Boron deficiency caused poor anther, pollen development, floret fertility and low grain set (Cheng and Rerkasem, 1993; Rerkasem et al., 1993; Huang et al., 2000). Rerkasem et al. (1997) also found that B influenced the male reproductive organs. Boron deficiency caused floret sterility which was mainly due to sterile pollen. They also reported that pollen of wheat required 8 to 10 mg B $kg^{-1}$ in order to avoid grain yield losses from sterility. In this study, there was high correlation between pollen viability and seed set, but pollen viability was more responsive to B application than seed set in both varieties. Many factors adversely affect the formation of seed in sunflowers, such as B deficient, water deficit and high temperature. In many cases, B deficiency is associated with the induction of pollen sterility and low seed set and its impact may be exacerbated by environmental factors. Therefore, pollen viability would be a better parameter for determination of optimum B application levels.

At the highest level of B application (12.50 kg B $ha^{-1}$), the pollen viability was decreased in sunflower variety S473 (Figure 1). This could be due to toxic effect as a result of excess B (Chatterjee and Nautiyal, 2000). However, it was not the direct toxic effect of B on pollen viability, because B concentration and uptake in S473 decreased at the highest rate of B application. The reduction of B uptake at high rate of B application might be due to the toxic effects of B in the roots which could prohibit the nutrient uptake and plant growth (Lukaszewski and Blevins, 1996). This corresponds with the results of Oyinlola (2007), who reported that B application at high levels (12 kg B $ha^{-1}$) resulted in toxicity symptoms, low plant growth and yield reduction of sunflower grown in sandy loam soil. In addition, soil application of B has been reported to improve the harvest index, seed set, seed yield and oil content in sunflower (Rerkasem, 1986; Oyinlola, 2007; Al-Amery et al., 2011).

The results also exhibited the genotypic differences in B sensitivity which S473 is more sensitive than Pacific 77 to B applications. With added B, S473 responded to higher levels of B than Pacific 77. In addition, the results indicated that S473 variety had better pollen viability and seed set than Pacific 77 on soils with optimal B, butworse on soils with low and high levels of B.

## Conclusion

The results of this study indicated that B application at planting date could increase pollen viability and percent seed set of sunflower. However, the application at high rate (more than 11.3 kg B $ha^{-1}$) resulted in no further benefit, but tended to decrease B uptake, pollen viability and seed set. From the result of the regression analysis, it could be concluded that, in this area, the optimum levels of B application for sunflower variety "Pacific 77" is 5.6 to 11.3 kg B $ha^{-1}$, while for variety "S473" is 5.7 to 10.4 kg B $ha^{-1}$.

## ACKNOWLEDGEMENT

The authors are grateful to Suranaree University of Technology for the facilities and Suranaree University of Technology research fund for financial support.

### REFERENCES

Al-Amery MM, Hamza JH, Fuller MP (2011). Effect of boron foliar application on reproductive growth of sunflower (*Helianthus annuus* L.). Int. J. Agron. doi: 10.1155/2011/230712. p. 5

Asad A, Blamey FPC, Edwards DG (2002) Dry matter production and boron concentrations of vegetative and reproductive tissues of canola and sunflower plants grown in nutrient solution. Plant Soil 243:243-252.

Asad A, Blamey FPC, Edwards DG (2003). Effects of boron foliar applications on vegetative and reproductive growth of sunflower. Ann Bot. 92:565-570.

Bingham FT (1982). Boron. In: Page AL (ed) Methods of soil analysis, Part 2: Chemical and mineralogical properties. American Society of Agronomy, Madison, WI, USA. pp. 431-448.

Blamey FPC, Zollinger RK,Schneiter AA (1997). In:Sunflower technology and production. SchneiterAA (ed) Sunflower production and culture. American Society of Agronomy, Madison, WI, USA. pp. 595-670.

Blevins DG,Lukaszewski KM (1998). Boron in plant structure and function. Plant Physiol. 49:481-500.

Cakmak I, Römheld V (1997). Boron deficiency-induced impairments of cellular functions in plants. Plant Soil 193:71-83.

Chatterjee C, NautiyalN (2000) Developmental aberrations inseeds of boron deficient sunflower and recovery. J. Plant Nutr. 23:835–841.

Cheng C, Rerkasem B (1993). Effects of boron on pollen viability in wheat. Plant Soil 155-156:313-315.

Dell B,Longbin H (1997). Physiological response of plants to low boron. Plant Soil .193:103-120.

Goldberg S, Glaubig RA (1985) Boron absorption on aluminum and iron oxide minerals. Soil Sci. Soc. Am. J. 49:1374-1379.

Huang L, Pant J, Dell B, Bell RW (2000). Effects of boron deficiency on anther development and floret fertility in wheat (*Triticumaestivum* L. 'Wilgoyne'). Ann. Bot. 85:493-500.

Lukaszewski KM, Blevins DG (1996). Rooth growth inhibition in boron-deficient or aluminum-stressed squash may be a result of impaired ascorbate metabolism. Plant Physiol. 112:1135-1140.

Oyinlola EY (2007). Effect of boron fertilizer on yield and oil content of three sunflowercultivars in the Nigerian Savana. J. Agron. 6:421-426.

Rerkasem B (1986). Boron deficiency in sunflower and green gram at Chiang Mai. J. Agric. 2(2):163-172.

Rerkasem B, Lordkaew S, Dell B (1997). Boron requirement for reproductive development in wheat. In: Ando T, Fujita K, Mae T, Matsumoto H, Mori S,Sekiya J (eds) Plant nutrition for sustainable food production and environment. Kluwer Academic Publishers. pp. 69-73.

Rerkasem B, Netsangtip R, Lordkaew S, Cheng C (1993). Grain set failure in boron deficiency wheat. Plant Soil 155-156:309-312.

Schneiter AA, Miller JF (1981) Description of sunflower growth stages. Crop Sci. 21:901–903.

Shivanna K, Rangaswamy NS (1992). Pollen biology: A laboratory manual. New York, USA.

Soil Survey Staff (2010). Keys to soil taxonomy. Natural Resources Conservation Service, USDA.

Yermiyahu U, Keren R, Chen Y (2001). Effect of composted organic matter on boron uptake by plants. Soil Sci. Soc Am. J. 65:1436-1441.

Zuur AF, Leno EN, Meesters, EHWG (2009). A Beginner's Guide to R. Springer, New York, USA.

2

# A novel strategy for the identification of 73 *Prunus domestica* cultivars using random amplified polymorphic DNA (RAPD) markers

**Mingliang Yu[1], Jianqing Chu[2,3], Ruijuan Ma[1], Zhijun Shen[1] and Jinggui Fang[2,3]**

[1]Institute of Horticulture, Jiangsu Academy of Agricultural Sciences, Nanjing 210014, P. R. China.
[2]College of Horticulture, Nanjing Agricultural University, Tongwei Road 6, Nanjing 210095, P. R. China.
[3]Jiangsu Fruit Crop Genetics Improvement and Seedling Propagation Engineering Center, Nanjing 210095, P. R. China.

**Despite the usefulness of DNA marker techniques, various DNA markers have not been widely applied for practical cultivar identification. We developed a novel strategy based on DNA molecular fingerprints from the genotyped plant individuals, following which a cultivar identification diagram (CID) was manually generated and used as referable information for quick plant and/or seed sample identification. Based on this, we used random amplified polymorphic DNA (RAPD) markers to identify a total of 73 plum cultivars of different origins. The cultivars could be clearly separated by the fingerprints of 9 RAPD primers. Experimental verification also indicated that the CID generated is referable and workable in the identification of any two or more plum cultivars studied, which remains the main advantage of this CID constructed manually over the phylogenetic trees from cluster analysis used in most reports on plant identification using DNA markers. Furthermore, fewer primers can be used to distinguish all cultivars using this approach. This new strategy developed and employed in plum cultivar identification may be applied in the plum industry to identify and separate plant and seed samples using DNA makers.**

**Key words:** Plum, Cultivar identification diagram, random amplified polymorphic DNA (RAPD) markers.

## INTRODUCTION

Plum (*Prunus* sp.) genus is taxonomically diverse and adapted to a broad range of climatic and edaphic conditions (Ramming and Cociu, 1991; Salesses et al., 1993), and contains more than 30 species (Weinberger, 1975) that are diploid ($2n = 2x = 16$) to hexaploid ($2n = 6x = 48$) (Rehder, 1954) in nature. A very large number of plum cultivars are known worldwide (Blazek, 2007). Accurate and rapid identification of plum varieties is therefore necessary for both breeders and commercial companies. When compared to some other fruit crops, plum has not received much attention from geneticists, cytogeneticists and molecular biologists. Reports on plum cultivar identification using molecular markers are limited (Heinkel et al., 1998; Rohrer et al., 2004; Shimada et al., 2006), most of those reports employ cluster analysis of the banding patterns. Although, these phylogenetic tree based dendrograms could give the genetic diversity levels and separate the plant individuals, they are not able to make easy and referable identification of plum cultivars. Therefore, developing a strategy that can make reliable, easy, and referable identification of plum cultivars is vital for the nursery industry and growers to protect plant patents and provide genetically uniform plants.

Molecular markers are advantageous in that they are not affected by the environment and can provide a powerful tool for proper characterization of cultivars. Recently, various DNA-based markers have been developed and used in studies on genetic diversity, fingerprinting and origins of cultivars in different fruits (Cheng and Huang, 2009; D'Onofrio et al., 2009; Elidemir and Uzun, 2009; Fang et al., 2005; Melgarejo et al., 2009;

*Corresponding author: E-mail: mly1008@yahoo.com.cn.

Papp et al., 2010). Among several markers available, the Random Amplified Polymorphic DNA (RAPD) (Williams et al., 1990) marker is useful in analysis of cultivars due to simplicity, efficiency, easy operation, and non-requirement of any previous sequence information. If optimization of the RAPD technique by choosing 11 nt primers and strict screening of PCR annealing temperature is done before the technique is employed in fingerprinting of plants, RAPD can be the technique of choice. So far, RAPD markers have been used in the cultivar identification and genetic relationship analysis of a number of fruit species, such as apricot-*Prunus armeniaca* (Ercisli et al., 2009), cherry- *Prunus cerasus* (Demirsoy et al., 2008), pistachio- *Pistacia vera* (Javanshah et al., 2007), pomegranate- *Punica granatum* L. (Hasnaoui et al., 2010), strawberry- *Fragaria ananassa* Duch (Wang et al., 2007). However, in practice fruit plant identification using available powerful DNA markers has not been done yet in efficient, recordable, and easy way due to limitations in analysis strategies of those DNA fingerprints called cluster analysis. The cluster analyses have not made the cultivar or species separation efficient in practice and thus utilization of DNA markers in plant and crop seed identification remains a non-popular practice (Hasnaoui et al., 2010).

The objective of this study was to develop a new method that can make the identification of plum cultivars as a practicable, efficient, recordable, and referable approach as possible using a cultivar identification diagram (CID) generated for the 73 selected plum cultivars. The CID may be employed like that of the periodic table of elements with advantages of highly referable and of use, workability, and flexibility by newer addition of cultivars upon availability of data. In addition, CID will provide valuable information and theoretical basis for identification of cultivars, genetic diversity analysis and genetic improvement of crops at molecular level besides essential requirement in granting of protection to new varieties through DUS (Distinctness, Uniformity and Stability) testing (Lu et al., 2009).

## MATERIALS AND METHODS

### Plant materials and genomic DNA extraction

A total of 73 plum genotypes (Table 1) were used in this study. Total genomic DNA of each genotype was extracted from young leaves using the modified cetyltrimethylammonium bromide (CTAB) method (Murray and Thompson, 1980; Bousquet et al., 1990). The extracted DNA was diluted to a final concentration of 30 ng $\mu L^{-1}$ with 1×TE buffer and stored at -20°C pending use.

### Amplification of RAPD markers

The reaction mixture (final volume 15 µl) contained 1.5µl 10 buffer, 1.2 µl $MgCl_2$(25 mM), 1.8 µl dNTP (2.5 mM), 1.2 µl primer (1.0 µM), 0.08µl rTaq Polymerase Dynazyme (5 U/µl) and 30 ng of genomic DNA. Amplification reactions were performed based on the standard protocol of Williams et al. (1990) with minor modification. The PCR was carried out in a Autorisierter Thermocycler (Eppendorf, Hamburg, Germany), programmed as follows: Pre-denaturation for 5 min at 94°C; then 42 cycles each consisting of a denaturation step for 30 s; an annealing step for 1 min at annealing temperature (Table 2); an extension step for 2 min at 72°C. Amplification was terminated by a final extension of 10 min at 72°C.

### RAPD analysis

Out of 54 11 nt RAPD primers, reproducible polymorphic bands were developed by 9 primers (Table 2). The 54 primers had been earlier designed and used for assorted experiments. The PCR products were detected on 1.3% (w/v) agarose gels in 1×TAE (0.04 M *Tris*-acetate, 0.001 M EDTA pH 8.0) buffer at 100 V. The gels were stained with 0.5 µg/ml of ethidium bromide and photographed under ultraviolet light. Polymorphic bands among the cultivars were observed from the photographs. In order to have reproducible, accurate and clear banding patterns, every amplification was repeated at least thrice separately.

### Data analysis

Only the clear and unambiguous bands in the photographic prints of gels were manually chosen and scored for each cultivar by each primer. When some cultivars had a specific band in the fingerprint generated from one primer, they could be separated singly, and those cultivars sharing the same banding pattern were separated into the same sub-group, while the others were separated into another sub-group. On this basis, all the plum cultivars were step by step completely separated from each other with more primers being employed.

### Test of use and workability of the cultivar identification diagram (CID)

Several plum cultivars, which were randomly chosen from the inter- and intra-groups, were used to verify the utilization and workability of CID showing the separation of 73 cultivars. The corresponding primers to be used for the separation of each group were easily picked out from the diagram.

## RESULTS

### Cultivar identification

To determine the suitability of the RAPD technique in identifying the plum cultivars, fifty four 11 nt primers were employed and the annealing temperatures for each primer were screened based on the quality and reproducibility of banding patterns 11 nt. Finally, all the 73 plum cultivars were successfully identified by the joint use of several 11 nt primers (Table 2). A notable example of the RAPD pattern, obtained with primer Y12, is shown in Figure 1A. The RAPD primer Y12 was the first to be used to amplify the 73 plum cultivars. The electrophoresis results showed that forty-nine plum cultivars (The lane numbers correspond to the codes in Figure 2, refer Table 2 for cultivar names) generated uniform, clear and reproducible bands, which were absent in the other

**Table 1.** 73 plum cultivars used in this study.

| No. | Cultivar | Origin | | No. | Cultivar | Origin | |
|---|---|---|---|---|---|---|---|
| | | Province | Country | | | Province | Country |
| 1 | Youyi | Liaoning | China | 38 | Beijingwanhong | Beijing | China |
| 2 | Bulin | unknown | America | 39 | Xiguali | Hebei | China |
| 3 | Haolaiwu | unknown | America | 40 | Qiuxiaojie | unknown | China |
| 4 | Qiyuehong | Hebei | China | 41 | Kelsey | unknown | America |
| 5 | Faguohong | Liaoning | China | 42 | Guangfunai | unknown | unknown |
| 6 | Chuandaojiuhong | unknown | Japan | 43 | Moerteli | unknown | America |
| 7 | Dashizaosheng | unknown | Japan | 44 | Taiyangli | unknown | Japan |
| 8 | 95-6 | Liaoning | China | 45 | Kaiseman | unknown | America |
| 9 | Aoli | unknown | America | 46 | Misili | unknown | America |
| 10 | Xiangjiaoli | Liaoning | China | 47 | American-hongxinli | unknown | America |
| 11 | Gaixiandali | unknown | America | 48 | Conghua-sanhuali | Guangdong | China |
| 12 | Changli-15 | Jilin | China | 49 | Aodaliya-14 | unknown | America |
| 13 | Hongbaoshi | unknown | America | 50 | Hongmenli | unknown | America |
| 14 | Heibaoshi | unknown | America | 51 | Liwang | unknown | Japan |
| 15 | Muhuangli | Heilongjiang | China | 52 | Xianfeng | unknown | America |
| 16 | Zaoyan | unknown | Japan | 53 | Heihupo | unknown | America |
| 17 | Angenuo | unknown | America | 54 | Taoli | unknown | China |
| 18 | Qiulizi | Liaoning | China | 55 | Suoruisi | unknown | Italy |
| 19 | Dalimei | Heilongjiang | China | 56 | Laluoda | unknown | America |
| 20 | Dongbeili | Dongbei | China | 57 | Meiguili | Yunnan | China |
| 21 | Taoyeli | unknown | China | 58 | Dazili | Hebei | China |
| 22 | Xiaosuli | Liaoning | China | 59 | Liyanghuangli | Jiangsu | China |
| 23 | Xiaoheli | Beijing | China | 60 | Yueguangli | unknown | Japan |
| 24 | Suilinghong | Heilongjiang | China | 61 | Hongliangjin | unknown | Japan |
| 25 | Fali | Yunnan | China | 62 | Kaersai | unknown | America |
| 26 | Fenghuali | Zhejiang | China | 63 | Hongmeili | unknown | America |
| 27 | Guiyang | unknown | China | 64 | Dahongli | Guangdong | China |
| 28 | Xiaoganyuhuangli | Shandong | China | 65 | Weikexun | unknown | America |
| 29 | Liheli | unknown | China | 66 | Zuili | Zhejiang | China |
| 30 | Meiguodali | Beijing | China | 67 | Hubeili | Hubei | China |
| 31 | Owent | unknown | America | 68 | Aozhakeshouxiang | unknown | America |
| 32 | Wuxiangli | Huabei | China | 69 | Shengmeigui | unknown | unknown |
| 33 | Meiguihuanghou | unknown | America | 70 | Jiaqingzi | Jiangsu | China |
| 34 | Taihouli | unknown | unknown | 71 | Shandong-yuhuangli | Shandong | China |
| 35 | Hongbulin | unknown | America | 72 | Aodeluoda(Green) | unknown | America |
| 36 | Aodeluoda(Red) | unknown | America | 73 | Haoyun | unknown | America |
| 37 | Furongli | Fujian | China | | | | |

cultivars. The fragment size of the specific band was about 650 bp (Figure 1A). When this special band was selected for cultivar identification, the cultivars could be separated into two groups. Another primer Y22 was chosen to differentiate the two groups of plum cultivars respectively, but there were no polymorphic bands, meaning that the 73 cultivars in the two groups could not be differentiated using the primer Y22 and hence, the primer Y4 (Figure not shown) was chosen to further distinguish cultivars within the two groups. The first and second group could be successfully separated into several secondary groups by the use of primer Y4 (Figure 2). Primer Y40 was used to amplify all of the secondary groups where 11 plum cultivars were directly identified from each other and the other smaller groups could also be further separated into many more sub-groups (Figure 1B). A further example of such separation is the RAPD pattern obtained with primer Y41 which had several polymorphic bands, meaning that the cultivars in this group could be successfully differentiated from each other (Figure 1C). After following this trend and eventually utilizing all the 9 primers, the original group of

**Table 1.** 11 nt primers chosen for further fingerprinting of 73 plum genotypes.

| Primer | Nucleotide sequence (5'–3') | Annealing temperature (°C) |
|---|---|---|
| Y-4 | GTTTCGCTCCT | 44.8 |
| Y-12 | CTGCTGGGACC | 40.4 |
| Y-24 | GGACCCAACCC | 44.4 |
| Y-30 | GTGTGCCCCAC | 44.4 |
| Y-34 | AAGCCTCGTCT | 44.4 |
| Y-39 | AGCGTCCTCCA | 44.4 |
| Y-40 | AGCGTCCTCCT | 42.8 |
| Y-41 | AGCGTCCTCCG | 44.4 |
| Y-60 | ACCCCCGACTC | 42.8 |

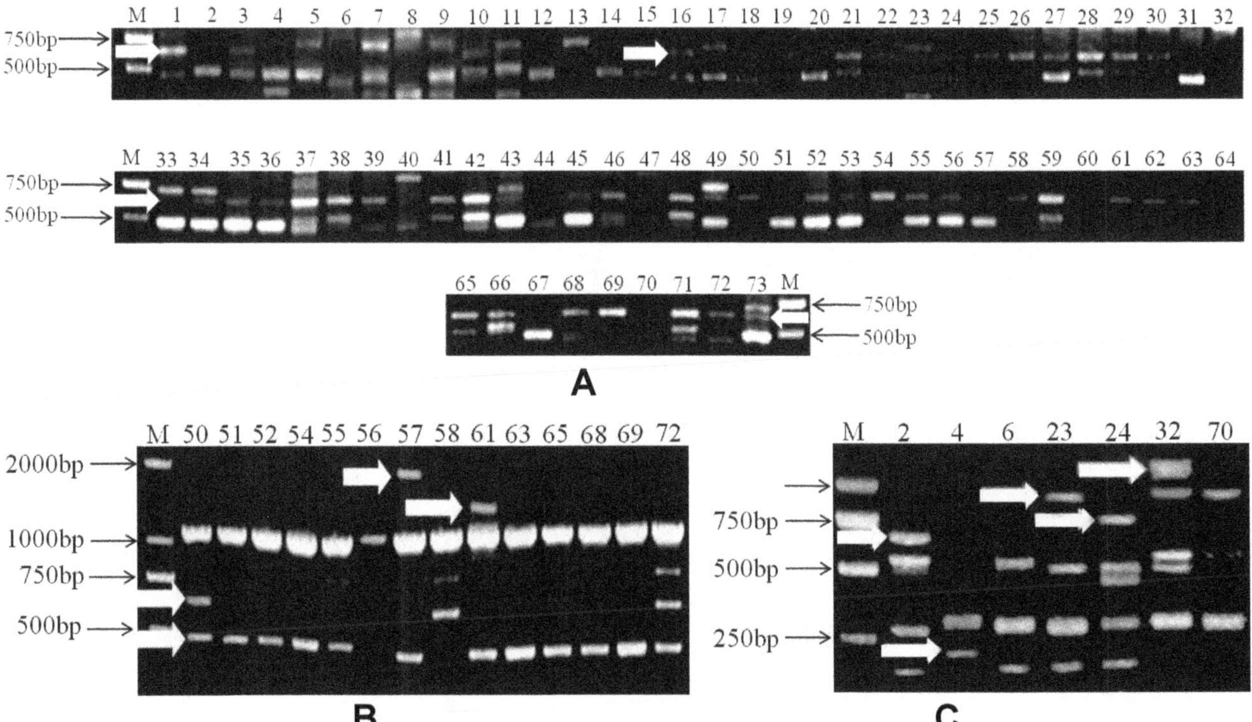


**Figure 1.** A: RAPD patterns of 73 plum genotypes obtained with primer Y12. Horizontal arrows indicate the specific bands. The lane numbers correspond to the codes in Table 1. M: DNA size marker. B and C: RAPD profiles obtained with RAPD primers. Horizontal arrows indicate the specific bands. The lane numbers correspond to the code in Table 1. M: DNA size marker. B obtained with the prime Y40, C obtained with the primeY41.

73 plum cultivars could be completely differentiated from each other (Figure 2).

### Test of the utilization and workability of the diagram in cultivar identification

An important aim of this study was to learn how to use the RAPD marker to distinguish the 73 plum cultivars. In addition, we endeavored to generate a referable plum cultivar identification diagram (CID) for identification of cultivars in future nursery industry practice and cultivar-right-protection. Four plums 'Conghua-sanhuali' and 'Kaersai' from the first and 'Owent', and 'Hubeili' from second group were chosen and used to verify the scientific aspects of this method using primers Y12 and Y4 (Figure 2). The PCR results clearly showed that the four plum cultivars could be identified by three specific bands as anticipated in Figure 2. Firstly, primer Y12 was used to amplify genomic sequence of the four plums cultivars (Figure 3A), where a specific band (~ 650 bp) was used for cultivar identification. The cultivars 'Conghua-sanhuali' and 'Kaersai' were separated into the same group due to a specific band that was absent in

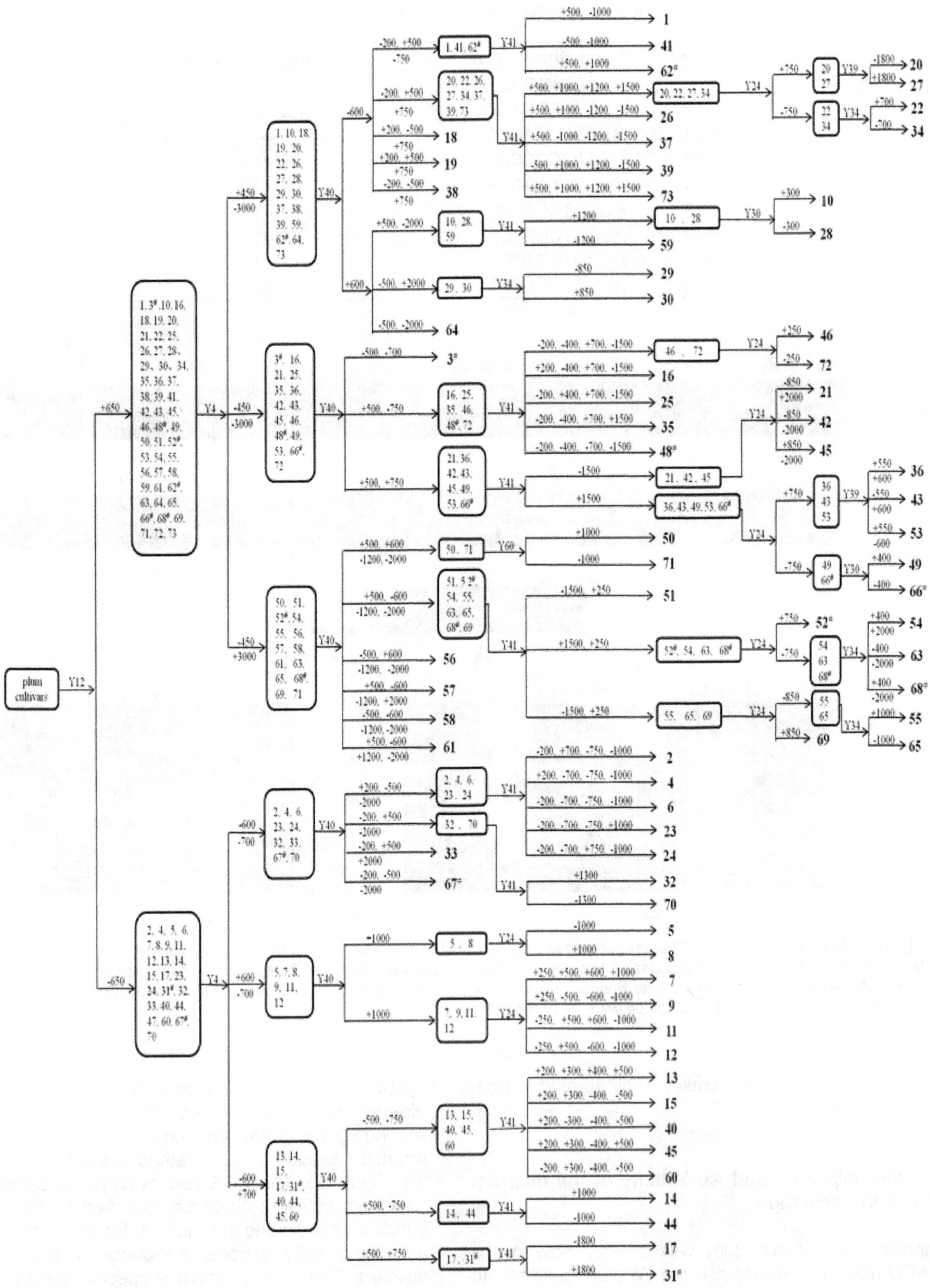


**Figure 2.** Classification of 73 plum cultivars based on nine 11 nt RAPD primers. The numbers on the branches show the size of the band in bp; (+) band present; (−) band absent; '#' this cultivar was used for validation. Terminal branch numbers in bold indicate the cultivar was uniquely identified.

'Owent' and 'Hubeili', which were clustered into some other group. The genomic sequences of these two groups were amplified by, primer Y4 (Figures 3B and C) that produced a specific ~700 bp band for identification of 'Owent' and 'Hubeili', and one another band (~ 450 bp) was used for identification of 'Conghua-sanhuali' and 'Kaersai'. Figures 3D and E further exemplify how this principle to separate cultivars works.

## DISCUSSION

DNA-based molecular markers have acted as versatile tools in various fields like taxonomy, genetic engineering, marker assisted selection (MAS), cultivar identification and variability studies. These classes of markers are found in abundance and are more precise, thus provide an opportunity for direct comparison of genetic materials, as well as not being affected by different environmental conditions or the developmental stage of plants (Reddy et al., 2002). Despite this, DNA markers have not been easily used in genotyping of plants. In fact, the situation is more serious than anticipated, with the question of whether DNA markers can be well and easily used in the identification of plant varieties yielding a negative response from many scientists. No efficient approaches have been developed to use DNA markers easily and efficiently in plant cultivar identification apart from the use of phylogenetic clusters or some fingerprints. Apparently, the clusters formed in phylogenetic trees cannot tell which information can be referable for identification of plant samples, while fingerprinting cannot present all the fingerprints of many cultivars together for identification. These weaknesses could probably be attributed to the fact that no analysis could connect the information of DNA fingerprints with cultivars in an easy, clear and readable way. The new approach developed in this study can use DNA markers efficiently to distinguish the cultivars as desired. It has the advantages of less cost, timeliness and objectivity among others. This strategy can realize the power of DNA markers in plant cultivar identification activities and can use the polymorphic bands of each primer gradually to distinguish every species and individual plant, from which a cultivar identification diagram (CID) can be finally constructed for further use on these cultivars.

Botanical classification of species and cultivars within plum is sometimes controversial, partly because of the ease of interspecific hybridization, which creates numerous intermediate types, and blurs the limits between taxa (Blazek, 2007). This therefore, makes it difficult to avoid the situations of homonym or synonym within the materials, and buttresses the need to identify plum species and cultivars for conservation studies and use of germplasm resources as well as plant variety protection. The ability to distinguish cultivars could be greatly enhanced by using appropriate molecular markers (Iezzoni and Brettin, 1998; Heinkel et al., 2000). The optimized RAPD was a preferred technique selected to make the identification in this study even more efficient and easy. However, the most important aim of this study was not just how to use RAPD marker to distinguish the 73 plum cultivars, which focuses on the utilization of DNA fingerprints in identifying plant cultivars, but also to develop a new strategy to properly utilize DNA marker in the separation of plum cultivars. This methodology could also be considered as a universal strategy to use in distinguishing cultivars and seed samples in other plant species.

By deployment of the CID strategy, only nine 11 nt RAPD primers were used to distinguish all the 73 selected plum cultivars in this study. The method is very convenient and fast for the user. Although a single RAPD primer cannot distinguish all plum cultivars simultaneously, this method represents a substantial increase in efficiency over previous studies. In addition, it reveals new evidence on the rapid identification of plum cultivars. The informative CID (Figure 2) of the plum cultivars can tell us which primers can be used to separate which plum cultivars. Basically, any two or more plum cultivars can be distinguished by the use of one special primer. For example, the plum cultivars 'Owent' and 'Hubeili' can be distinguished by the use of primer Y4 according to Figure 3B. If the result of PCR amplification shows a special band and the fragment size is about 700 bp, the cultivar can be judged to be 'Owent', otherwise the cultivar is 'Hubeili'. The same principle can be used to distinguish any other two plum cultivars. If more new plum cultivars are released in production, the set of 9 primers selected in this study can be used to run the DNA samples of the new cultivars and the PCR banding patterns can let us know where to position the new cultivars in the CID. If they cannot be separated from the 73 already identified using the 9 primers, additional new primers can be found and used to separate, and then position the cultivars on the CID. By contrast, Shimada et al. (2006) required 20 primers in a RAPDs study to distinguish genetic diversity of 42 plum varieties. It appears that not much work needs to be done and this exercise to generate a larger CID of plum cultivars, which is definitely a significant resource for the plum industry.

Although the method may not accurately reflect genetic relationships among the cultivars, in theory the genetic distance between cultivars separated by the first primer is far greater than the distance between cultivars separated by the last primer. This method is definitely a great addition to plant cultivar identification for cultivar-right-protection and early identification.

This is a first report on using RAPD primers in sequence to identify plum cultivars. In order to verify the reliability of this theory, experimental verification which is an absolute necessity was done and it gave satisfactory results. Therefore, this experiment suggests the possibility of utilizing the DNA markers even in plant

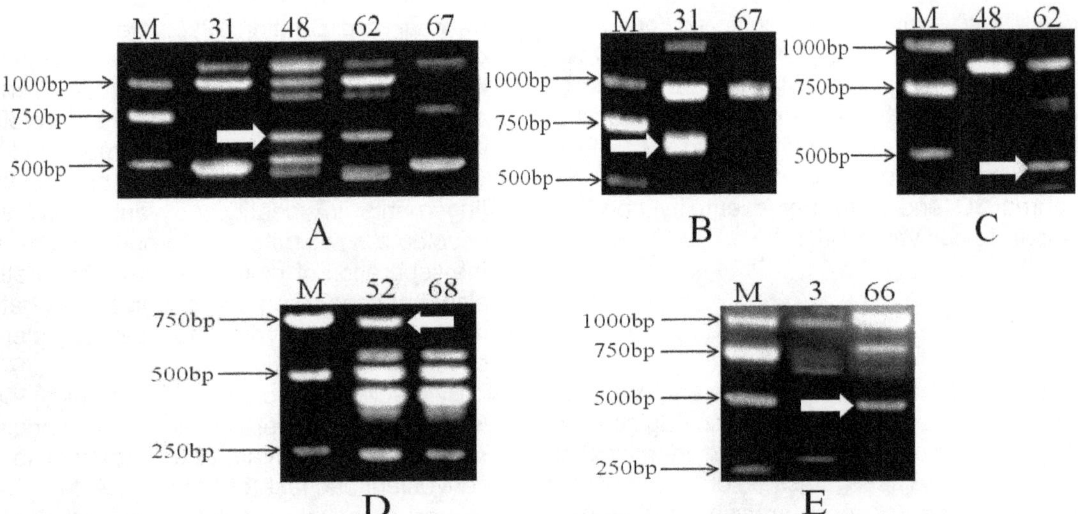


**Figure 3.** RAPD profiles obtained with RAPD primers. Horizontal arrows indicate the specific bands. The lane numbers correspond to the code in Table 1. M: DNA size marker. A obtained with the Y12 primer, B and C obtained with the Y4 primer, D obtained with the Y24 primer, E obtained with the Y40 primer.

species which have a highly heterozygous genome, without requiring a genetic linkage map and/or any DNA sequence information to distinguish the cultivars. It seems to be an effective technique for convenient development of selection markers in fruit trees. In addition, these polymorphic bands may be developed into special molecular markers for the cultivars identification in future. The amazing results of identification using this new strategy is that a readable and referable CID can be constructed and used in the identification of the related plant species in a manner similar to the use of a periodic table of elements in providing the basic information of chemical elements. We believe that as research on this method progresses, we can use this technique and other molecular makers to develop a table for each species, whether plants or other organisms, which in turn can provide us with the information needed to separate the cultivars as desired. There is also need to test the different markers such as SCAR, SSR and others to ascertain suitability since some markers are sometimes unstable.

In conclusion, this method is rapid, simple, and produces reliable results, since it was possible to demonstrate that a standard set of primers can be used to distinguish between a large number of plum cultivars.

## REFERENCES

Blazek J (2007) A survey of thegenetic resources used in plum breeding. Acta Hort. 734:31-45.

Bousquet J, Simon L, Lalonde M (1990). DNA amplification from vegetative and sexual tissues of tree using polymerase chain reaction. Can. J. For Res. 20:254-257.

Cheng ZP, Huang HW (2009). SSR fingerprinting Chinese peach cultivars and landraces (*Prunus persica*) and analysis of their genetic relationships. Sci. Hort. 120:188-193.

Demirsoy L, Demir T, Demirsoy H, Kacar YA, Okumus A (2008). Identification of some sweet cherry cultivars grown in Amasya by RAPD markers. Acta Hort. 795:147-152.

D'Onofrio C, Lorenzis G, de Giordani T, Natali L, Scalabrelli G, Cavallini A (2009). Retrotransposon-based molecular markers in grapevine species and cultivars identification and phylogenetic analysis. Acta Hort. 827:45-52.

Elidemir AY, Uzun I (2009). Assessment of genetic diversity of some important grape cultivars, rootstocks, and wild grapes in Turkey using RAPD markers. Acta Hort. 827:275-278.

Ercisli E, Agar G, Yildrim N, Esitken A, Esitken A, Orhan E (2009). Identification of apricot cultivars in Turkey (*Prunus armeniaca* L.) using RAPD markers. Romanian Biotechnol. Lett. 14: 4582-4588.

Fang J, Qiao Y, Zhang Z, Chao CT (2005). Genotyping fruiting-mei (*Prunus mume* Sieb. Et Zucc) cultivars using AFLP. Hort. Sci. 40:325-328.

Hasnaoui N, Messaoud M, Jemni C, Mokhtar T (2010). Molecular Polymorphisms in Tunisian pomegranate (*Punica granatum* L.) as revealed by RAPD fingerprints. Diversity 2:107-114.

Heinkel R, Hartmann W, Stösser R (2000). On the origin of the plum cultivars "Cacaks Beauty', 'Cacaks Best', 'Cacaks Early' and 'Cacaks Fruitful' as investigated by the inheritance of random amplified polymorphic DNA (RAPD) fragments. Sci. Hort. 83:149-155.

Heinkel R, Hartmann W, Stuttgart-Hohenheim RS (1998). Parental analysis of four 'Cacaks'. Zwetschensorten. Erwerbsobstbau. 40:147-149. (In German)

Iezzoni AF, Brettin TS (1998). Utilization of molecular genetics in cherry. Acta Hort. 498: 53-62.

Javanshah A, Tajabadipour A, Mirzaei S (2007). Identification of a new phenotype (Siah Barg) of pistachio (*Pistacia vera* L.) with shiny-blackish green leaves using RAPD assay. Int. J. Agric. Biol. 9:307-310.

Lu X, Liu L, Gong Y, Zhao L, Song X, Zhu X (2009). Cultivar identification and genetic diversity analysis of broccoli and its related species with RAPD and ISSR markers. Sci. Hort. 122:645-648.

Melgarejo P, Martcnez JJ, HerncLndez Fca, Martcnez R, Legua P, Oncina R, Martcnez-Murcia A (2009). Cultivar identification using 18S-28S rDNA intergenic spacer-RFLP in pomegranate (*Punica granatum* L.). Sci. Hort. 120:500-503.

Murray GC, Thompson WF (1980). Rapid isolation of high molecular weight DNA. Nucl. Acid .Res. 8:4321- 4325.

Papp N, Szilvassy B, Abranko L, Szabo T, Pfeiffer P, Szabo Z, Nyeki J, Ercisli S, Stefanovits-Banyai E, Hegedus A (2010). Main quality attributes and antioxidants in Hungarian sour cherries: identification of genotypes with enhanced functional properties. Int. J. Food Sci. Technol. 45:395- 402.

Ramming DW, Cociu V (1991). Plum (*Prunus*). In: Moore JV, Ballington JR (eds) Genetic resources of temperate fruit and nut crops. Acta Hort. 290:239-288.

Reddy PM, Saral N, Siddiq EA (2002). Inter simple sequence repeat (ISSR) polymorphism and its applications in plant breeding. Euphytica 128:9-17.

Rehder A (1954). Manual of cultivated trees and shrubs. Second edition. Dioscorides Press, Portland, Oregon.

Rohrer JR, Ahmad R, Southwick SM, Potter D (2004). Microsatellite analysis of relationships among North American plums. Plant Syst. Evol. 244:69-75.

Salesses G, Grasselly C, Renaud R, Claverie J (1993). Lesporte-greffe des espèces fruitières à noyau du genre Prunus. In: Gallais A, Bannerot H (eds) Améliorationdes espèces cultivées. INRA Editions, Paris. pp. 605-619.

Shimada T, Hamaya H, Haji T, Yamaguchi M, Yoshida M (2006) Genetic diversity of plums characterized by random amplified polymorphic DNA (RAPD) analysis. Euphytica 109:143-147.

Wang Z, Zhang Z, Li H, Gao X, Du G, Tan C (2007). Identification of strawberry cultivars by RAPD and SCAR markers. Acta Hort. Sinica. 34:591-596.

Weinberger JH (1975). Plums, In: J. Janick J. N. Moore (Eds), Advances in fruit breeding. Purdue Univ. Press, West Lafayette, IN, U.S.A. pp. 336-347.

Williams JGK, Kubelik AR, Livak KJ, Rafalski A, Tingey SV (1990). DNA polymorphisms amplified by arbitrary primers are useful as genetic markers. Nucleic Acids Res. 18:6531-6535.

3

# Effects of leaf harvest on crude protein and mineral contents of selected early maturing lines of lablab (*Lablab purpureus*)


**B. M. Baloyi[1], V. I. Ayodele[1] and A. Addo-Bediako[2]***

[1]Department of Soil Science, Plant Production and Agricultural Engineering, University of Limpopo, Private Bag X1106, Sovenga, 0727, South Africa.

[2]Department of Biodiversity, University of Limpopo, Private Bag X1106, Sovenga, 0727, South Africa.



**The crude protein and mineral elements of the tropical legume *Lablab purpureus* leaves were determined in eight different lines to evaluate the effect of leaf harvest on the contents of crude protein and mineral elements. Leaves of the eight lablab lines were harvested at 6, 8 and 10 weeks after planting (WAP) for the study. The crude protein and mineral contents were found to be higher at the early harvest stages than at a later stage. At 15 WAP, leaves were harvested from defoliated plants and non-defoliated plants to determine the effect of leave harvest on crude protein and mineral contents. There was no significant difference in crude and mineral contents between leaves from defoliated plants and non-defoliated plants.Thus leaf harvest at the vegetative stage did not compromise the crude protein and mineral element contents of lablab as a vegetable crop. The results show leaves can be harvested at early stage and yet have little or no effect on the crude protein and mineral contents in leaves at a later growth (e.g. flowering) phase.**



**Key word:** Crude protein, Lablab lines, leaf harvesting, mineral elements.


## INTRODUCTION

*Lablab purpureus* is a widespread food crop especially in Africa and Asian. It produces edible leaves, pods and seeds. The tenderer leaves are usually consumed by humans and the older leaves for forage. Hay from the whole plant (if cut at a young, leafy stage) is nutritionally comparable to alfalfa, although somewhat less digestible. The consumption of legume products has been proved to reduce the risk of a number of chronic diseases (Gossalau and Chen, 2004; Gundgaard et al., 2003).

Lablab is a prolific food crop and thrives on relatively soils of low fertility. It improves the land's nitrogen content through the action of the highly active beneficial bacteria in the root nodules. The plant is simple to establish and easy to manage under subsistence conditions. It gives high yields and resists droughts that usually affect leguminous crops. It is commonly preferred due to the high biomass (forage) yields in drought conditions (Murphy and Colucci, 1999) prevailing in these regions than cowpeas (*Vigna unguiculata* L.). It can be grown alone, inter-planted with field crops, or included in crop rotations as it can be used as a cover crop.

Lablab has a high nutritional value with the crude-protein contents of 20 to 28%. In addition, amino acids are moderately well balanced, with especially high lysine content (6 to 7%), which means that lablab seeds complement cereal diets well. The seeds, in addition to contributing relatively good quality protein, are also a good source of energy. Among legumes, lablab is one of the best sources of iron (155 mg/100 g of leaves dry weight) ((Deka and Sarkar, 1990; Norton and Poppi, 1995). Lablab leaves are known not to contain tannins (Murphy and Colucci, 1999), making them a good feed for monogastric animals (Agishi, 1991). Lablab hay may improve and increase live weight and milk yield of cattle


*Corresponding author. E-mail: abe.addo-bediako@ul.ac.za.

(Beckmann and Clements, 2002)

Harvesting of the leaves for human consumption is becoming common in many areas where the crop is grown. Nevertheless, the effect of defoliation on the overall performance of the plants is not known. The response of plants to defoliation depends on the intensity or extent, frequency and timing of foliage removal (Salisbury and Ross, 1992; Saidi et al., 2010). Leaf harvesting procedures have the potential to reduce the yield of essential components of the crop (Rahman et al., 2008).

The aim of the study was to determine and compare crude protein and mineral elements contents in different lablab lines at different harvesting stages. In addition, the study was to ascertain the most appropriate stage to harvest the leaves in order to derive optimum crude protein and mineral elements for human and livestock consumption.

## MATERIALS AND METHODS

The field trial was conducted at the Horticultural Skills Centre, University of Limpopo, South Africa (23°53'10"S, 29°44'15"E) during 2010/11 summer growing season. A randomized complete block design (RCBD)with four replications consisting of factorial combinations of two leaf harvest regimes (no leaf harvest, leaf harvest) and five entries (CPI60795, Q6880B, CPI52554, CPI52506, CPI52513, CPI81364, CQ3620, and CQ52552) was used. Each experimental plot was 3 × 4 m (12 m²). Two or three lablab seeds were sown at 30 cm between plants and 60 cm between rows.

The seedlings were thinned to one, two weeks after planting. Mechanical weed control with hand hoe was also regularly done first at three weeks after planting (WAP) and subsequently once during the vegetative stage and close to crop maturity. Supplemented irrigation with horse pipe was carried out during the trial.

Young leaves were harvested from five plants tagged at the middle rows per plot. Data were collected from six weeks after planting at two-weekly interval (6, 8 and 10), and then flowering at 15 WAP. The fully frown but succulent leaves were harvested since the older leaves of legumes are not used by humans as vegetable (Karikari and Molatakgosi, 1999). Harvested leaves were washed with distilled water and dried in ventilated oven at 65°C to a constant weight for mineral content. Dried leaves were then stored in brown paper bag at room temperature and were sent to the laboratory for chemical analysis.

The chemical analysis of leaf samples was conducted at Cedara Feed Laboratory in Kwa Zulu-Natal, South Africa. The dried leaves were ground into powder using a milling machine and then sieved through 20 mesh sieves. Proximate analysis was carried out using the methods recommended by Association of Official Analytical Chemists International (AOAC, 1990). The following parameters were determined: Crude protein, calcium (Ca), iron (Fe), potassium (K), zinc (Zn), phosphorus (P), sodium (Na) and magnesium (Mg). All analyses were carried out duplicate and reported as mean values on a dry weight basis.

The samples were analysed based on nitrogen analysis utilizing the Kjeldahi system according to AOAC for crude protein content. The crude protein was calculated using a nitrogen conversion factor of 6.25 (AOAC, 1990). The Na and K content were determined by flame photometry and P was determined calorimetrically with a Jemway 6100 spectrophotometer. The other mineral elements were determined after wet digestion with a mixture of nitric acid, sulphuric acid and hydrochloric acid using Atomic Absorption Spectrophotometer (AAS Model SP9) (Dada and Oworu, 2010).

All data were subjected to analysis of variance (ANOVA), using SAS software programme (2001). The differences between means were assessed and compared using the least of significant difference (LSD) and significant level of 0.05.

## RESULTS

### Crude protein

The analysis of variance showed a significant difference in crude protein among the lablab lines at 6 WAP (the first harvest). The mean crude protein (CP) ranged from 24.7% in Q6880B to 30.3% in CP152513at 6 WAP. At 8 WAP (second harvest), the mean crude protein values were higher than those at 6 WAP and ranged from 28.0% in Q6880B to 33.5% in CQ3620.However, the values decreased at 10 WAP (third harvest) and ranged from 22.8% in Q6880B to 28.5% in CP160795 (Table 1).

At 15 WAP, lower crude protein contents were observed in lablab leaves as compared to the earlier stages. The grand mean crude protein at 6, 8 and 10 WAP (vegetative stage) were 27.7, 30.1, and 26.9% respectively, while at 15 WAP the grand mean was 22.7%. At 15 WAP, there was no significant difference in crude protein between leaves from defoliated plants and those of the control (non-defoliated plants) ($p > 0.05$). Thus, leaf harvest had no significant effect on the crude protein contents of the lablab leaves.

### Mineral content

There were variations in the mineral content among the different harvesting stages. The highest Ca content of 1.43% was recorded in CP160795 at 6 WAP. Mg content was highest (0.34%) was in CP181364 at 6 WAP. K highest content of 2.41% was in CP181364 at 8 WAP. Na highest content of1.84% was in CQ3620 at 10 WAP. The highest P content of 0.65%% was found in CP152513 at 8 WAP (Table 1).

The highest Mn content of 139.5 ppm was in CQ3620 at 6 WAP. Cu content was highest (9 ppm) in CP152506 at 8 WAP. Fe highest content of 427.5 ppm was in CP160795 at 6 WAP and the highest Zn content of 156.5 ppm was recorded in CQ3620 at 6 WAP (Table 2). All the mineral contents with the exception of Na content were highest at either 6 or 8 WAP. The mean Fe and Ca contents were significantly higher at 6 WAP than the other two stages ($p<0.05$); however, P, and Zn were higher at 8 WAP than other harvesting stages, though the difference was not significant. Meanwhile, Na content was significantly higher at 10 WAP than at 6 WAP and 8 WAP ($p<0.05$). There were significant differences in interaction between Lablab lines and harvesting stages in K and Na.

**Table 1.** Crude protein and mineral contents of lablab leaves on 100%DM basis at three harvesting stages (6, 8 and 10 WAP).

| Lines | Crude protein (%) | | | Ca (%) | | | Mg (%) | | | K (%) | | | Na (%) | | | P (%) | | |
|---|---|---|---|---|---|---|---|---|---|---|---|---|---|---|---|---|---|---|
| | Harvest stages (WAP) | | | | | | | | | | | | | | | | | |
| | 6 | 8 | 10 | 6 | 8 | 10 | 6 | 8 | 10 | 6 | 8 | 10 | 6 | 8 | 10 | 6 | 8 | 10 |
| CPI60795 | $25.2^{bc}$ | 29.5 | 28.5 | 1.43 | $0.90^{a}$ | $0.90^{ab}$ | $0.26^{ab}$ | $0.20^{ab}$ | $0.20^{b}$ | $2.0^{bc}$d | $2.06^{ab}$ | $1.90^{a}$ | 0.03 | $0.01^{ab}$ | $0.03^{c}$ | 0.42 | 0.58 | $0.40^{ab}$ |
| Q6880B | $24.7^{c}$ | 28.0 | 22.8 | 1.13 | $0.82^{ab}$ | $0.90^{bc}$ | $0.25^{ab}$ | $0.20^{ab}$ | $0.20^{b}$ | 1.80d | $2.07^{ab}$ | $1.40^{c}$ | 0.03 | $0.01^{ab}$ | $0.01^{c}$ | 0.43 | 0.57 | $0.30^{c}$ |
| CPI52554 | $26.8^{abc}$ | 29.6 | 28.2 | 1.15 | $0.69^{ab}$ | $0.70^{bc}$ | $0.22^{b}$ | $0.20^{ab}$ | $0.20^{b}$ | $1.86^{cd}$ | $2.20^{a}$ | $2.00^{a}$ | 0.02 | $0.02^{a}$ | $0.02^{c}$ | 0.50 | 0.63 | $0.50^{a}$ |
| CPI52506 | $29.6^{ab}$ | 28.2 | 28.1 | 1.17 | $0.69^{ab}$ | $0.90^{ab}$ | $0.28^{ab}$ | $0.20^{b}$ | $0.20^{b}$ | $2.20^{a}$ | $1.89^{ab}$ | $1.90^{ab}$ | 0.03 | $0.01^{ab}$ | $0.01^{c}$ | 0.55 | 0.59 | $0.50^{a}$ |
| CPI52513 | $30.3^{a}$ | 31.1 | 28.0 | 1.25 | $0.60^{b}$ | $0.8^{bc}$ | $0.24^{ab}$ | $0.20^{b}$ | $0.20^{b}$ | $2.23^{a}$ | $1.98^{ab}$ | $1.80^{ab}$ | 0.02 | $0.01^{ab}$ | $0.02^{c}$ | 0.57 | 0.65 | $0.50^{ab}$ |
| CPI81364 | $28.9^{abc}$ | 31.0 | 26.9 | 1.41 | $0.82^{ab}$ | $1.16^{a}$ | $0.34^{a}$ | $0.30^{a}$ | $0.30^{a}$ | $1.9^{bc}$d | $2.41^{a}$ | $1.80^{ab}$ | 0.03 | $0.01^{ab}$ | $1.81^{a}$ | 0.49 | 0.61 | $0.50^{ab}$ |
| CQ3620 | $30.1^{a}$ | 33.5 | 27.8 | 1.33 | $0.58^{b}$ | $0.80^{bc}$ | $0.28^{ab}$ | $0.20^{b}$ | $0.20^{b}$ | $2.0^{abc}$ | $1.49^{b}$ | $1.80^{ab}$ | 0.02 | $0.01^{ab}$ | $1.84^{a}$ | 0.56 | 0.64 | $0.40^{ab}$ |
| CQ52552 | $26.1^{abc}$ | 29.5 | 24.9 | 0.83 | $0.63^{ab}$ | $0.61^{c}$ | $0.22^{b}$ | $0.20^{b}$ | $0.20^{b}$ | $2.08^{ab}$ | $1.89^{ab}$ | $1.60^{bc}$ | 0.01 | $0.01^{ab}$ | $1.63^{b}$ | 0.51 | 0.51 | $0.40^{bc}$ |
| Mean | 27.7 | 30.1 | 26.9 | 1.21 | 0.72 | 0.84 | 0.26 | 0.22 | 0.21 | 2.01 | 2.00 | 1.80 | 0.02 | 0.01 | 0.67 | 0.50 | 0.59 | 0.44 |
| LSD (0.05) | 4.7 | ns | ns | ns | 0.29 | 0.26 | 0.12 | 0.06 | 0.05 | 0.20 | 0.67 | 0.24 | ns | 0.01 | 0.15 | ns | ns | 0.09 |
| Significance | | | | | | | | | | | | | | | | | | |
| L | | ** | | | ** | | | ** | | | ** | | | ns | | | ** | |
| HS | | ** | | | ** | | | ** | | | ** | | | ns | | | ** | |
| L X HS | | ns | | | ns | | | ns | | | ** | | | ** | | | ns | |

L = Lablab lines; HS = harvesting stages; means followed by the same letter(s) within a column are not significantly different at $p < 0.05$ among the *Lablab* lines. ** F-test significant at $p < 0.05$. $^{NS}$, not significant.

Similar to the crude proteins, at 15 WAP there were no significant differences in Na and Zn contents of leaves from defoliated plants and those from non-defoliated plants among the lablab lines ($p > 0.05$). Although in most lines, higher mineral contents were recorded in leaves from non-harvested plants than those from harvested plants, the differences were insignificant (Table 3).

## DISCUSSION

### Crude protein

There were obvious differences in variation and distribution of crude protein among the lablab lines. It implied that the genotypic variations provided opportunities to select materials with high contents of crude protein. In the present experiment, the crude protein contents in CPI52513, CPI81364 and CQ3620 were high and could be selected after considering other factors. The crude protein content is found to be higher at the early stages of growth (vegetative phase) than at the later stage (15 WAP), indicating importance to harvest the leaves at early stages to derive most of the crude protein. This supports the finding of Miller-Cebert et al. (2009) in canola leaves where significantly higher protein content was found at pre-bolting stage than rosette and blooming growth stages. The lower protein content of leaves at the final harvest (15 WAP) for both harvested and non-harvested plants is an indication that there is a decline in protein content (nutrient) with the age of the plant.

### Mineral content

There was a genetic variation in mineral composition among lablab lines as observed in the crude protein content. This variation has also been reported in other crops such as cassava (Ravindran and Rajaguru, 1988; Dada and Oworu, 2010) and kenaf (Hossain et al., 2011).

At the vegetative phase (6, 8 and 10 WAP) generally there was a decline in mineral contents with successive leaf harvest. This suggests that

**Table 2.** Mineral contents of lablab leaves on 100% DM basis at three harvest stages (6, 8 and 10 WAP).

| Lablab line | Mn (ppm) | | | Cu (ppm) | | | Fe (ppm) | | | Zn (ppm) | | |
|---|---|---|---|---|---|---|---|---|---|---|---|---|
| | Harvest stages (WAP) | | | | | | | | | | | |
| | 6 | 8 | 10 | 6 | 8 | 10 | 6 | 8 | 10 | 6 | 8 | 10 |
| CPI60795 | 73.0 | 69.5 | 127.0$^{a}$ | 7.5 | 7.0 | 7.5$^{ab}$ | 427.5 | 221.0$^{a}$ | 174.5$^{ab}$ | 54.0 | 60.0 | 64.5$^{ab}$ |
| Q6880B | 105.5 | 92.5 | 76.5$^{b}$ | 8.0 | 8.0 | 6.5$^{ab}$ | 364.0 | 218.5$^{ab}$ | 137.0$^{ab}$ | 93.0 | 103.5 | 56.0$^{ab}$ |
| CPI52554 | 59.0 | 49.0 | 51.0$^{b}$ | 7.0 | 7.0 | 6.5$^{ab}$ | 370.5 | 217.5$^{ab}$ | 132.0$^{ab}$ | 55.5 | 47.5 | 47.5$^{ab}$ |
| CPI52506 | 57.5 | 88.5 | 96,0$^{ab}$ | 6.5 | 9.0 | 8.0$^{a}$ | 365.5 | 186.0$^{bcd}$ | 206.5$^{a}$ | 45.5 | 102.5 | 88.5$^{ab}$ |
| CPI52513 | 105.5 | 68.5 | 68.0$^{b}$ | 6.5 | 6.5 | 6.0$^{b}$ | 335.5 | 162.5$^{d}$ | 127.0$^{b}$ | 63.5 | 51.5 | 32.5$^{b}$ |
| CPI81364 | 58.5 | 52.0 | 50.5$^{b}$ | 7.5 | 7.5 | 7.0$^{ab}$ | 420.0 | 171.0$^{cd}$ | 145.0$^{ab}$ | 67.5 | 68.0 | 50.5$^{ab}$ |
| CQ3620 | 139.5 | 58.5 | 75.0$^{b}$ | 8.0 | 7.5 | 6.0$^{b}$ | 354.5 | 204.0$^{abc}$ | 167.5$^{ab}$ | 156.5 | 97.0 | 89.0$^{a}$ |
| CQ52552 | 61.5 | 114.0 | 134.5$^{a}$ | 6.0 | 7.0 | 6.0$^{b}$ | 338.5 | 166.5$^{d}$ | 198.0$^{ab}$ | 52.0 | 79.0 | 89.0$^{a}$ |
| Mean | 82.5 | 74.1 | 84.8 | 7.1 | 7.4 | 6.7 | 372.0 | 193.4 | 161.0 | 73.4 | 76.1 | 62.3 |
| LSD (0.05) | ns | ns | 46.1 | ns | ns | 1.8 | ns | 34.9 | 77.6 | ns | ns | 56.3 |
| Significance | | | | | | | | | | | | |
| L | | ns | | | ** | | | ns | | | ** | |
| HS | | ** | | | ** | | | ** | | | ** | |
| L × HS | | ns | | | ns | | | ns | | | ns | |

L, Lablab lines; HS, harvesting stages; Means followed by the same letter(s) within a column are not significantly different at $p< 0.05$ among the *Lablab* lines; ** F-test significant at $p < 0.05$; $^{NS}$, not-significant.

**Table 3.** Effects of leaf removal on crude protein and mineral element contents of lablab at final harvest (15 WAP).

| Lablab line | Treatment | CP | Ca | Mg | K | Na | P | Mn | Cu | Fe | Zn |
|---|---|---|---|---|---|---|---|---|---|---|---|
| | | | | % | | | | | ppm | | |
| CP160795 | LH | 22.6 | 2.59$^{a}$ | 0.40$^{ab}$ | 1.49$^{ab}$ | 0.03 | 0.24$^{b}$ | 131.5$^{a}$ | 6.50$^{ab}$ | 653.0$^{a}$ | 100.0 |
| | NLH | 22.6 | 2.71$^{a}$ | 0.36$^{bcd}$ | 1.52$^{ab}$ | 0.07 | 0.25$^{b}$ | 122.5$^{a}$ | 6.00$^{ab}$ | 571.0$^{ab}$ | 79.50 |
| Q6880B | LH | 20.9 | 2.26$^{c}$ | 0.39$^{abc}$ | 1.28$^{b}$ | 0.02 | 0.23$^{b}$ | 111.0$^{a}$ | 6.50$^{ab}$ | 378.0$^{b}$ | 105.0 |
| | NLH | 21.2 | 2.07$^{d}$ | 0.39$^{abc}$ | 1.53$^{ab}$ | 0.02 | 0.29$^{ab}$ | 107.5$^{a}$ | 7.00$^{ab}$ | 460.0$^{a}$ | 96.0 |
| CP152554 | LH | 22.8 | 2.34$^{b}$ | 0.35$^{bcd}$ | 1.56$^{ab}$ | 0.02 | 0.29$^{ab}$ | 131.5$^{a}$ | 6.50$^{ab}$ | 529.0$^{a}$ | 102.5 |
| | NLH | 23.5 | 2.20$^{c}$ | 0.32$^{bcd}$ | 1.51$^{ab}$ | 0.02 | 0.29$^{ab}$ | 119.0$^{a}$ | 6.50$^{ab}$ | 525.5$^{a}$ | 91.5 |
| CP152506 | LH | 21.9 | 2.30$^{b}$ | 0.40$^{ab}$ | 1.63$^{ab}$ | 0.01 | 0.30$^{ab}$ | 131.0$^{a}$ | 6.00$^{ab}$ | 475.0$^{a}$ | 124.5 |
| | NLH | 23.3 | 2.30$^{b}$ | 0.47$^{a}$ | 1.52$^{ab}$ | 0.02 | 0.28$^{b}$ | 120.5$^{a}$ | 7.50$^{a}$ | 569.5$^{a}$ | 140.0 |
| CP152513 | LH | 23.2 | 2.34$^{b}$ | 0.37$^{bcd}$ | 1.56$^{ab}$ | 0.02 | 0.28$^{b}$ | 103.0$^{a}$ | 6.00$^{ab}$ | 515.5$^{a}$ | 71.0 |
| | NLH | 23.4 | 2.36$^{a}$ | 0.34$^{bcd}$ | 1.44$^{ab}$ | 0.02 | 0.26$^{b}$ | 121.5$^{a}$ | 6.00$^{ab}$ | 534.0$^{a}$ | 73.0 |
| CP181384 | LH | 22.4 | 2.23$^{c}$ | 0.40$^{ab}$ | 1.44$^{ab}$ | 0.02 | 0.26$^{b}$ | 96.0$^{b}$ | 5.50$^{b}$ | 383.5$^{b}$ | 81.5 |
| | NLH | 23.8 | 2.76$^{a}$ | 0.47$^{a}$ | 1.53$^{ab}$ | 0.02 | 0.22$^{b}$ | 99.0$^{b}$ | 7.50$^{a}$ | 656.0$^{a}$ | 112.5 |
| CQ3620 | LH | 22.8 | 2.20$^{d}$e | 0.37$^{bcd}$ | 1.41$^{ab}$ | 0.02 | 0.27$^{b}$ | 125.5$^{a}$ | 6.50$^{ab}$ | 608.5$^{a}$ | 114.5 |
| | NLH | 22.9 | 2.25$^{cd}$ | 0.34$^{bcd}$ | 1.55$^{ab}$ | 0.02 | 0.28$^{b}$ | 140.0$^{a}$ | 6.50$^{ab}$ | 664.0$^{a}$ | 152.5 |
| CQ52552 | LH | 23.7 | 2.07$^{d}$e | 0.32$^{bcd}$ | 1.52$^{ab}$ | 0.02 | 0.30$^{ab}$ | 126.0$^{a}$ | 5.50$^{ab}$ | 493.0$^{a}$ | 95.5 |
| | NLH | 22.5 | 1.81e | 0.35$^{bcd}$ | 1.74$^{ab}$ | 0.02 | 0.36$^{a}$ | 151.5$^{a}$ | 6.00$^{ab}$ | 550.0$^{a}$ | 132.5 |
| Grand mean | | 22.9 | 2.73 | 0.36 | 1.51 | 0.02 | 0.27 | 121.1 | 6.38 | 535.3 | 104.5 |
| LSD (0.05) | | ns | 0.42 | 0.09 | 0.07 | ns | 0.09 | 46.68 | 1.98 | 267.7 | ns |
| CV (%) | | 8.60 | 6.29 | 11.14 | 12.23 | 40.12 | 15.54 | 18.09 | 14.60 | 23.48 | 53.67 |

LH, Leaf harvested; NLH, No leaf harvested; ns, not significant; Means followed by the same letter within a column are not significantly different at $p< 0.05$among the *Lablab* lines.

leaves harvested at the early stage of lablab growth contained more of the minerals than those harvested at a later stage of the vegetative phase. Thus the nutritive value of legumes declined as the plant matures (Deka and Sarkar, 1990). However, there were no significant differences among the harvesting stages.

At 15 WAP, there was no consistent trend in mineral content; Ca, Mg, Mn, Fe and Zn contents were slightly higher than the contents recorded at the vegetative phase, while K, Na, P and Cu contents declined slightly from the vegetative phase. In cassava, a decrease in Na, K, Ca and P composition of cassava leaf with an increase in the age of the crophas been reported (Dada and Oworu, 2010).It has also been reported in *Amaranthus* that the nutritional composition of leaves declines with the age of the plant (Modi, 2007). These suggest that absolute nutrient potential derivable from the leaf of the crop may not be fully exploited at the maturity phase.

## Conclusion

The nutritional value of lablab leaves does not only depend on the variety (line) but also the stage of the leaf harvest. Leaves harvested at the early stage had higher protein and some of the mineral contents than those harvested at a later growing stage (flowering).There were no significant changes in crude protein and mineral contents between leaves from plants with harvested leaves and those with non-harvested leaves. Lablab with its high crude protein and mineral content at both vegetative and flowering phases could be employed more often in tropical and sub-tropical agricultural production systems, to improve nutrition, boost food security, and foster rural development.

## ACKNOWLEDGEMENTS

The authors are grateful for the contributions of the Australian Tropical Forage Genetic resources Centre, Commonwealth Scientific and Industrial Research Organization (CSIRO) in providing the initial lablab germplasms in South Africa, Limpopo Department of Agriculture for providing the seeds and that of Australian Centre for International Agricultural Research (ACIAR) for financial support.

.

### REFERENCES

Agishi EC (1991). A bibliography *on Lablab purpureus.*Plant Science Division Working Document No.A6. International Livestock Centre for Africa (ILCA), Addis Ababa, Ethiopia.

AOAC (1990). Association of Official Analytical Chemists.Official Methods of Analysis.International, Arlington, VA.

Beckmann R, Clements B (2002). Legume offers a ray of hope in South Africa. In: A love of legumes. First Edition: Beckmann R and Clements b (ed), Pulb. Partners, Kenya, pp. 11-17.

Dada OA, Oworu OO (2010). Mineral and nutrient leaf composition of two cassava (*Manihotesculentus*Crantz) cultivars defoliated at varying phonological phases. Nat. Sci. Biol. 2:44-48.

Deka RK,SarkarCR (1990). Nutrient composition and anti-nutritional factors of *Dolichoslablab* L seeds. Food Chem. 38:239-246.

Gossalau A, Chen KY (2004). Nutraceuticals apoptosis and disease prevention. Nutrition 20:95-102.

Gundgaard J, Nielsen JN, Olsen J, Sorensen J (2003). Increased intake of fruit and vegetables: Estimation of impact in terms of life expectancy and healthcare costs. Public Health Nutr. 6:25-30.

Hossain MD, Hanafi MM, Jol H, Jamal T (2011). Dry matter and nutrient partitioning of kenaf (*Hibiscus cannabinus*L.) varieties grown on sandy bris soil. Aust. J. Crop. Sci. 5:654-659

Karikari SK, Molatakgosi G (1999). Response of cowpea (*Vignaunguiculata* (L) Walp) varietiesto leaf harvesting in Botswana. UNISWA J. Agric. 8:9-11.

Miller-Cebert RL, Sistani NA, Cebert E (2009). Comparative protein and foliate content among canola cultivars and other cruciferous leafy greens. J. Food Agric. Environ. 7:46-49.

Modi AT (2007). Growth temperature and plant age influence on nutritional quality of*Amaranthus*leaves and seed germination capacity. Water SA 33(3):368-375.

Murphy AM, Colucci PE (1999). A tropical forage solution to poor quality diets: A review of *Lablab purpureus*. Livest.Res.Rural Dev. (http://www.cipav.org.co/lrrd/lrrd11/2/colu.htm).

Norton BW, Poppi DP (1995). Composition and Nutritional Attributes of Pasture Legumes. In. Tropical Legumes in Animal Nutrition; D'Mello, J P F. and C Devendra (Eds), CAB International:Wallingford: UK. pp 45-60

Ravindran V, Rajaguru H (1988).Effect of stem pruning on cassava root yield and leaf yield.Srilankan. J. Agric. Sci. 25:32-37.

Rahman SA, Ibrahim U, Ajoji FA (2008). Effect of defoliation at different growth stages on yield and profitability of cowpea (*Vignaunguiculata* (L.)Walp.).Elec. J. Env. Agric. Food Chem. 7:3248-3254.

Saidi M, Itulya FM, AguyohJN, Mshenga PM, Oworu G (2010). Effects of cowpea leaf harvesting initiation time on yields and profitability of a dual–purpose sole cowpea and cowpea-maize intercrop. Elec. J. Env. Agric. Food Chem. 9:1134-1144.

Salisbury FB, Ross CW (1992).Plant physiology ($4^{th}$ Ed.).Wadsworth, Belmont, California, p.184.

SAS Institute (2001). The SAS system for windows, v 8.2. SAS Inst., Cary, NC.

# 4

# Screening of elite material against major diseases of safflower under field conditions

**S. V. Pawar[1], Utpal Dey[1], V. G. Munde[1], Hulagappa[2] and Anamika Nath[3]**

[1]Department of Plant Pathology, Marathwada Agricultural University, Parbhani, Maharashtra, India.
[2]Department of Plant Pathology, University of Agricultural Sciences, Dharwad, Karnataka, India.
[3]Department of Plant Breeding and Genetics, Mahatma Phule Krishi Vidyapeeth, Rahuri, Maharashtra, India.

**A field experiment with three replications was conducted at the All India Coordinated Research Project (AICRP) on oilseeds at Marathwada Agricultural University (MAU), Parbhani, Maharashtra on the screening of different elite material against major diseases. Significant differences in resistance to all the diseases were found in the elite material tested. Among the 46 elite lines, 21 and 33 elite lines registered highly resistant reaction against *Alternaria* leaf spot and to root rot, respectively while 15 lines registered resistant reaction to wilt. This study concludes that screening elite lines for resistance to diseases is an important step in developing varieties/hybrids with improved resistance to different diseases.**

**Key words:** Safflower, diseases, resistance.

## INTRODUCTION

Safflower (*Carthamus tinctorius* L.) occupies prominent place in the agricultural wealth and economy of India. It belongs to family *Compositae* and believes to be native of Afganistan. The word *Carthamus* is arabic word *quartum* (means the colour of dye obtained from florets). It is described as "*Kusumbha*" in ancient Sanskrit literature. Other Indian names, like *Kusum*, *Karrad* (Hindi), *Kusumpuli* (Bengali), *Kusumbo* (Gujrathi), *Kardi*, *Kurdi* (Marathi), *Sendurakam* (Tamil), *Kusuma* (Telgu), *Kusube*, *Kusume* (Kannada), *Kusumba* (Punjabi) seem to have been derived from "*Kusumbha*". Presently the most common name being "*Kusum*" or "*Kardi*". It is a rich source of proteins and edible oil and so many farmers plant it. It is known to suffer from many fungal, bacterial and viral diseases at different stages of crop growth (Bhale et al., 1998). Seed is the costliest input in safflower cultivation and is highly prone to losses in germination and vigour due to seed mycoflora. Safflower plant is also prone to infection by several seed-borne fungi (Ramesh and Avitha, 2005). Seeds also act as carrier in transmission of pathogens and thereby cause economic threat to safflower cultivation. Considering the economic losses in this present investigation attempts were therefore made to as certain this spectrum of fungal flora associated with the seeds of safflower elite materials.

### MATERIALS AND METHODS

The seeds of elite materials of safflower were received from Directorate of Oil Seeds Research, Hyderabad. These elite lines were screened in the field under artificial epiphytotic conditions for various diseases during monsoon season of 2012 at AICRP on oilseeds, Marathwada Agricultural University, Parbhani, Maharashtra. The screening of elite material against major diseases was done in three replications. Forty six elite materials were screened in the field under artificial epiphytotic conditions during *monsoon,* 2012. The test lines were sown in a randomised block design with the Gross plot size being single row of 3.0 m. Distance between rows was 30 cm and plant to plant distance was kept 15 cm as closer distance favours disease development. *Alternaria* susceptible genotype Manjira, Rhizoctonia and Fusarium susceptible genotype Nira were sown after every fifth row of test

*Corresponding author. E-mail: utpaldey86@gmail.com

**Table 1.** 0-9 disease scale for *Alternaria*.

| Disease incidence (0-9) scale | Disease incidence (%) | Reaction |
|---|---|---|
| 0 | No symptoms | Immune |
| 1 | <1 | Resistant |
| 3 | 1-10 | Moderately resistant |
| 5 | 11-25 | Tolerant |
| 7 | 26-50 | Susceptible |
| 9 | >50 | Highly susceptible |

**Table 2.** Disease rating scale (Mayee and Datar, 1986).

| Disease incidence (0-9) scale | Disease incidence (%) | Reaction |
|---|---|---|
| 0 | No wilting | Immune |
| 1 | <1 | Resistant |
| 3 | 1-10 | Moderately resistant |
| 5 | 11-20 | Tolerant |
| 7 | 21-50 | Susceptible |
| 9 | 51 and above | Highly susceptible |

material. Recommended agronomic practices and insect pest control measures were followed as per the package of practices of University of Agricultural Sciences, Dharwad, Karnataka (Anonymous, 2003). Further, the elite materials were categorized as highly resistant, resistant, moderately resistant, susceptible and highly susceptible based on 0 to 9 disease scale for *Alternaria*. (Table 1). Percent disease score was calculated as per the standard area diagram developed by Mayee and Datar (1986). For recording the disease intensity (*Fusarium* wilt and *Macrophomina* rot) under field condition, 0 to 9 disease rating scale developed by Mayee and Datar (1986) was used (Table 2). For this purpose five leaves located at the bottom, five in the middle and five at the top of the plant were chosen and scored as per scale given subsequently.

## RESULTS AND DISCUSSION

Continuous efforts to locate resistant sources and their utilisation in resistance breeding programme are imperative to manage the diseases in the long run. Screening was therefore undertaken to evaluate a large number of elite line collections against major diseases during *monsoon* 2012. The lines were evaluated based on 0 to 9 disease rating scale. The reaction of the different lines is presented in Table 3. Significant variations in disease severity index (0 to 9 scale) for major diseases of safflower were observed in various lines. Of the 46 elite line collections evaluated, only 21 lines, *viz.*, IVT-11-11, IVT-11-14, IVT-11-15, IVT-11-16, IVT-11-17, IVT-11-18, IVT-11-19, IAHT-I-11-01, IAHT-I-11-02, IAHT-I-11-03, IAHT-I-11-04, IAHT-I-11-05, IAHT-I-11-06, IAHT-I-11-07, IAHT-I-11-09, AVHT-II-11-01, AVHT-II-11-02, AVHT-II-11-04, AVHT-II-11-05, AVHT-II-11-01, HUS 305 (Resistant check) registered resistant reaction. These 21 lines were identified as resistant. Four lines were found to be susceptible to *Alternaria* leaf spot.

Results (Table 4) revealed that among the 46 elite lines evaluated only 14 lines, *viz.*, IVT-11-02, IVT-11-03, IVT-11-04, IVT-11-11, IVT-11-14, IVT-11-15, IVT-11-16, IVT-11-19, IVT-11-20, IVT-11-21, IAHT-I-11-06, AVHT-II-11-02, AVHT-II-11-05, AVHT-II-11-01 and HUS 305 (resistant check) registered resistant reaction, Twenty nine lines were identified as moderately resistant, and two lines were found to be susceptible to wilt.

The results (Table 5) of this present study indicated that among the 46 elite lines evaluated only 33 lines, *viz.*, IVT-11-01, IVT-11-05, IVT-11-07, IVT-11-08, IVT-11-11, IVT-11-12, IVT-11-13, IVT-11-15, IVT-11-16, IVT-11-18, IVT-11-19, IVT-11-20, IVT-11-21, IVT-11-22, IVT-11-23, IVT-11-24, IVT-11-25, IVT-11-26, IVT-11-27, IVT-11-28, IVT-11-29, IAHT-I-11-01, IAHT-I-11-02, IAHT-I-11-03, IAHT-I-11-04, IAHT-I-11-05, IAHT-I-11-06, IAHT-I-11-09, AVHT-II-11-01, AVHT-II-11-02, AVHT-II-11-03, AVHT-II-11-04 and HUS 305 (resistant check) registered highly resistant reaction, seven lines were identified as resistant, five lines were moderately resistant and one line was found to be susceptible to root rot.

These findings will help to develop a new set of agronomically desirable disease-resistant hybrids to enhance and sustain safflower productivity. This present study revealed that out of the 44 lines tested; only 32 lines registered high level of resistance (HR) and recorded least disease rating of 1.0, while susceptible check Manjira exhibited maximum rating scale of 4.0. This suggests that the disease development was highly satisfactory and the categorization of materials into different classes is appropriate. Thus, it can be emphasized from the results that the identified highly resistant lines hold excellent promise for resistance against major diseases of safflower and can be used for

**Table 3.** Disease severity on selected on elite material against *Alternaria* leaf spot caused by *Alternaria carthami.*

| S/N | Entries | Mean | Reaction |
|---|---|---|---|
| 1 | IVT-11-01 | 20 | MR |
| 2 | IVT-11-02 | 17.5 | MR |
| 3 | IVT-11-03 | 20 | MR |
| 4 | IVT-11-04 | 20 | MR |
| 5 | IVT-11-05 | 29.5 | S |
| 6 | IVT-11-06 | 24 | MR |
| 7 | IVT-11-07 | 22.5 | MR |
| 8 | IVT-11-08 | 22.5 | MR |
| 9 | IVT-11-09 | 30 | S |
| 10 | IVT-11-10 | 17.5 | MR |
| 11 | IVT-11-11 | 7.5 | R |
| 12 | IVT-11-12 | 17.5 | MR |
| 13 | IVT-11-13 | 12.5 | MR |
| 14 | IVT-11-14 | 10 | R |
| 15 | IVT-11-15 | 5 | R |
| 16 | IVT-11-16 | 10 | R |
| 17 | IVT-11-17 | 10 | R |
| 18 | IVT-11-18 | 9 | R |
| 19 | IVT-11-19 | 10 | R |
| 20 | IVT-11-20 | 18.5 | MR |
| 21 | IVT-11-21 | 12.5 | MR |
| 22 | IVT-11-22 | 30 | S |
| 23 | IVT-11-23 | 15 | MR |
| 24 | IVT-11-24 | 20 | MR |
| 25 | IVT-11-25 | 22.5 | MR |
| 26 | IVT-11-26 | 17.5 | MR |
| 27 | IVT-11-27 | 15 | MR |
| 28 | IVT-11-28 | 25 | MR |
| 29 | IVT-11-29 | 17.5 | MR |
| 30 | IAHT-I-11-01, | 6.5 | R |
| 31 | IAHT-I-11-02 | 3.5 | R |
| 32 | IAHT-I-11-03 | 10 | R |
| 33 | IAHT-I-11-04 | 5 | R |
| 34 | IAHT-I-11-05 | 6.5 | R |
| 35 | IAHT-I-11-06 | 4.5 | R |
| 36 | IAHT-I-11-07 | 5.5 | R |
| 37 | IAHT-I-11-08 | 12.5 | MR |
| 38 | IAHT-I-11-09 | 10 | R |
| 39 | AVHT-II-11-01 | 5 | R |
| 40 | AVHT-II-11-02 | 5 | R |
| 41 | AVHT-II-11-03 | 15 | MR |
| 42 | AVHT-II-11-04 | 7.5 | R |
| 43 | AVHT-II-11-05 | 7.5 | R |
| 44 | AVHT-II-11-01 | 9 | R |
| 45 | Manjira | 30 | S |
| 46 | HUS 305 | 7.5 | R |

developing hybrids and composites in future programme of breeding for disease resistance.

**Table 4.** Disease severity on selected on elite material against wilt caused by *Fusarium oxysporum* f.sp. *carthami.*

| S/N | Entries | Mean | Reaction |
|---|---|---|---|
| 1 | IVT-11-01 | 16.77 | MR |
| 2 | IVT-11-02, | 8.66 | R |
| 3 | IVT-11-03 | 5.71 | R |
| 4 | IVT-11-04 | 9.75 | R |
| 5 | IVT-11-05 | 12.88 | MR |
| 6 | IVT-11-06 | 12.75 | MR |
| 7 | IVT-11-07 | 15.74 | MR |
| 8 | IVT-11-08 | 14.55 | MR |
| 9 | IVT-11-09 | 37.3 | S |
| 10 | IVT-11-10 | 11.83 | MR |
| 11 | IVT-11-11 | 8.94 | R |
| 12 | IVT-11-12 | 11.57 | MR |
| 13 | IVT-11-13 | 13.97 | MR |
| 14 | IVT-11-14 | 8.44 | R |
| 15 | IVT-11-15 | 3.58 | R |
| 16 | IVT-11-16 | 8.91 | R |
| 17 | IVT-11-17 | 15.07 | MR |
| 18 | IVT-11-18 | 12.03 | MR |
| 19 | IVT-11-19 | 8.39 | R |
| 20 | IVT-11-20 | 8.75 | R |
| 21 | IVT-11-21 | 6.39 | R |
| 22 | IVT-11-22 | 11.71 | MR |
| 23 | IVT-11-23 | 16.75 | MR |
| 24 | IVT-11-24 | 15.26 | MR |
| 25 | IVT-11-25 | 13.81 | MR |
| 26 | IVT-11-26 | 16.33 | MR |
| 27 | IVT-11-27 | 11.41 | MR |
| 28 | IVT-11-28 | 15.16 | MR |
| 29 | IVT-11-29 | 13.38 | MR |
| | IAHT (Trial) | | |
| 30 | IAHT-I-11-01 | 13.55 | MR |
| 31 | IAHT-I-11-02 | 13.54 | MR |
| 32 | IAHT-I-11-03 | 11.13 | MR |
| 33 | IAHT-I-11-04 | 12.22 | MR |
| 34 | IAHT-I-11-05 | 10.03 | MR |
| 35 | IAHT-I-11-06 | 5.705 | R |
| 36 | IAHT-I-11-07 | 13.89 | MR |
| 37 | IAHT-I-11-08 | 19.90 | MR |
| 38 | IAHT-I-11-09 | 13.58 | MR |
| | AVHT (Trial) | | |
| 39 | AVHT-II-11-01 | 14.03 | MR |
| 40 | AVHT-II-11-02 | 5.46 | R |
| 41 | AVHT-II-11-03 | 14.14 | MR |
| 42 | AVHT-II-11-04 | 11.08 | MR |
| 43 | AVHT-II-11-05 | 6.06 | R |
| 44 | AVHT-II-11-01 | 5.64 | R |
| 45 | Manjira | 31 | S |
| 46 | HUS 305 | 7.1 | R |

**Table 5.** Disease severity on selected on elite material against root rot caused by *Rhizoctonia bataticola*.

| S/N | Entries | Mean | Reaction |
|---|---|---|---|
| 1 | IVT-11-01 | 0 | HR |
| 2 | IVT-11-02 | 5.78 | R |
| 3 | IVT-11-03 | 5.45 | R |
| 4 | IVT-11-04 | 6.62 | R |
| 5 | IVT-11-05 | 0 | HR |
| 6 | IVT-11-06 | 12.75 | MR |
| 7 | IVT-11-07 | 0 | HR |
| 8 | IVT-11-08 | 0 | HR |
| 9 | IVT-11-09 | 10.94 | R |
| 10 | IVT-11-10 | 5.91 | R |
| 11 | IVT-11-11 | 0 | HR |
| 12 | IVT-11-12 | 0 | HR |
| 13 | IVT-11-13 | 0 | HR |
| 14 | IVT-11-14 | 13.58 | MR |
| 15 | IVT-11-15 | 0 | HR |
| 16 | IVT-11-16 | 0 | HR |
| 17 | IVT-11-17 | 17.63 | MR |
| 18 | IVT-11-18 | 0 | HR |
| 19 | IVT-11-19 | 0 | HR |
| 20 | IVT-11-20 | 0 | HR |
| 21 | IVT-11-21 | 0 | HR |
| 22 | IVT-11-22 | 0 | HR |
| 23 | IVT-11-23 | 0 | HR |
| 24 | IVT-11-24 | 0 | HR |
| 25 | IVT-11-25 | 0 | HR |
| 26 | IVT-11-26 | 0 | HR |
| 27 | IVT-11-27 | 0 | HR |
| 28 | IVT-11-28 | 0 | HR |
| 29 | IVT-11-29 | 0 | HR |
| 30 | IAHT-I-11-01 | 0 | HR |
| 31 | IAHT-I-11-02 | 0 | HR |
| 32 | IAHT-I-11-03 | 0 | HR |
| 33 | IAHT-I-11-04 | 0 | HR |
| 34 | IAHT-I-11-05 | 0 | HR |
| 35 | IAHT-I-11-06 | 0 | HR |
| 36 | IAHT-I-11-07 | 8.6 | R |
| 37 | IAHT-I-11-08 | 14.22 | MR |
| 38 | IAHT-I-11-09 | 0 | HR |
| 39 | AVHT-II-11-01 | 0 | HR |
| 40 | AVHT-II-11-02 | 0 | HR |
| 41 | AVHT-II-11-03 | 0 | HR |
| 42 | AVHT-II-11-04 | 0 | HR |
| 43 | AVHT-II-11-05 | 8.03 | R |
| 44 | AVHT-II-11-01 | 13.52 | MR |
| 45 | Manjira | 31.5 | S |
| 46 | HUS 305 | 0 | HR |

Awadhiya (1992) identified *A. carthami, Fusarium moniliforme, Botrytis cinerea, Macrophomina phaseolina, Stachybotrys* spp. and *Oedocephalum* spp. from seeds of 50 safflower cultivars in states of Maharashtra, Karnataka, Andhra Pradesh and Madhya Pradesh in India. *A. carthami* was the only pathogen found in all varieties tested. In studies of healthy, discoloured, wrinkled and deformed seeds of five varieties (APRR 2, HUS 304, JSF 1, NS 99-A and SF 364) no particular association of the pathogen with the condition of seed was found. Chavan and Kakde (2009) isolated nine fungal species from safflower cultivars. Among these, *Aspergillus* spp. showed dominance, followed by *Fusarium* spp. and *Alternaria* spp. Bhima variety showed maximum susceptibility to fungi and got infected by *Aspergillus niger. A. flavus, Fusarium oxysporum, Alternaria dianthicola* and *Alternaria dianthi* while C1L and C1B varieties were least susceptible to fungi.

This study confirms that differences in resistance to major diseases exist in germplasm of safflower. The resistant nature of elite lines observed in present field trials confirmed the reports by Singh et al. (1987), Borkar and Shinde (1988), Zad (1992), Khanam (1993) and Ismail et al. (2004). These findings suggest that it is possible to improve an existing elite line through further selection and screening of the progenies of the parental line.

## REFERENCES


Anonymous (2003). Package of Practices for Improved Cultivation of Crops. Publication centre, Directorate of Extension, University of Agricultural Science, Dharwad. p. 26.

Awadhiya GK (1992). Seed borne pathogenic mycoflora of safflower. Crop Res. 5(2):344-347.

Bhale MS, Bhale Usha, Khare MN (1998). Diseases of important oilseed crops and their management. In: Khurana SM Paul (ed.) Pathological problems of economic crop plants and their management, Scientific Publishers(India), Jodhpur. pp. 251-279.

Borkar SG, Shinde R (1988). Alternaria leaf spot of safflower. Seed Res. 16:126-127

Ramesh CH, Avitha KM (2005). Presence of external and internal mycoflora of sunflower seeds. J. Mycol. Plant Pathol. 35(2):362-363.

Chavan AM, Kakde RB (2009). Detection of fungal load on abnormal oil-seeds from Marathwada region. Bioinfolet 6:149-150.

Ismail M, Irfan Ul-Haque M, Riaz A (2004). Seed-borne mycoflora of safflower (*Carthamus tinctorius* L.) and their impact on seed germination. *Mycopatho*logy 2:51-54.

Khanam M (1993). Seed-borne fungi associated with safflower and their effect on germination. Sarhad J. Agric. 9:153-156.

Mayee CD, Datar VV (1986). Phytopathometry. Technical Bull-I, MAU, Parbhani. pp. 88-89.

Singh SN, Agarwal SC, Khare MN (1987). Seedborne mycoflora of safflower, their significance and control. Seed Res. 15:190-194.

Zad SJ (1992). Safflower seed-borne diseases. Mededelingen van de Faculteit Landbouwwetenschappen, Universiteit Gent 57:161-163

# Allelopathic potential of sunflower and caster bean on germination properties of dodder (*Cuscuta compestris*)

**M. Seyyedi, P. Rezvani Moghaddam, R. Shahriari, M. Azad and E. Eyshi Rezaei**

Ferdowsi University of Mashhad, Faculty of Agriculture, P. O. Box 91775-1163, Mashhad, Iran.

**Allelopathic impacts of two crops, sunflower (*Helianthus annus* L.) and caster bean (*Ricinus communis* L.) were evaluated against dodder (*Cuscuta compestris*) germination properties. Different plants residue, plant parts (root, shoot, leaf and whole plant), various concentrations of aqueous extract and decay durations were employed as study factors under completely randomized design (CRD) with factorial arrangement by three replications in this study. The results indicated that germination percentage, germination rate, emergence rate, seedling length of dodder was sharply influenced by sunflower and caster bean residue application. Aqueous extract of sunflower inhibited dodder seed germination more efficiently in comparison with caster bean especially in higher concentrations. Moreover, shoot aqueous extract allelochemicals showed substantial potential to the inhibition of dodder germination in contrast with other parts of plants. However, fresh leaf solid residue indicated great potential to dodder germination suppression. Increasing aqueous extract concentration significantly inhibited dodder germination and emergence under controlled conditions. In conclusion, dodder germination can be controlled by sunflower and caster bean allelochemicals. Therefore, allelopathic potential of these two plants can be consider as a sustainable approach in integrated management systems.**

**Key words:** Allelochemicals, aqueous extract, decay duration, *Cuscuta compestris*.

## INTRODUCTION

Dodder (*Cascuta campestris*) is an annual holoparasitic higher plant in the Convolvulaceae family (Mishra et al., 2007). Dodder seedlings are thin, long, delicate, rootless and leafless (Weinberg et al., 2003). Potatoes, tomatoes and sugar beets are most important crops highly influenced by parasitic impact of dodder (Nadler-Hasasr and Rubin, 2003). Dodder life cycle is entirely dependent on host for water supplying, assimilates and minerals (Mishra et al., 2007). There is no large number of chemical control patterns to avoiding from parasitic impact of dodder. In addition, environmental concerns about synthetic herbicides application become rising (Lanini and Kogan, 2005).

In general, allelopathy refers to any direct or indirect beneficial or harmful effects produced by plants which influence the growth and developments of other plants (Narwal, 2010). Allelopathic plants compete with other plants, by producing different secondary metabolic components such as alkaloids and glycosides and introducing them to the soil rhizosphere of plants (Jarchow and Cook, 2009; Morris et al., 2009; Weston, 2005). All plant species produce a variety of natural compounds that may be released into the environment as exudates, leachate, or volatile. Therefore, allelopathy may be a widespread occurrence (Morris et al., 2009). Allelochemicals which produce by allelopathic plants showed directly negative influences on seed germination and plant growth of other plants even in low concentrations (Kupidłowska et al., 2006). Consequently, the inhibitory effects of allelochemicals might be used against weeds as a controlling tool for decreasing the

*Corresponding author. E-mail: eh_ey145@stu-mail.um.ac.ir.

weed emergence in field conditions (Xuan et al., 2005; Belz, 2007).

The allelochemicals derived from many crops such as sunflower (*Helianthus annus*) and caster bean (*Ricinus communis*) and could prevent some broad and narrow leaf weeds growth (Bhowmik and Inderjit, 2003; Khanh et al., 2005). Anjum and Bajwa (2005) studied the effects of bioactive annuionone from aqueous extracts of sunflower leaves on growth of five weeds including *Chenopodiun album* L., *Coronopis didymus* L., *Medicago polymorpha* L., *Rumex dentatus* L. and *Phalaris minor*, they reported this extract can be used as a natural herbicides. Moreover, some studies demonstrated that Sunflower residues decreased growth of different weeds such as *Cyamopsis tetragonoloba*, *Pennisetum americanum* and *S. biocolor* (Batish et al., 2002). They concluded that due to decomposing tissue of sunflower by soil microorganisms, some allelochemicals such as phenolics were released and inhibited the growth of those weeds (Batish et al., 2002).

Jamil et al. (2009) reported that sunflower water extract can be used for controlling wild oats (*Avena fatua*) and canary grass (*Phalaris minor*). Ricin is one of the plant toxins which derived from caster been seeds has been used as an organic herbicide for weed controlling (Doan, 2004; Aslani et al., 2007). It was well documented that there was allelopathic activity in different plant parts of caster bean (Ilavarasan et al., 2006). On the other hand, some studies reported that caster bean extract can be used for insect controlling (Upasani et al., 2003). This study was built on assessment of the allelopathic effects of sunflower and caster bean on germination and emergence of dodder inhibition.

## MATERIALS AND METHODS

Evaluation of sunflower and caster bean parts allelopathic impacts on germination and emergence properties of dodder were conducted in three separate experiments. These experiments were performed under laboratory and greenhouse conditions. All experiments were carried out by using completely randomized designs (CRD) with factorial arrangement by three replications.

### Experiment one

Three factors employed in this experiment, first factor was two levels of allelopathic plant residue species (sunflower and caster bean), the second factor was different parts of sunflower and caster bean included root, stem, leave and whole plant without inflorescence and the third factor was concentrations of aqueous extract at 10 levels (0, 2, 3, 4, 5, 6, 7, 8, 9 and 10%). Sunflower and caster bean were collected at the end of flowering and beginning with seed filling stage, from research station of faculty of agriculture, Ferdowsi University of Mashhad, Iran. Parts of plants were dried in shade for seven days and then grinded separately. Stock solution was prepared through 10 g of each plant samples powder mixing with 100 ml distilled water. The prepared solutions were shaken for three days at 25 to 30°C and subsequently filtered through a double Whatman (No.2) filter paper. Obtained 10% (w/v) extracts of different parts were diluted to gain study solutions.

Dodder seeds were collected from a potato farm in 2009, and then treated with 98% sulfuric acid-scarified for 20 min (Nadler-Hassar and Rubin, 2003). Various concentrations of aqueous extracts (4 ml) which gained from different parts were added to each sterilized Petri dishes contained a filter paper and then 20 dodder seeds were sown in Petri dishes. The Petri dishes were placed in dark germinator at 30°C (Benvenuti et al., 2005). The daily counting of germinated seeds started 24 h after sowing and it continued until germination process is completed (13 days). Seedlings with 2 mm length of radicle were termed "germinated" (Benvenuti et al., 2005). Germination rate was calculated by:

$$GR = \sum_{i=1}^{n} \frac{ni}{di} \quad (1)$$

Where *ni* is number of germinated seeds in first day of counting, and *di* is first day of counting.

### Experiment two

Similar to first experiment, three factors included two different allelopathic plants residue (sunflower and caster bean), four parts of sunflower and caster bean (root, stem, leaves, and total plant without inflorescence) and various aqueous extract concentrations of them at 5 levels (0, 2.5, 5, 7.5 and 10%) were employed in this experiment. Aqueous extracts were prepared according to experiment one instruction. In addition, 10 seeds of dodder were sown in plastic pots (10 cm × 15 cm × 8 cm) after acid-scarifying dodder seeds. In that case, different aqueous extract concentrations were added to each pot.

### Experiment three

The third experiment was performed by three factors same as pervious experiments; two different allelopathic plants residue (sunflower and caster bean), sunflower and caster bean parts at 4 levels (root, stem, leaf and whole plant without inflorescence) and decay durations at 7 levels (0, 15, 30, 45, 60, 75 and 90 days decay) were employed in this experiment. All plant parts were separately chopped (in 2 mm) and added to the soil by 5% (w/w) in each pot at the same time. Afterward, 10 dodder seeds which scarified previously were sown in each pot.

The pots of experiments two and three were kept in 25 to 30°C room temperature. The irrigation of pots was carried out by distilled water every day. Daily counting of emerged seeds in each pot in experiments two and three were recorded 24 h after sowing and it continued until seeds emergence is fixed. The rate of emergence for second and third experiments was calculated by:

$$ER = \sum_{i=1}^{n} \frac{ni}{di} \quad (2)$$

Where *ni* is number of emergence seeds in first day of counting, *di* is first day of counting.

### Study measurements

Germination percentage and rate, and length of dodder seedlings were measured in pot/Petri dish levels.

### Statistical analysis

In order to evaluate the treatments impacts on study parameters,

**Table 1.** Effects of different allelopathic plant species, plant parts and aqueous extract concentrations on germination percentage, germination rate, and seedling length of dodder.

| Plant species | Germination percentage (%) | Germination rate (seed per day) | Seedling length (mm) |
|---|---|---|---|
| Sunflower | 55[b] | 2.9[b] | 43[b] |
| Caster bean | 61[a] | 3.6[a] | 47[a] |
| LSD | 1.9 | 0.1 | 5 |
| Plant part | | | |
| Root | 62[a] | 3.6[a] | 59[a] |
| Shoot | 50[c] | 3.0[c] | 34[c] |
| Leaf | 57[b] | 3.1[bc] | 43[b] |
| Whole plant | 62[a] | 3.3[b] | 44[b] |
| LSD | 2.7 | 0.22 | 3.7 |
| **Aqueous extract concentration (%)** | | | |
| 0 | 83[a] | 4.4[a] | 56[ab] |
| 2 | 65[bc] | 4.2[a] | 60[a] |
| 3 | 59[d] | 3.5[bc] | 47[cd] |
| 4 | 62[cd] | 3.8[b] | 43[de] |
| 5 | 50[ef] | 2.7[d] | 51[bc] |
| 6 | 67[b] | 3.7[b] | 40[fe] |
| 7 | 52[e] | 2.6[d] | 40[fe] |
| 8 | 60[d] | 3.2[c] | 37[f] |
| 9 | 37[g] | 1.7[e] | 39[fe] |
| 10 | 46[f] | 2.4[d] | 36[f] |
| LSD | 4.3 | 0.34 | 5.8 |

Similar letters in each column show non-significant differences according to Duncan's Multiple Range Test at 5% level of probability.

analysis of variance (ANOVA) was performed as standard procedure for factorial randomized block designs. The t-test was used to find significant differences among treatments. The significant differences between treatments were compared by Duncan's multiple range tests at 5% probability level.

## RESULTS

### First and second experiments

#### *Allelopathic plant species, part and aqueous extract concentration*

Different allelopathic plants, plant parts and aqueous extract concentrations of allelopathic plants showed significant impact (*$P > 0.05$*) on germination properties of dodder seeds in Petri dish level (Table 1).

Sunflower residue inhibited dodder seed germination more efficiently in comparison with caster bean residue. Germination percentage and rate of dodder was 55% and 2.9 seed per day whenever sunflower residue was applied in Petri dish level (Table 1).

In addition, sunflower residue application cause sharply decreases in seedling length of dodder in contract to caster bean residue in Petri dish (43 mm) level (Table 1). However, plant species did not indicate significant effect on dodder seed germination percentage and emergence rate on pot level (Table 2). Seedling length of dodder was higher under sunflower residue in pot level (Table 2).

Application of shoot residue significantly decline the germination percentage (50% in Petri dish and 42% in pot levels), rate (3 seed per day in Petri dish and 0.76 seed per day in pot level) and seedling length (34 mm in Petri dish level) in contrast to other parts of study plants in both Petri dish and pot levels (Table 1 and 2).

Various parts of allelopathic plants did not showed significant impact on seedling length in pot level (Table 2).

The results were shown that different aqueous extract concentrations of allelopathic plants directly influenced germination percentage, rate and seedling length of dodder.

Increment of aqueous extract concentration penetratingly decrease study parameters in both Petri dish and pot levels (Table 1 and 2). Lowest values of germination percentage (46 and 37% in Petri dish and 30% in pot levels), germination and emergence rate (2.4 and 1.7 seed per day in Petri dish and 0.41 seed per day in pot level) and seedling length (36 mm in Petri dish and 36 mm in pot levels) was obtained on highest concentration of aqueous extract (Table 1 and 2).

**Table 2.** Effects of various allelopathic plant species, plant parts and aqueous extract concentrations on germination percentage, emergence rate, and seedling length of dodder.

| Plant species | Germination percentage (%) | Emergence rate (seed per day) | Seedling length (mm) |
|---|---|---|---|
| Sunflower | 48[a] | 0.88[a] | 55[a] |
| Caster bean | 50[a] | 0.91[a] | 47[b] |
| LSD | 4.7 | 0.09 | 6.2 |
| Plant part | | | |
| Root | 53[a] | 0.97[a] | 54[a] |
| Shoot | 42[b] | 0.76[b] | 54[a] |
| Leaf | 49[a] | 0.87[ab] | 49[a] |
| Whole plant | 54[a] | 0.97[a] | 47[a] |
| LSD | 6.6 | 0.13 | 8.8 |
| | Aqueous extract concentration (%) | | |
| 0 | 90[a] | 2.13[a] | 74[a] |
| 2.5 | 44[b] | 0.69[b] | 46[bc] |
| 5 | 44[b] | 0.64[b] | 54[b] |
| 7.5 | 40[b] | 0.59[b] | 45[bc] |
| 10 | 30[c] | 0.41[c] | 36[c] |
| LSD | 7.4 | 0.15 | 9.9 |

Similar letters in each column show non-significant differences according to Duncan's Multiple Range Test at 5% level of probability.

### *Interactive effects of plant residue species, part and aqueous extract concentration*

The results showed tooth response of dodder germination properties to application of aqueous extract of sunflower and caster bean which were extracted from different parts of plant residue. Effects of applied treatments indicated more influence on dodder germination in Petri dish level (Figure 1 and 2).

In general, sunflower aqueous extract gradually decrease the germination percentage, emergence and germination rate and seedling length of dodder in both Petri dish and pot levels especially under higher concentrations of aqueous extract (Figures 1 and 2).

Uppermost decrease in dodder germination and seedling length were obtained under higher extract concentrations gained from sunflower shoot in both Petri dish and pot levels except seedling length parameter in pot levels (Figures 1 and 2) .

## Third experiment

### *Plant species, part and decay duration*

Plant species did not show significant effect on germination percentage and seedling rate on third experiment, but sunflower residue (0.59 seed per day) considerably decrease emergence rate of dodder in comparison with caster bean residue (0.70 seed per day) Table 3). All study parameters significantly influenced ($P > 0.05$) by various parts of allelopathic plants (Table 3). Utmost decrease in germination percentage (20%), emergence rate (0.30 seed per day) and seedling length (44 mm) of dodder was gained in application of leaf residue (Table 3). Summations of decay duration gradually decrease germination percentage (28% in 0 days of decay duration), emergence rate (0.34 seed per day in 30 days decay duration) and seedling length (53 mm in days decay duration) (Table 3).

### *Interactive effects of plant residue species, part and decay duration*

Residue decay duration increment showed direct impact on germination properties of dodder seeds. Caster bean residues showed momentous efficiency in prevention of dodder germination in comparison with sunflower residue especially under lower decay duration in the third experiment (Figure 3). Leaf residue of sunflower and caster bean showed highest prevention of dodder seeds in lower decay durations. However, shoot residue indicated maximum germination inhibition under higher decay durations (Figure 3).

## DISCUSSION

The results evidently demonstrated that sunflower and caster bean residues significantly inhibited seed germination and seedling elongation of dodder under controlled conditions. High inhibition of dodder germination was

**Table 3.** Impacts of various allelopathic plant species, plant parts and decay duration of plant materials on germination percentage, emergence rate, and seedling length of dodder.

| Plant species | Germination percentage (%) | Emergence rate (seed per day) | Seedling length (mm) |
|---|---|---|---|
| Sunflower | 38[a] | 0.59[b] | 69[a] |
| Caster bean | 41[a] | 0.70[a] | 76[a] |
| LSD | 3.6 | 0.07 | 7.6 |
| Plant part | | | |
| Root | 58[a] | 0.97[a] | 85[a] |
| Shoot | 35[c] | 0.55[c] | 79[a] |
| Leaf | 20[d] | 0.30[d] | 44[b] |
| Whole plant | 46[b] | 0.75[b] | 82[a] |
| LSD | 5.1 | 0.1 | 10 |
| **Decay duration (day)** | | | |
| 0 | 28[d] | 0.59[c] | 58[c] |
| 15 | 37[c] | 0.67[bc] | 59[c] |
| 30 | 22[d] | 0.34[d] | 53[c] |
| 45 | 24[d] | 0.36[d] | 48[c] |
| 60 | 47[b] | 0.70[bc] | 94[b] |
| 75 | 48[b] | 0.77[b] | 115[a] |
| 90 | 72[a] | 1.09[a] | 84[b] |
| LSD | 6.7 | 0.13 | 14 |

Similar letters in each column show non-significant differences according to Duncan's Multiple Range Test at 5% level of probability.

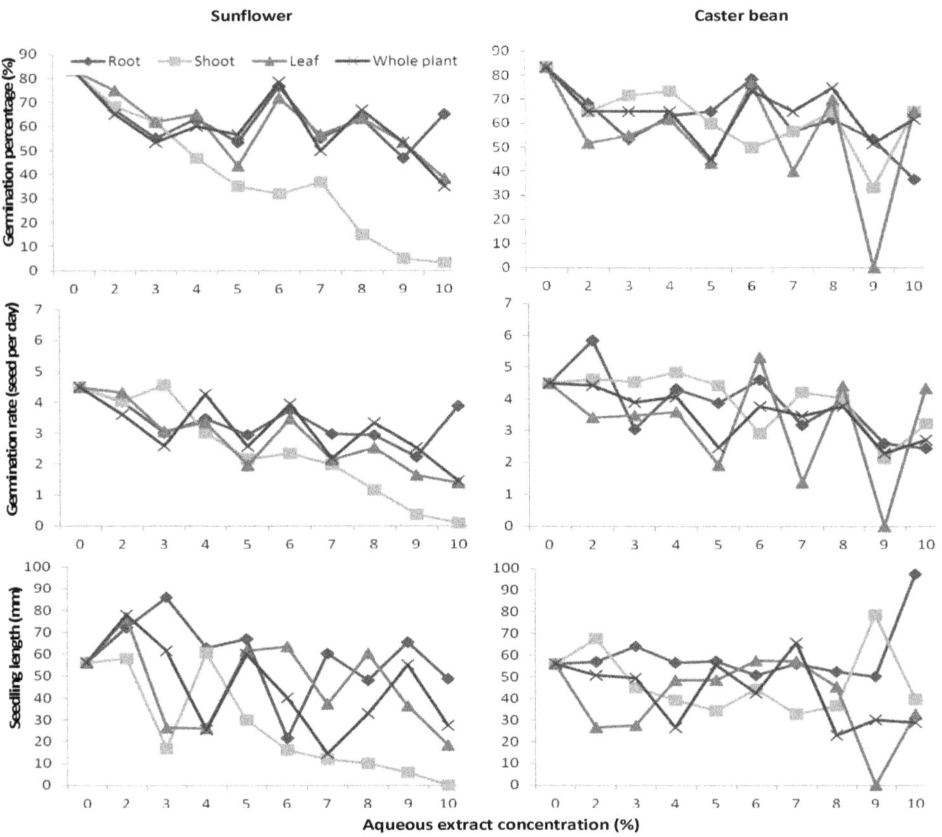


**Figure 1.** Interactive effects of plant residue species, part and aqueous extract concentration on germination percentage, rate and seedling length of dodder in first experiment.

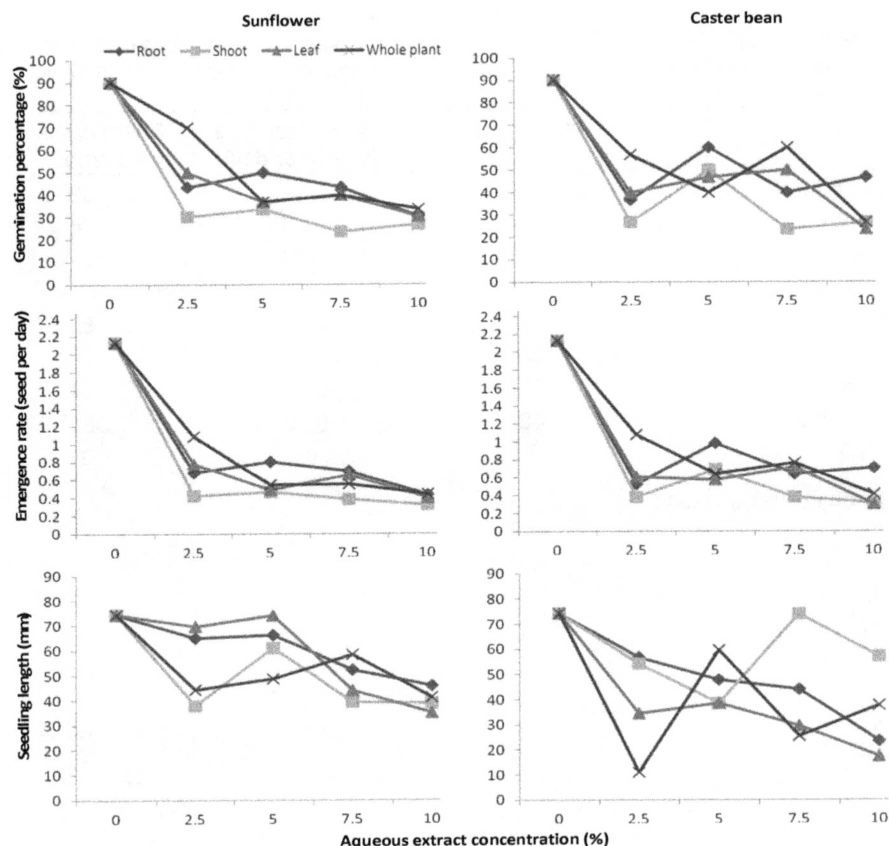


**Figure 2.** Interactive effects of plant residue species, part and aqueous extract concentration on germination percentage, emergence rate and seedling length of dodder in second experiment.

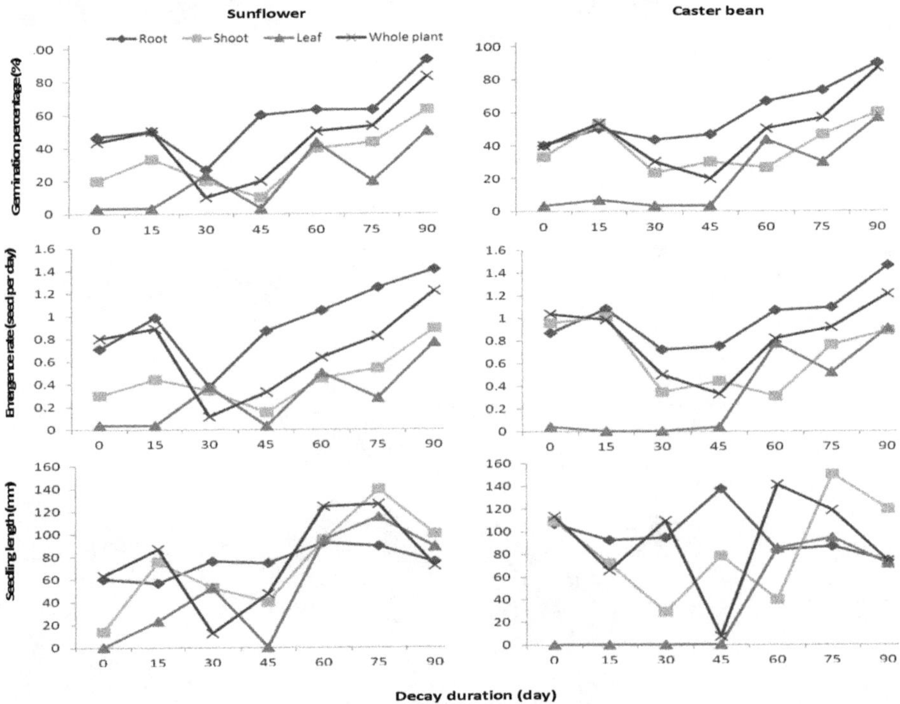


**Figure 3.** Interactive effects of plant residue species, part and decay duration on germination percentage, emergence rate and seedling length of dodder in third experiment.

obtained whenever aqueous extract of sunflower residues applied as allelochemical in comparison with aqueous extract of caster bean in both Petri dish and pot levels. The present study results confirmed sunflower allelopathic inhibition impact on germination of weeds. Anjum and Bajwa (2007) found sunflower aqueous extract significantly restrain *R. dentatus* germination in field condition.

Chemical studies on sunflower extract represented that this species is a wealthy source of phenolic compounds and terpenoids, particularly sesquiterpene lactones with a broad range of biological activities such as allelopathy (Macias et al., 2000). Highest decrease in dodder germination was gained by shoot residue aqueous extract but leaf residue showed maximum decline in dodder germination whenever fresh solid caster been residue was applied. Machado (2007) indicated that shoot residues extract of different allelopathic plants prevented downy brome (Bromus *tectorum* L.) growth efficiently in comparison with leaf residues extract. Differences in shoot, leaf and root extract effects may point toward the existence of different allelochemicals or concentrations of allelochemicals in roots and shoots.

Germination declines of dodder was raise appreciably when higher concentrations of sunflower and caster bean were applied. It shows that general effect of allelopathic plants was extremely quantity dependent (Skulman et al., 2004). Application of sunflower extract in higher concentrations inhibited seed germination of mustard (*Sinapis alba*) under controlled conditions (Bogatek et al., 2006). In conclusion, sunflower and caster bean residues showed great potential for dodder germination control but relative effectiveness of different plant parts such as shoot, leaf and root extracts and their concentration and decay duration are important in creating weed control strategies.

## REFERENCES

Anjum T, Bajwa R (2005). A bioactive annuionone from sunflower leaves. Photochemistry 66:1919-1921.

Anjum T, Bajwa R (2007). Field appraisal of herbicide potential of sunflower leaf extract against *Rumex dentatus*. Field Crops. Res. 100:139-142.

Aslani MR, Maleki M, Mohri M, Sharifi K, Najjar-Nezhad V, Afshari E (2007). Castor bean (*Ricinus communis*) toxicosis in a sheep flock. Toxicon 49:400-406.

Batish DR, Tung P, Singh HP, Kohli RK (2002). Phytotoxicity of sunflower residues against some summer season crops. J. Agron. Crop Sci. 188:19-24.

Belz RG (2007). Allelopathy in crop/weed interactions - An update. Pest. Manage. Sci. 63:308-326.

Benvenuti S, Dinelli G, Bonetti A, Catizone P (2005). Germination ecology, emergence and host detection in *Cuscuta campestris*. Weed. Res. 45:270-278.

Bhowmik PC, Inderjit B (2003). Challenges and opportunities in implementing allelopathy for natural weed management. Crop. Prot. 22:661-671.

Bogatek R, Gniazdowska A, Zakrzewska W, Oracz K, Gawronski SW (2006). Allelopathic effects of sunflower extracts on mustard seed germination and seedling growth. Biol. Plantarum 50:156-158.

Doan LG (2004). Ricin: Mechanism of toxicity, clinical manifestations, vaccine development. A review. J. Toxicol. Clin. Toxicol. 42:201-208.

Ilavarasan R, Mallika M, Venkataraman S (2006). Anti-inflammatory and free radical scavenging activity of Ricinus communis root extract. J. Ethnopharmacol. 103:478-480.

Jamil M, Cheema ZA, Mushtaq MN, Farooq M, Cheema MA (2009). Alternative control of wild oat and canary grass in wheat fields by allelopathic plant water extracts. Agron. Sust Devel. 29:475-482.

Jarchow ME, Cook BJ (2009). Allelopathy as a mechanism for the invasion of *Typha angustifolia*. Plant Ecol. 204:113-124.

Khanh TD, Chung MI, Xuan TD, Tawata S (2005). The exploitation of crop allelopathy in sustainable agricultural production. J. Agron. Crop Sci. 191:172-184.

Kupidłowska E, Gniazdowska A, Stpien J, Corbineau F, Vinel D, Skoczowski A, Janeczko A, Bogatek R (2006). Impact of sunflower (*Helianthus annuus* L.) extracts upon reserve mobilization and energy metabolism in germinating mustard (*Sinapis alba* L.) seeds. J. Chem. Ecol. 32:2569-2583.

Lanini WT, Kogan M (2005). Biology and management of Cucuta in crops. Ciencia. Investigaci6n. Agraria. 32:165-179.

Macias FA, Galindo JCG, Molinillo JMG, Castellano D (2000). Dehydrozaluzanin C: a potent plant growth regulator with potential use as a natural herbicide template. Phytochemistry 54:165–171.

Machado S (2007). Allelopathic potential of various plant species on Downy brome: implications for weed. Agron J. 99:127–132.

Mishra JS, Moorthy BTS, Bhan M, Yaduraju NT (2007). Relative tolerance of rainy season crops to field dodder (*Cuscuta campestris*) and its management in Niger (Gulzotia abyssinica). Crop Prot. 26:625-629.

Morris C, Grossl PR, Call CA (2009). Elemental allelopathy: processes, progress and pitfalls. Plant Ecol. 202:1-11.

Nadler-Hassar T, Rubin B (2003). Natural tolerance of *Cuscuta campestris* to herbicides inhibiting amino acid biosyntesis. Weed Res. 43:341-347.

Narwal SS (2010). Allelopathy in ecological sustainable organic agriculture. Allelopathy J. 25:51-72.

Skulman BW, Mattice JD, Cain MD, Gbur EE (2004). Evidence for allelopathic interference of Japanese honeysuckle (Lonicera japonica) to loblolly and short leaf pine regeneration. Weed Sci. 52:433–439.

Upasani SM, Kotkar HM, Mendki PS, Maheshwari VL (2003). Partial characterization and insecticidal properties of Ricinus communis L. foliage flavonoids. Pest. Manag Sci. 59:1349-1354.

Weinberg T, Lalazar A, Rubin B (2003). Effects of bleaching herbicides on field dodder (*Cuscuta campestris*). Weed. Sci. 51:663-670.

Weston LA (2005). History and current trends in the use of allelopathy for weed management. Hortic. Technol. 15:529-534.

Xuan TD, Shinkichi T, Khanh TD, Chung IM (2005). Biological control of weeds and plant pathogens in paddy rice by exploiting plant allelopathy: An overview. Crop Prot. 24:197-206.

# Effect of fire on flowering of *Hyparrhenia hirta* (L.) Stapf ($C_4$), *Merxmuellera disticha* (Nees) Conert ($C_3$) and *Themeda triandra* Forsskal ($C_4$) on the Signal Hill, Cape Town, South Africa

**M. H. Ligavha-Mbelengwa and R. B. Bhat**

Department of Botany, University of Venda, Thohoyandou, Limpopo 0950, South Africa.

**Flowering in *Merxmuellera disticha* was more strongly stimulated by fire in summer. Flowering tillers produced between 9 and 11 months after fire were significantly more abundant than those produced 21 to 24 months thereafter. Growth in terms of size of individuals was considered insignificant for analysis. Neighbored *Hyparrhenia hirta* showed, to a certain degree, an out-of-season growth, production and reproduction, whilst *M. disticha* and *Themeda triandra* on the same plots did not. The xylem water potential of *H. hirta* [-10.9(±0.29)], *M. disticha* [-14.6(±0.80)] and *T. triandra* [-13.8(±0.29)] on one stand differed significantly from one another ($p < 0.0001$), and that of *H. hirta* [-19.7(±0.81] and *M. disticha* [-34.5(±1.26)] on the other stand also differed significantly from each other ($p < 0.0001$). One way analysis of variance (ANOVA) thereof produced the following (F-ratio = 13.893; degrees of freedom = 29, $p < 0.0001$), and for *H. hirta* versus *M. disticha*, (F-ratio = 97.605; degrees of freedom = 23, $p < 0.0001$). Multiple range analysis tests revealed a significant difference between xylem water potentials of *H. hirta* from that of *M. disticha* and *T. triandra* but not between *M. disticha* and *T. triandra*.**

**Key words:** South Africa, Signal Hill, fire, flowering plants, impact.

## INTRODUCTION

Veld burning has unquestionably been a feature of the African landscape since time immemorial. Periodic or planned burning of the veld is thought to preserve pasture grass communities (Scott, 1970). The beneficial effect of veld burning may be merely the removal of excess cover, since where the veld has been grazed short or excess grass had been removed by mowing, there was no need to burn (Scott, 1951). Many studies in various parts of Southern Africa, such as the Eastern Cape, Natal, and the North-eastern Transvaal have confirmed the necessity for regular/frequent fires for the maintenance of grassland communities (Acocks, 1966; Rethman and Booysen, 1986; Tainton et al., 1970; Downing, 1974; Friedel and Blackmore, 1988). But there are studies by the scientists who have reported the persistence of grasses on unburned treatment (Bowman et al., 1988). In general, however, fire appears to be a key factor in maintaining South African grasslands. The role of fire in preventing bush encroachment has also been demonstrated.

(Trollope, 1974). To date, many studies on fire have shown it to be the most important agent for improving grassland ecosystems (Tolsma et al., 2010; Smith and Nelson, 2011). Grass species richness improved when burnt annually, and decreased when veld was protected from fire for at least 5 years (Yeaton et al., 1988). Season of veld burning might also play an important part in maintaining South African grasslands. Rethman and Booysen (1986) found that production of *Themeda triandra* and its regrowth potential were significantly influenced by the season of defoliation. Opperman and Roberts (1978) on their study of *T. triandra*, *Elyonurus argenteus* and *Heteropogon contortus* suggested an optimum disturbance (burning, grazing, cutting, or drought) time to prevent damage to the shoot apices because that may have a detrimental effect on leaf, shoot and seed production. System processes such as nutrient recycling and water relations have been proved to be fire dependent (Raison, 1979; Van Wilgen and Le Maitre, 1981; Stock and Lewis, 1986).

It was the aim of this project of the study to compare the effect of fire on growth and flowering of *Hyparrhenia hirta* ($C_4$), *Merxmuellera disticha* ($C_3$) and *T. triandra* ($C_4$) in 9 months, 21 months and 15 years after burning the veld on Cape Town's Signal Hill.

## MATERIALS AND METHODS

### Study site

The study area was Signal Hill, Cape Town, South Africa (Altitude: 300 to 350 m, latitude: 33°54' and longitude: 18°23'). The study site was chosen for the presence of *H. hirta*, *M. disticha* and *T. triandra* associations, the accessibility and proximity of the site to a representative weather station (Kloofnek), which, however, only keeps limited weather records.

Experimental sites were established on both the east-and west-facing slopes. Both the two areas on the two slopes (east- and west-facing slopes) considered for analysis were each about 500 × l500 m, and the plots used for the different facets of this study were randomly located in the two areas. All sample sites were located on the mid- and upper-elevations of the sites.

The climate of the area is typically Mediterranean. For temperature, precipitation and relative humidity, data for City Hospital station-Cape Town was considered suitable to represent Signal Hill. Most of the annual total falls in winter and little, falls in spring and summer. Local topography clearly plays an important role in influencing temperature and precipitation. Soils at the two sites seemed heterogeneous because of discrete monospecific and mixed stands of grasses and shrubs on both the slopes.

A survey of fire effects was carried out after a wildfire which burned an area approximately 50 × 150 m of the topmost section of Signal Hill's west-facing slope. The fire burned in late February, 1988 (Clive May -Kloofnek station supervisor - personal communication). Other plots on the east-facing slope studied here are all known to have had their last fire in 1975.

The Signal Hill plant communities comprise grasses, shrubs and herbs, but only the localized patches dominated by grass (*H. hirta*, *M. disticha* and *T. triandra*) were the focus for this study.

The role of fire in the 1988 burn was assessed by counting the number of flowering tillers 11 months (on January, 1989) after fire, and another count of the same individuals was performed on December, 1989 and January/February, 1990. Two plots, a monospecific stand of *M. disticha*, and a mixed stand of *H. hirta* and *T. triandra* were used. At least 20 individuals were monitored. Wire rods with numbered tags were used to mark the individuals monitored. It was not possible to compare *M. disticha* versus itself between the 1988 burn and the 1975 one, because *M. disticha* on the 1975 burn failed to produce any flowering tillers. Flowering tillers were also counted on *H. hirta* and *T. triandra* in the 1988 burn. Growth in terms of individual species' sizes and leaf greening were slightly observed for both the 1975 and 1988 burn. The data thereof were not analyzed.

Xylem water potential measurements of the three grass species on two unaltered plots in the 1975 burn (east-facing slope) were done on the following 2 days: 05/10/1988 and 21/02/1989. The aim was to use the measurements to explain the differences in the vigor of growth and reproduction of the three grass species.

## RESULTS

Flowering in *M. disticha* was more strongly stimulated by fire in the summer of 1988/1989 than that of 1989/1990 (Wilcoxon test by ranks: $Z = 3.214$, $p < 0.005$, total pairs = 25). Flowering tillers produced between 9 and 11 months after fire were significantly more abundant than those produced 21 to 24 months thereafter.

*H. hirta* and *T. triandra* species on a 1988 burn showed leaf greening flowering longer than those on the 1975 burn. Growth in terms of size of individuals was considered insignificant for analysis. Neighbored *H. hirta* on a 1975 burn showed, to a certain degree, an out-of-season growth, production and reproduction, whilst *M. disticha* and *T. triandra* on the same plots did not. The xylem water potential of *H. hirta* [-10.9(±0.29)], *M. disticha* [-14.6(±0.80)] and *T. triandra* [-13.8(±0.29)] on one stand differed significantly from one another ($p < 0.0001$), and that of *H. hirta* [-19.7(±0.81)] and *M. disticha* [-34.5(±1.26)] on the other stand also differed significantly from each other ($p < 0.0001$). One way analysis of variance (ANOVA) thereof (1975 burn) produced the following (F-ratio = 13.893; degrees of freedom = 29, $p < 0.0001$), and for *H. hirta* versus *M. disticha*, (F-ratio = 97.605; degrees of freedom = 23, $p << 0.0001$). Multiple range analysis tests revealed a significant difference between xylem water potentials of *H. hirta* from that of *M. disticha* and *T. triandra* but not between *M. disticha* and *T. triandra*.

Normal annual flowering for *H. hirta* ($C_4$), and *T. Triandra* ($C_4$) (to a lesser degree relative to that of *H. hirta*) was observed in the 1975 burn. There was insignificant or no flowering for *M. disticha* ($C_3$) in the same aged veld.

## DISCUSSION

Flowering was more abundant in a 1 year burn than in 2 years burn for all species. An increase in flowering frequency following fire has been noted on, the Florida Lake Wales Ridge for a number of grass species (*Aristida stricta*, *Panicum abscissum* and *Andropogon* spp.). Late spring and summer fire stimulated a vigorous flowering response, whereas winter fires encouraged only a vegetative response (Abrahamson, 1984). Frequent burning has been reported to stimulate flowering (Daubenmire, 1968; Rowley, 1970; Vogel, 1973; Dickingson and Dodd, 1976; Christensen, 1981; Gill, 1981; Gunderson et al., 1983; Whelan, 1986). Within species, both timing of flowering (Curtis and Partch, 1950; Gill and Ingwerson, 1976; Abrahamson, 1984), and number of flowering stems produced (Burton, 1944; Stone, 1951; Kucera and Ehrenreich, 1962) change following the fire. Though there are reports of a number of studies on flowering phenologies in fire-dominated habitats (Parrish and Bazzaz, 1979; Anderson and Schelfhout, 1980; Tepedino and Stanton, 1980; Rabinowitz et al., 1981), the significance of fire upon flowering of coexisting species is still not yet fully established.

The fact that all the three grass species on the 1988 burn produced out-of-season flowering tillers, regardless of whether neighbored or non-neighbored, suggests a possible post-fire improvement in soil nutrients (Christensen and Muller, 1975).

Growth improvement and out-of-season leaf greening observed on 1 year burn, compared to 2 and 15 years after fire, signifies the importance of fire on, these aspects. Many $C_3$ and $C_4$ plants have been reported to show both an out-of-season growth and leaf greening as a consequence of fire (Perce and Cowling, 1984). Some studies (White, 1983; Bowman et al., 1988) have shown that grass cover is dependent upon frequent burning. Seasonal variation of fire effects (Gill, 1981; Henderson et al., 1983; Lovel et al., 1983; Snyder, 1986, in Platt et al., 1988) has been reported. Foliage increase on burned plots and retrogression on unburned plots was witnessed. The current study has demonstrated the role of fire as "space creator", and as a stimulant of plants' production and reproduction on Signal Hill.

Elimination of *T. triandra* from pasture has been linked to the absence of fire, and repeated defoliations (Downing, 1974); thus, suggesting periodic burning for maintaining *T. triandra* vigor. On Signal Hill, unburned tussocks of *T. triandra* have become moribund. Downing (1974) found a similar phenomena, and suggested such tussocks would die eventually if fire could be withheld for long enough. Moribund T. *triandra* tussocks on Signal Hill will eventually die out if no effective management (for example, periodic burning) could be exercised. Other grass communities of Signal Hill and elsewhere might be similarly affected.

*H. hirta* on unaltered plots on a 1975 burn continued to grow and flower, whilst their *M. disticha* and *T. Triandra* counterparts did not. This might be linked to *H. hirta* having shown highest xylem water potentials relative to the other two species. However, xylem water potentials in this instance were measured once on each of the two plots. Neither sizes nor distances of neighbors of the measured individual species were considered. The number of neighboring plants per pressure bombed individual was also not recorded. This might have affected the xylem water potentials of the monitored plants.

The fact that the two $C_4$ grass species still flowered in a 1975 burn, whilst their *M. disticha* counterparts did not, appears to suggest a "generalist" strategy by them ($C_4$) of coping with both warm and cool temperature.

Whether February was the appropriate time for burning the three grass species is in doubt because numerous herbs and shrubs proliferated, and a year after fire most of the spaces formerly occupied by grasses were dominated by herbs and shrubs. In this case (Western Cape), the suitable time for burning could be winter because it is rainy and hence wet, thus, fire would not easily exceed predetermined boundaries. Seeds and re-sprouters will have enough moisture for germination and establishment. Flowering could also be encouraged under sufficient soil moisture content. However, it could not be verified which season is best for veld burning.

In conclusion, fire was found to enhance growth, production and reproduction of both $C_3$ and $C_4$ plants. The $C_3$ species differ, however, in being completely dependent on fire for flowering and seed production. In other words, $C_3$ species does not flower at all in old unburnt veld, whereas $C_4$ species do. It would, therefore, be interesting to determine whether $C_4$ species successfully produce seed in older veld and whether this seed is capable of establishment. The further investigation on the importance of low versus high fire intensity will throw more light of the impact of fire on flowering plants.

## REFERENCES


Abrahamson WG (1984). Species responses to fire on the Florida. Lake Wales Ridge. Am. J. Bot. 71(1):35-43.

Acocks JPH (1966). Non-selective grazing as a means of veld reclamation. Proc. Grassld. Soc. S. Afr. 1:33-39.

Anderson RC, Schelhout S (1980). Phenological pattern among tallgrass prairie plants and their implications for pollinators' competition. Am. Mid. Nat. 104:253-263.

Bowman MJS, Wilson BA, Hooper RJ (1988). Responses of Eucalyptus forest and woodland to four regimes at Munmarlary, northern territory, Australia. J. Ecol. 76:215-232.

Burton GW (1944). Seed production of several southern grasses as influenced by burning and fertilization. Am. Soc. Agro. J. 36:523-529.

Christensen NL (1981). Fire regimes in southeastern ecosystems. In: Mooney H.A., Bonnickson T.M., Christensen N.L., Lotan J.E., Reiners W.A. (eds). Fire regimes and ecosystems properties. USDA Forest Serv. Gen Tech. Rep Wo-26:112-136.

Christensen NL, Muller CH (1975). Effects of fire on factors controlling plant growth in *Adenostoma*chaparral. Ecol. Mono. 45:29-55.

Curtis JT, Partch ML (1950). Some factors affecting flower productionin *Andropogonqerardii*. Ecology 31:488-489.

Daubenmire R (1968). Ecology of fire in grasslands. Adv. Ecol. Res. 5:209-266.

Dickingson CE, Dodd JL (1976). Phenological pattern in the shortgrass prairie. Am. Mid. Nat. 96:367-378.

Downing BH (1974). Reactions of grass communities to grazing and fire in the sub-humid lowlands of Zululand. Proc. Grassland Soc. S. Afr. 9:33-37.

Friedel MH, Blackmore AC (1988). The development of veld assessment in the Northern Transvaal Savanna. I. Red Turfveld. J. Grassland. Soc. S. Afr. 5(1):20-37.

Gill AM (1981) Adaptive responses of Australian vascular plant species to fires. In: Gill A.M., Groves R.H., Noble I.R. (eds) Fire and the Australian Biota. Australian Academy of Science, Canberra. pp. 243-271.

Gill AM, Igwerson F (1976). Growth of *Xanthorrheaaustralis* R. Br. in relation to fire. J. Appl. Ecol. 13:195-203.

Gunderson L, Taylor D, Craig J (1983). Fire effects on flowering and fruiting patterns of understorey plants in pinelands of Everglades National Park, National Park Service, South Florida Research Center, Report SFRC-83/04. p. 36.

Henderson RA, Lovell DL, Howell EA (1983). The flowering responses of 7 grasses to seasonal timing of prescribed burns in remnant Wisconsin prairie. In: Brewer R. (ed) Proc. 8th N. Am. Prairie Conf. W. Michigan University. pp. 7-10.

Kucera CL, Ehrenreich JH (1962). Some effects of annual, burning oncentral Missouri prairie. Ecol. 43:334-336.

Lovell DL, Henderson RA, Howell EA (1983). The response of forb species to seasonal timing of prescribed burns in remnant Wisconsin prairie. In: Brewer R. (Ed) Proc. 8th North American Prairie Conf., w. Michigan University. pp. 11-15.

Opperman DPJ, Roberts BR (1978). Die fenologieseontwikkeling van *Themedatriandra*, *Elyorunusargenteus* and *Heteropoqoncontortus*onderveldtoestande in die sentralleOranje-Vrystaat. Proc. Grassland. Soc. S. Afr.13:135-140.

Parrish JAD, Bazzaz FA (1979). Difference in pollination niche relationships in early and late successional plant community. Ecology 60:597-610.

Perce SM, Cowling RM (1984). Phenology of fynbos, renosterveld and subtropical thicket in the South Eastern Cape. S. Afr. J. Bot. 3:1-16.

Platt WJ, Evans GW, Davis MM (1988). Effects of fire season on flowering herbs of forbs and shrubs in longleaf pine forests. Oecologia 76:353-363.

Rabinowitz D, Rapp JK, Sork VL, Rathke BJ, Reese GA,Weaver JC (1981). Phenological properties of wind- and insect- pollinated prairie plants. Ecology 62:49-56.

Raison RJ (1979). Modification of the soil environment by vegetation fires, with particular reference to nitrogen transformations: A review. Plant Soil 51:73-108.

Rethman NFG, Booysen PDev (1986). The influence of time of defoliation on the vigour of a Tall Grassveld sward in the next season. Proc. Grassland. Soc. S. Afr. 3:91-94.

Rowley J (1970). Effects of burning and clipping on temperature, growth and flowering of narrow-leaved snow tussock. N.Z. J. Bot. 8:264-282.

Scott JD (1951). A contribution to the study of the problems of the Drakensberg Conservation Area. South African Department of Agricultural Science Bulletin. p. 324.

Scott JD (1970). Pros and Cons of eliminating veld burning. Proc. Grassland. Soc. S. Afr. 5:23- 26.

Scott JD (1951). A contribution to the study of the problems of the Drakensberg Conservation Area. South African Department of Agricultural Science Bulletin. p. 324.

Snyder JR (1986). The impact of wet season and dry season prescribed fires on Miami Rock Ridge Pineland, Everglades National Park, South Florida Research Center Report SFRC-86/06. p. 106.

Smith M, Nelson BW (2011). Fire favors expansion of bamboo-dominated forests in the south–west Amazon. J. Trop. Ecol. 27:59-64.

Stock WD, Lewis OAM (1986). Soil nitrogen and the role of fire as a mineralizing agent in a South African fynbos ecosystem. J. Ecol. 74:317-328.

Stone EC (1951). The stimulative effect of fire on the floweringof the golden Brodiola (*BrodioeaixiodesWats.* var. *luqensJeps.*) Ecology 32:534-537.

Tainton NM, Booysen P De v, Scott JD (1970). Response of tall grassveld to different intensities, seasons and frequencies of clipping. Proc. Grassland Soc. S. Afr. 5:32-41.

Tepedino VJ, Stanton NL (1980). Spatiotemporal variation in phenology and abundance in resources on shortgrass prairie. Great Basin Nat. 40:197-215.

Tolsma AD, Tolhurst KG, Read SM (2010). Effects of fire, post-fire defoliation, drought and season on regrowth and carbohydrate reserves of alpine snow grass *Poafawcettiae* (Poaceae). Austr. J. Bot. 58(3):157-168.

Trollope WSW (1974). Role of Fire in Preventing Bush Encroachment in the Eastern Cape. Proc. Grassland Soc. South Afr. 9:67-72.

Van Wilgen BW, LeMaitre DC (1981). Preliminary estimates of nutrient levels in fynbos vegetation and the role of fire in nutrient cycling. S. Afr. For. J. 119:24-28.

Vogel RJ (1973). Fire in the southeastern grasslands. Proc. Ann. Tall Timbers Forest Ecol. Conf. 12:175-198.

Whelan RJ (1986). Seed dispersal in relation to fire. In: Murray D.R. (ed} Seed dispersal. Academic Press, San Diego, Cal. pp. 237-271.

White AS (1983). The effect of thirteen years of annual prescribed burning on a *Quercus ellipsoidalis*community in Minnesota. Ecology 64(5):1081-1085.

Yeaton RI, Frost S, Frost PGH (1988). The structure of a grass community in *Burkeaafricana* savanna during recovery from fire. S. Afr. J. Bot. 54(4):367-371.

# 7

# Effect of indolebutyric acid (IBA) and naphthaleneacetic acid (NAA) plant growth regulaters on Mari gold (*Tagetes erecta* L.)

**Zia Ullah[1], Sayed Jaffar Abbas[1,2], Nisar Naeem[1], Ghosia Lutfullah[2], Taimur Malik[1], Malik Atiq Ullah Khan[1] and Imran Khan[1]**

[1]Biotechnology Center, Agricultural Research Institute, Tarnab, Peshawar, Pakistan.
[2]Center of Biotechnology and Microbiology, University of Peshawar, Pakistan.

**This experiment was conducted for the optimization of auxin [indolebutyric acid (IBA) and naphthaleneacetic acid (NAA)] required for the regeneration of Mari gold. Significant differences were observed for bloom flowers. The maximum branches per plant (6.8) were observed at 400 ppm. Maximum flower size (10.7) was exhibited at 100 ppm IBA concentration and minimum flower size (2.8) at 300 ppm. The 200 and 400 ppm IBA concentration showed maximum leaf size (4.0). The maximum leaves per plant (44.0) were observed at 100 ppm IBA concentration while 200 ppm showed minimum leaves per plant (6.0). At 100 ppm IBA concentration maximum plant height (9.40 cm) was observed and minimum plant height (5.92 cm) was recorded at 400. 200 and 300 ppm concentration of IBA have no significant effect on roots per plants (37.0) while 400 ppm has maximum effect on roots per plant (84.4). Root size increased (6.50) at 100 ppm IBA concentration, while increasing the IBA concentration decreases the root size (4.10) respectively. Maximum value (7.2) was observed at 100 ppm IBA for non-bloom flower and minimum value was (3.2) at 300 ppm. Maximum branches per plant (5.0) were recorded at 400 ppm NAA concentration followed by 300 ppm (4.0) while minimum value (3.2) was observed at 200 ppm. The three concentration of NAA 100, 300 and 400 ppm are significantly different from control for flower size and there is no difference between control and 200 ppm concentration of NAA. By dipping the seedling of Mari gold in NAA maximum leaf size (3.20) was recorded at 200 ppm concentration, while minimum leaf size was observed at 400 ppm. Increased in leaves per plant were observed with increase in NAA concentration. Maximum plant height (7.80) was recorded at 100 and 200 ppm NAA concentration while minimum value was (5.0) at 300 ppm concentration. By dipping the seedling in higher concentration of NAA showed the maximum increase in roots per plant (123.2) and root size (6.8). 300 ppm showed maximum value for non-bloom flower and 200 ppm showed the minimum value (3.2).**

**Key words:** Gold, indolebutyric acid (IBA), naphthaleneacetic acid (NAA, auxins.

## INTRODUCTION

Marigold (*Tagetes erecta* L.) belongs to family Asteraceae and grown as an ornamental crop for loose flowers, as a landscape plant, and as a source of pigment for poultry feed. Control of flowering is one of the most important practical aspects in application of plant growth regulators. There are many examples of utilization of plant growth hormones to regulate the flowering in aromatic plant (Shukla and Farooqi, 1990), for example, application of Ethrel (2-Chloroethyl phosphonic acid), Naphthalene acetic acid (NAA) and Kinetin improved flowering in Jasminum. Kaplan (1960) established that Marigold is a native of central and South America, especially Mexico. From Mexico it spread to different parts of the world during the early part of the 16th century. Both leaves and flowers are equally important from the medicinal point of view. Leaf paste is used externally against boils and carbuncles. Marigold is adaptable to different types of soil conditions and thus can be grown successfully in a wide variety of soils. However, a deep, fertile, friable and well drained soil (pH 7.00 to 7.5) having good water holding capacity, is the most desirable. Loose flowers are sold in the market which is mainly used in making garlands. The flower is also used as cut flowers arrangement. Furthermore, Marigolds are grown for beautification as landscape plants due to its variable height and various colors of flowers. It is highly suitable as a bedding plant, in an herbaceous border and is also ideal for newly planted shrubbery to provide color and fill spaces. French Marigold is ideal for rockeries, edging, hanging baskets and window boxes. Auxin is well known to stimulate the rooting of cuttings (Hartmann et al., 2002). However, it has been found in various studies that rooting percentage and rooting time was poor; of the two auxins tried for root regeneration IBA was more responsive than NAA The most widely used auxin for commercial rooting is IBA (Nickel, 1990). Two synthetic materials, indole-3-butyric acid (IBA) and naphthalene acetic acid (NAA), were even more effective than the naturally occurring or synthetic IAA for rooting. Today, IBA and NAA are still the most widely used auxins for rooting stem cuttings and for rooting tissue-culture-produced micro cuttings Zimmerman and Wilcoxon (1935). It has been repeatedly confirmed that auxin is required for initiation of adventitious roots on stems and indeed, it has been shown that divisions of the first root initial cells are dependent upon either applied or endogenous auxins (Gaspar et al., 1988; Stromquist and Hansen, 1980). IBA has been found to occur naturally. The formation of root primordium cells depends on the endogenous auxins in the cutting and on a synergic compound such as a diphenol. These substances lead to the synthesis of ribonucleic acid (RNA), which act upon root primordium initiation (Hartmann et al., 2002). The application of some plant growth retardants, together with auxin, has been used to improve the rooting capacity of cuttings in some species (Davis and Haissig, 1990; Pan and Zhao, 1994). Plant growth regulators have gained wide acceptance for optimizing the yield of plants by modifying growth, development and stress behaviour (Shukla and Farooqi, 1990). Synthetic plant growth regulators, such as auxins, cytokinins and various growth retardants when applied exogenously to the plant, influence various aspects of plant development and biosynthesis of its important components (Shukla and Farooqi, 1990; Kewalanand and Pandy, 1998). Marigold requires a mild climate for luxuriant growth and flowering. Marigold seedlings are easily transplanted and established in field without much mortality. At the time of transplanting, they should be stocky and bear three to four true leaves.

In the present study, the seedlings of Mari gold were treated by plant growth regulators IBA and NAA. The aim of this study was to test the potential effect of plant growth regulators on the Marigold plants and to select its optimum concentration.

#### MATERIALS AND METHODS

In this study a commercial cultivar *T. erecta* of Marigold was used as plant material and 2 different auxins Alpha-naphtalene acetic acid (NAA), and 3-indole butyric acid (IBA) were used as plant growth regulators. Four different concentrations e.g. 100, 200, 300 and 400 ppm were used as treatments of each plant growth regulators. The stem cuttings of the Marigold plants were treated with each concentration of each plant growth regulators for 1 min and then simultaneously transferred in the pots. These treatments were compared with the control which did not apply any growth regulator. The experiment was conducted in Completely Randomized Design (CRD) and each treatment was replicated 5 times. The recommended agronomic practices were applied equally to all the plants in the pots.

## RESULTS AND DISCUSSION

### Effect of different concentration of IBA on Marigold cuttings

Through statistical analysis it was observed that significant differences occurred between treatments containing IBA and control, while non-significant differences was observed among three different higher concentration of IBA (200, 300 and 400 ppm) (Table 1) used for bloom flower by dipping the root of Marigold cuttings. The longest flowering period was at 400 ppm although it had no positive effect on flower yield. In fact, late pruning time (near to flowering time) caused longer flowering period that was possibly due to late flower bud information; these results was supported by Saffari et al., (2004) and Mesen (1993) who observed inhibition in shooting with increased concentration of IBA in other species. It was observed that reduction in time of flowering not only affected seed yield but also the color of flowers. The maximum branches per plant (6.8) was observed by the treated seedling with 400 ppm IBA concentration while minimum branches per plant (3.6) was recorded by 300 ppm IBA (Table 1). These results showed that an adequate time is necessary in which branches can grow longer enough having more flower buds resulting in higher yield. Similar result was also reported by Paul et al. (1995). Maximum flower size (10.7) was exhibited at 100 ppm IBA concentration and minimum flower size (2.8) at 300 ppm positive effect on flowering (Table 1). At 100 ppm the flowers were normal and had abundant pollen with increased seed set. Other concentrations and treatment times were not so helpful and had either reduced number of flowers or aborted pollen with no or reduced seed set; these results are

**Table 1.** Effect of different concentration of IBA on bloom flower, branches plant$^{-1}$, flower size leaf size and leaves plant$^{-1}$.

| Treatment | Bloom flower | Branches/plant | Flower size | Leaf size | Leaves/plant |
|---|---|---|---|---|---|
| Control | $0.4^{b}$ | $3.2^{b}$ | $2.4^{b}$ | $2.8^{ab}$ | $27.4^{b}$ |
| 100 ppm | $1.4^{a}$ | $5.0^{ab}$ | $10.7^{a}$ | $3.6^{ab}$ | $44.0^{a}$ |
| 200 ppm | $1.6^{a}$ | $4.2^{b}$ | $3.8^{b}$ | $4.0^{a}$ | $6.0^{c}$ |
| 300 ppm | $1.6^{a}$ | $3.6^{b}$ | $2.8^{b}$ | $2.6^{b}$ | $9.2^{c}$ |
| 400 ppm | $1.6^{a}$ | $6.8^{a}$ | $5.2^{b}$ | $4.0^{a}$ | $30.2^{b}$ |
| $LSD_{0.05}$ | 0.72 | 2.11 | 5.35 | 1.30 | 8.58 |

*Means followed by a common letter in the respective column do not differ by $LSD_{0.05}$.

**Table 2.** Effect of different concentration of IBA on plant height, root plant$^{-1}$, root size and non-bloom flower.

| Treatment | Plant height | Roots/plant | Root size | Non-bloom flower |
|---|---|---|---|---|
| Control | $5.92^{b}$ | $20^{c}$ | $3.60^{d}$ | $4.80^{b}$ |
| 100 ppm | $9.40^{a}$ | $56.8^{b}$ | $6.50^{a}$ | $7.20^{a}$ |
| 200 ppm | $7.80^{ab}$ | $37.4^{bc}$ | $4.10^{cd}$ | $4.20^{b}$ |
| 300 ppm | $7.00^{ab}$ | $37.0^{bc}$ | $5.20^{bc}$ | $3.20^{b}$ |
| 400 ppm | $5.92^{a}$ | $82.4^{a}$ | $6.0^{ab}$ | $4.60^{b}$ |
| $LSD_{0.05}$ | 2.72 | 22.67 | 1.13 | 2.13 |

*Means followed by a common letter in the respective column do not differ by $LSD_{0.05}$.

supported by Ozel et al. (2006). The 200 and 400 ppm IBA concentration showed maximum leaf size (4.0), while decreasing in IBA concentration the leaf size was also decreased (Table 1). The maximum leaves per plant (4.0) were observed by treating the seedling with 100 ppm IBA concentration while 200 ppm showed minimum leaves per plant (6.0). The effect of different IBA concentrations, leaf size and propagation media on rooting ability of leafy stem cuttings of *Milicea excelsa* was investigated by Ofori (1996). At 100 ppm IBA concentration, maximum plant height (9.40 cm) was observed and minimum plant height (5.92 cm) was recorded at 400 ppm. 200 and 300 ppm concentration of IBA have no significant effect on roots per plants (37.0) while 400 ppm has maximum effect on roots per plant (84.4) (Table 2). Root size increased (6.50) at 100 ppm IBA concentration while increasing the IBA concentration decreases the root size (4.10) respectively. Our results are in agreement with Felker and Clarke (1981), Klass et al. (1987), Ofori et al. (1996), Tchoundjeu and Leaky (1996), Mesen et al. (1997) and Berhe and Negash (1998) who reported differences in rooting frequency depending on the exogenous auxin or combination of auxins used, with IBA often giving the best results. However, Majeed et al. (2009) recorded the highest rooting rate (50%) for *Aesculus indica* cuttings treated with increasing the IBA concentration. Baul et al. (2008) also observed a similar trend in the vegetative propagation of *Stereospermum suaveolens* with cuttings treated with 0.2% IBA producing the longest root. Maximum value (7.2) was observed at 100 ppm IBA for non-bloom flower and minimum value was (3.2) at 300 ppm (Table 2).

### Effect of different concentration of NAA on Marigold cuttings

Through mean comparison there are significant differences for bloom flower among treatments containing NAA and control, while non-significant differences among the four treatments of NAA (100, 200, 300, 400 ppm) (Table 3). Maximum branches per plant (5.0) were recorded at 400 ppm NAA concentration followed by 300 ppm (4.0) while minimum value (3.2) was observed at 200 ppm. The three concentration of NAA 100, 300 and 400 ppm are significantly different from control for flower size and there is no difference between control and 200 ppm concentration of NAA (Table 3). NAA caused higher amount of flowers in plants rather than other treatments and also control (Table 4). Farooqi et al. (1993) reported the same result for Kinetin application on Damask rose in India. By dipping the seedling of Marigold in NAA maximum leaf size (3.20) was recorded at 200 ppm concentration, while minimum leaf size was observed at 400 ppm (Table 3). Increase in leaves per plant was observed with increase in NAA concentration. Waseem et al. (2007), who also found that the lowest concentration of NAA (0.5 mg L$^{-1}$), when used alone, showed its superiority over all the other concentration of NAA by producing the maximum number of shoots per explants, leaves and

**Table 3.** Effect of different concentration of NAA on bloom flower, branches plant$^{-1}$, flower size leaf size and leaves plant$^{-1}$.

| Treatment | Bloom flower | Branches/plant | Flower size | Leaf size | Leaves/plant |
|---|---|---|---|---|---|
| Control | $0.4^{b}$ | $3.2^{ab}$ | $2.4^{b}$ | $2.84^{ab}$ | $2.74^{a}$ |
| 100 ppm | $2.0^{a}$ | $3.2^{ab}$ | $6.6^{a}$ | $2.60^{ab}$ | $10.6^{b}$ |
| 200 ppm | $1.6^{a}$ | $2.2^{b}$ | $4.9^{ab}$ | $3.20^{a}$ | $9.00^{b}$ |
| 300 ppm | $2.0^{a}$ | $4.0^{a}$ | $6.5^{a}$ | $2.90^{ab}$ | $11.4^{b}$ |
| 400 ppm | $2.0^{a}$ | $5.0^{a}$ | $5.7^{a}$ | $2.30^{b}$ | $14.6^{b}$ |
| $LSD_{0.05}$ | 0.95 | 1.84 | 3.05 | 0.80 | 7.46 |

*Means followed by a common letter in the respective column do not differ by $LSD_{0.05}$.

**Table 4.** Effect of different concentration of NAA on plant height, root plant$^{-1}$, root size and non-bloom flower.

| Treatment | Plant height | Roots/plant | Root size | Non-bloom flower |
|---|---|---|---|---|
| Control | $5.92^{ab}$ | $20.0^{d}$ | $3.6^{b}$ | $4.8^{abc}$ |
| 100 ppm | $7.80^{a}$ | $60.2^{c}$ | $5.2^{ab}$ | $3.8^{bc}$ |
| 200 ppm | $7.80^{a}$ | $85.0^{bc}$ | $6.4^{a}$ | $3.2^{c}$ |
| 300 ppm | $5.00^{b}$ | $89.6^{b}$ | $6.4^{a}$ | $6.2^{a}$ |
| 400 ppm | $7.40^{a}$ | $123.2^{a}$ | $6.8^{a}$ | $5.6^{ab}$ |
| $LSD_{0.05}$ | 2.06 | 26.70 | 1.60 | 2.13 |

*Means followed by a common letter in the respective column do not differ by $LSD_{0.05}$.

nodes per shoot. Ali et al. (2005) also reported in Chrysanthemum that an increase of NAA in MS medium resulted in decreasing the multiplication rate. Maximum plant height (7.80) was recorded at 100 and 200 ppm NAA concentration while minimum value was (5.0) at 300 ppm concentration (Table 4). By dipping the seedling in higher concentration of NAA showed the maximum increase in roots per plant (123.2) and root size (6.8) (Table 4). The effect of auxin in promoting rooting of cuttings is well known ((Nanda, 1970; Hartmann et al.1997; Husen and Mishra 2001; Husen and Pal 2003;2006;2007), while very little information is available on the effectiveness of auxin in relation to the branch position, 300 ppm showed maximum value for non-bloom flower and 200 ppm showed the minimum value (3.2) (Table 4).

## REFRENCES


Ali N, Kafi, M, Mirmasoumi M. Babalar M (2005). Micropropagation of Damask Rose (Rosa damascene Mill) cvs. Azaran and Ghamsar. Int. J. Agric. Biol. 7:535–538.

Berhe D, Negash L (1998). Asexual propagation of Juniperus procera from Ethiopia: a contribution to the conservation of African pencil Cedar. For. Ecol. Manage. 4454:179–90.

Baul TK, Mezbahuddin M, Mohiuddin M (2008). Vegetative propagation and initial growth performance of *Stereospermum suaveolens* DC, a wild tropical tree species of medicinal value. New For. 37(3):375-283.

Davis TD, Haissig BE (1990). Chenical control of adventitious root formation in cuttings. Bull. Plant Growth Regul. Soc. Am. 18:1-17.

Farooqi AH, Sharma S, Naqvi AA, Khan A (1993) The effect of Kinetin on flower and oil production in *Rosa damascena.* J. Essent. Oil Res. 5:305–309.

Felker P, Clarke PR (1981). Rooting of mesquite (Prosopis) cuttings. J. Range Manage. 34:466–468.

Gaspar T, Hofinger M (1988). Auxin metabolism during adventitious rooting. In: Davis, T.D.; Haissig, B.E. and Sankhla, N. (Eds.). Adventitious root formation in cuttings. Portland: Dioscoride Press. pp. 61-69.

Hartmann HT, Kester DE, Davis JFT, Geneve RL (2002) Plant propagation: principles and practices. New Jersey: Prentice Hall.

Husen A, Pal M (2003). Clonal propagation of teak (*Tectona grandis* Linn. f.): effect of IBA application and adventitious root regeneration on vertically split cuttings. Silvae Gene. 52(3–4):173-176.

Husen A, Pal M (2006). Variation in shoot anatomy and rooting behaviour of stem cuttings in relation to age of donor plants in teak (*Tectona grandis* Linn. f.). New For. 31(1):57–73.

Husen A, Pal M (2007). Metabolic changes during adventitious root primordium development in *Tectona grandis* Linn. f. cuttings as affected by age of donor plants and auxin (IBA and NAA) treatments. New For. 33(3):309–323.

Kewalanand JN, Pandy CS (1998). Effect of plant growth regulators on the growth, herbage and oil yield of Japanese mint (*Mentha arvensis*) and it's economic there from. J. Med. Arom. Plant Sci. 20:725–730.

Klass S, Wright J, Felker P (1987). Influence of auxins, thiamine and fungal drenches on the rooting of *Prosopis alba* clone B2 V50 cuttings. J. Hort. Sci. 62:97–100.

Majeed M, Khan MA, Mughal AH (2009). Vegetative propagation of *Aesculus indica* through stem cuttings treated with plant growth regulators. J. Fores. Res. 20(2):171-173.

Mesen JF (1993). Vegetative propagation of Central American hardwoods. Ph.D Thesis, University of Edinburgh, Scotland.

Mesen JF, Newton AC, Leakey RRB (1997). Vegetative propagation of Cordia alliodora (Ruiz & Pavon) Oken: The effect of IBA concentration, propagation medium and cutting origin. For. Ecol. Manage. 92:45–54.

Nickel LG (1990). Plant growth regulators. Agricultural uses. Springer, New York. P. 4-5.

Ofori DA, Newton AC, Leakey RRB, Grace J (1996). Vegetative propagation of *Milicea excelsa* by leaf stem cuttings: effects of

auxin concentration, leaf size androoting medium. Forest Ecol. Manage.
84:39-48.

Ozel CA, Khawar KM, Mirici IS, Arslan O, Ozcan S (2006). Induction of *Ex Vitro* Adventitious Roots on Soft Wood Cuttings of Centaurea tchihatcheffii tchihatcheffii Fisch et. Mey usingIndole 3-Butyric Acid and α-Naphthalene Acetic Acid. Int. J. Agric. Biol. 8:1.

Pan R, Zhao Z (1994). Synergistic effects of plant growth retardants and IBA on the formation of adventitious roots in hypocotyl cuttings of mungbean. Plant Growth. Regul. 14:15-19.

Paul TM, Siddique MAA, John AQ (1995). Effect of severity and time of pruning on growth and flower production of *Rosa damascena* Mill. An important aromatic plant. Adv. Plant Sci. 8:28–32.

Saffari VR, Khalighi A, Lesani H, Babalar M, Obermaier JF (2004). Effects of Different Plant Growth Regulators and Time of Pruning on Yield Components of Rosa damascena Mill. Int. J. Agric. Biol. 6:6.

Shukla A, Farooqi AH (1990) Utilization of plant growth regulators in aromatic plant production. Curr. Res. Med. Arom. Plants 12:152–7.

Stromquist, LH, J. Hansen J (1980) Effects of auxin of irradiance on the rooting of cuttings of *Pinus sylvestris*. Physiol. Plant 49: 346-350.

Tchoundjeu Z, Leakey RRB (1996) Vegetative propagation of African mahogany: Effect of auxin, node position, leaf area and cutting length. New For. 11:125–36.

Waseem K, Khan MQ, Jaskani J, Khan MS (2007) Impact of different auxins on the regeneration of Chrysanthemum (*Dendranthema morifolium* L.) through shoot tip culture. Pak. J. Agric. Res. 20:51–57.

Zimmerman PW, Wilcoxon F (1935). Several chemical growth substanceswhich cause initiation of roots and other responses in plants. Contrib. Boyce. Thompson. Inst. 7:209-228.

# 8

# Prospects of clonal selection for enhancing productivity in Saffron (*Crocus sativus* L.)

**Gowhar Ali, Asif M. Iqbal, F. A. Nehvi, Sheikh Sameer Samad, Shaheena Nagoo, Sabeena Naseer and Niyaz A. Dar**

Saffron Research Station, Pampore Sher-e-Kashmir University of Agricultural Sciences and Technology of Kashmir, Shalimar, 191121, Jammu & Kashmir, India.

**Saffron a member of Iridaceae family is a perennial spice species. It is derived from the stigma of the flower of the saffron crocus (*Crocus sativus* L.), which is collected and dried to produce the spice. The main compounds that accumulated throughout stigma development in *C. sativus* L. were crocetin, its glucoside derivatives and picrocrocin, all of which increased as stigmas reached a fully developed stage. The volatile composition of *C. sativus* stigmas changed notably as stigmas developed with each developmental stage being characterized by a different volatile combination. In red stigmas, b-cyclocitral, the 7,8 cleavage product of b-carotene, was highly produced, suggesting the implication of both b-carotene and zeaxanthin in crocetin formation breeding of saffron (*C. sativus* L.), its position, urgency and topicality of the problem are considered in the present review. Clonal selection is proposed for genetic improvement of saffron in order to increase yield production and quality.**

**Key words:** Saffron, clonal selection, variability.

## INTRODUCTION

Saffron (*Crocus sativus* L.) belongs to the family Irridaceae and genus *Crocus*, of which about 80 spp. are so far known. Its cultivation in the world extends through 0 to 90°E longitude (Spain to Kashmir) and 30 to 45°N latitude (Persia to England). It is well known that saffron (*C. sativus* L.) is a very valuable irreplaceable spice with exceptional medicinal properties which has been cultivated since ancient times. Ancient history of saffron cultivation goes back to many thousands years ago. It was originated very likely in Greece and then distributed in the other Mediterranean and Near East countries (Turkey, Italy, Azerbaijan, Iran, Iraq, North of India, etc.). There are no wild forms and it exists solely in culture. Being triploid with chromosome number $2n=3x=24$ and basic number of $x=8$, saffron never bears seed and it is propagated exclusively in a vegetative way by corms. Its vegetation period starts in autumn, continues through winter and finishes in the middle of spring. Therefore its growth and development is very slow, facing the cold period of the year which results in low productivity of the plant. All over the world saffron is known as one cultivar, as descent of certain triploid sterile plant arisen once spontaneously in nature which was caught by sight of man and involved into cultivation (Mathew, 1977). Saffron, the dried red stigmas of *C. sativus*, has been used as flavouring and colouring agent since then and is currently considered the world's most expensive spice. Saffron is made up of a complex mixture of volatile and non-volatile compounds that contribute to its overall aroma and flavor (Moraga et al., 2009)

### Propagation

Cytological, DNA, and reproductive studies on the allied species of *C. sativus* such as *C. cartwrightianus*, *C.*

*Corresponding author. E-mail: asifquresh@gmail.com.

*thomasii*, *C. hadriaticus*, indicate a more likely parent of saffron may be *C. Cartwrightianus*or and *C. thomasii*. Both species are diploid with a karyotype similar to saffron. In addition, their pollen can fertilize the egg cell of saffron, giving rise to seeds which are viable, germinate and form new corms. Thus saffron can originate through fertilization of a normal reduced egg cell with an unreduced male gamete of the same *Crocus* species or by crossing between an egg cell and the male unreduced gamete of another species. The saffron is a sterile geophyte that produces annual renewal corms and is propagated only by them (Mathew, 1982). The physical properties of saffron corms are prerequisite to designing and developing harvesting, handling, sizing and sowing equipments of corms (Hassan-Beygy et al., 2010).

Cytological studies indicate that saffron is a sterile triploid (2n=3x=24) plant. The origin of saffron by allopolyploidy seems more probable considering the recent data on its karyotype and molecular biology. Many authors have hypothesized the origin of saffron by autotriploid (Mathew, 1977). On the basis of overall length and centromeric position, the somatic chromosomes could be assembled in seven triplets, one pair and one single chromosome. The karyotype consists of a series of chromosome triplets with the exception of one group of three chromosomes of which one is different from the others. Because of this unique chromosome in saffron genotype, selfing can be excluded (Chichiricco, 1984). Since *C. sativus* is sterile, the presence of this chromosome may be explained by the existence of chromosomal polymorphism at diploid level in the progenitors, most probably *C. Thomasii* or *C. cartwrightianus*.

Saffron has been propagated and still continues to be propagated vegetatively. There is a supposition that saffron as a clone can be scarcely changed genetically and its improvement is hardly possible through clonal selection (Dhar et al., 1988, Piqueras et al., 1999). Meanwhile, other suppositions exist as well. For example, Rzakuliyev (1959) investigating in Apsheron (Baku) specimens of saffron obtained from 6 regions (2 regions in Italy, France, Istanbul, Yalta and Mashtağa) during 3 vegetations in 1934 to 1937 concluded that it is possible to create a new high yielding cultivars of this plant on the basis of clonal selection. Kapinos (1965) studied the morphogenesis and cytoembryology of *C. sativus* under climatic conditions of Apsheron and came to the conclusion that this plant represents variegated blend of genetically heterogeneous forms – clones, and clonal selection on it would be very promising. Apparently, the lack of new cultivars of saffron at present may not be explained by the impossibilities of the improvement of this plant through clonal selection. To solve this problem, researchers need to try clonal selection for achieving improvement in saffron. Many researchers have addressed the problem of clonal selection of saffron, including a large group from Azerbaijan who have tackled this problem forthright (Rzakuliyev, 1959; Kapinos , 1965; Agayev 1993; Agayev, 1994b; Agayev et al. 2007). In 1981, Dhar et al. (1988) surveyed the natural populations of Kashmir saffron (*C. sativus* L.) recording the range of variation and tabulating perianth length, perianth width, length of the colored part of the stigma, flower number/stigma, fresh weight/stigma and dry weight/stigma. These authors together with Piqueras et al. (1999), however, assumed that saffron as a clone cannot be altered genetically to any major extent and that its improvement is hardly possible through clonal selection. In Kashmir, Munshi and Zargar (1991) identified an elite subpopulation developed from the progenies of corms selected from extensive saffron belts. Munshi (1992) derived information on coefficients of variation, heritability and genetic advance using data on ten yield components in 11 diverse saffron genotypes (mostly from Jammu and Kashmir with some exotic varieties) grown between 1986 and 1989. Significant genotypic differences were observed for all characteristics, except day to 100% flowering. The coefficients of variation were greatest for the number of daughter corms/mother corm and number of flower/space. The highest yield for dry saffron was obtained in 1987 by SKUAST (Sher-e-Kashmir University of Agricultural Sciences Technology of Kashmir) genotypes Nag c8708 and Bodi c8606 (3.57 and 3.30 kg/ha, respectively). GrilliCaiola et al. (2001) reported phenotypic differences in terms of flower size, tepal shape and color intensity. Pistil weight was found to show a wide range, revealing the possibility of saffron improvement through selection. Similar results of wide ranges of variability have also been reported by Zargar (2002) in Kashmir. The productivity, growth and quality attributes of ten saffron accessions of Birjand, Ghaen, Gonabad, Torbat-Haydarieh, Ferdows, Istahban, Kerman, Isfahan and Shahr-Kord were studied under natural environmental conditions at ShahrKord in Central Iran. Ehsanzadeh et al. (2004) concluded that the three latter accessions could be grown when satisfactory stigma yield is the goal. Picci (1987), however, found only small differences in yields from saffron grown in Abruzzi, Emilia and Sardinia. Indian experts, in accordance with the results of Iranian authors in similar studies, have confirmed the yield superiority (Table 1 and Figure 2) of ten genotypes (e.g. SMD-3, SMD-11, SMD-31, SMD-45, with 4.3, 4.2, 4.8, 7.6 kg/ha of dry pistil yield, respectively) (Nehvi et al., 2007a, b). A comparison of many hundreds and thousands of clones, each grown from one corm of the same weight, resulted in the identification of "superior" clones in terms of exceptionally large numbers of flowers and large (>10 g) corms. Based on the number of flowers and number of large corms, which are the two most economically important attributes of saffron, the clones were classified as extraordinary, superior, ordinary, inferior and declining clones. This background information illustrates that the

**Table 1.** Performance of elite saffron clones available with SKUAST-K (Gowhar et al., 2011).

| Genotype | Saffron yield (kg/ha) | No. of daughter corms | Average corm weight (g) | Crocin content (%) |
|---|---|---|---|---|
| SMD-3 | 6.30 | 6.26 | 5.37 | 15.49 |
| SMD-11 | 6.20 | 4.79 | 5.40 | 13.88 |
| SMD-31 | 6.8 | 6.32 | 5.66 | 14.81 |
| SMD-45 | 7.6 | 5.06 | 8.20 | 17.10 |
| SMD-52 | 5.4 | 4.55 | 7.63 | 13.89 |
| SMD-68 | 5.3 | 3.45 | 8.44 | 16.63 |
| SMD-79 | 5.4 | 2.91 | 8.49 | 16.91 |
| SMD-81 | 5.2 | 3.20 | 7.49 | 16.59 |
| SMD-211 | 6.06 | 4.12 | 5.18 | 15.57 |
| SMD-224 | 5.5 | 3.04 | 4.25 | 14.92 |

and importance of clonal selection with reference to improving the traits of cultivated saffron thus suggesting that in the existing plantations clonal selection of saffron is possible and promising.

Having an ancient history of cultivation, saffron apparently may contain a lot of genetically changed forms (clones) as the result of spontaneous mutations in somatic cells. The task is to find and study the individually, to separate the economically valuable forms and to bring them to the new industrial cultivars. Thus clonal selection of saffron can help us to isolate the genetically diverse superior clones which can be used as cultivars with higher number of flowers and higher quantity of large corms.

## HOW NEW CLONES OF SAFFRON DEVELOPED

Clone of saffron is the vegetative progeny of supposed initial triploid plant and population of saffron is mixture of different clones. During evolution, populations of saffron have been cultivated and propagated, and various mutant clones have appeared. These mutant clones never shared their gene pool with other plants. Each mutant clone evolved and multiplied independently. Today, the populations of saffron represent a conglomerate of very different clones that have arisen as a result of numerous changes during cultivation by man. Despite their different vitality potential, these clones do not compete, as they are cultivated by the man under the same identical conditions. The application of clonal selection in saffron means that clones with (many) excellent parameters should be selected for, even when they have certain undesirable characters. It is not possible to eliminate the identified defects due to the sterility of saffron. Therefore, it is essential that those clones designated for selection should necessarily have all positive attributes. Extraordinary and superior plants of saffron significant selection value can be separated from the general population of plants, tested again and then transformed, by propagation, into exclusively high-yielding cultivars. Two main difficulties of saffron breeding through clonal selection are as follow.

### Recognition of new clones

In any plantation, saffron is represented by plants existing in highly different "ages" of individuals connected with different sizes of corms underground. Above the ground, these plants differ in the size and number of their flowers (at the stage of flowering), also in their number and size of leaves. If some plants are sharply different from the others in certain characters, for the aim of breeding, a researcher cannot practically identify them. Naturally he does not see corms underground, and cannot elucidate the cause of the differences whether these differences are due to the age (size) of corms, or because of the genotypes of plants. So, genetically different (if existing) and similar plants will continue to grow together and not be subjected to selection.

### Multiplication of clones

Let us suppose that farmer recognizes certain plant(s) which could be used as a clone with good economic characters. Multiplication of such clone(s) would be a problem. One saffron corm with proper care produces an average of 4 corms of middle size during vegetative growth (one year). At such intensity of propagation, it could be brought to about 1 million corms after 17 years. This amount could be enough for planting on the area of 2 ha. It is clear that a farmer will never accomplish such an exhausting work of many years. Therefore the farmers would not pursue the aim to make new cultivar of saffron even if they have been lucky to find some clones with very highly expressed economically valuable characters. Concerning the researcher, in our opinion, selection of potential clone can be multiplied for development of new cultivar. Alternatively, superior selections of saffron could be propagated rapidly via *in vitro* technique (Homes et al.,

1987). Investigations in this direction are very promising. Unfortunately the experiments pursuing rapid corm propagation of saffron have not been successful so far and a few *in vitro* developed corms had been produced. Matured corms of *in vitro* origin in mass production had not been produced. Accordingly *in vitro* propagation protocols need to be refined and strengthened.

## METHODS OF CLONAL SELECTION OF SAFFRON

With an objective of breeding new cultivars of saffron with economically valuable traits, two approaches need to be followed;

1. Selection of superior clones in existing plantations,
2. Creating new valuable forms through induced mutations (experimentally).

### Selection of superior clones in existing plantations

Corms of saffron are selected from the different regions where the crop has been cultivated from ancient times and corms collected from different collection areas are kept separately. The peel is removed from all selected saffron corms and corms are classified into groups based on weight. The corms are divided in groups on weight groups, that is 3.0 to 3.9, 4.0 to 4.9, 5.0 to 5.9, 6.0 to 6.9, 7.0 to 7.9, 8.0 to 8.9, 9.0 to 9.9, 10.0 to 10.9, 11.0 to 11.9, 12.0 to 12.9 g, and so on. The corms of each weight group are planted separately in rows in the field at different places in pits; the corms were planted in rows separately in the field with one corm per hill (PIT). The distance between adjacent rows is 50 cm and between pits is 50 cm. The plants of each identified pit are inspected during flowering at 3$^{rd}$ and 4$^{th}$ generation after planting. The pits are labeled, indicating the number of flowers and dates (Agayev et al., 2009).

The traits noted are:

1. Pits with highest no of flowers.
2. Pits with ordinary number of flowers.
3. Pits with minimal (one to three) number of flowers
4. Pits with complete absence of flowers.

After four years of vegetative generation and complete study, all of the corms from all labeled pits are dug out and packed in proper packages of each pit separately. Each package is marked with the following data:

1. Name of the region
2. Number of flowers (registered at flowering time in 4$^{th}$ vegetative generation)
3. Any other data if noted

The corms of each package is weighed and divided into three groups

1. Corms weighing >10 (large corms)
2. Corms weighing <10 but > 5 (middle corms)
3. Corms weighing <5 (small corms)

These clones in packages will be objects of further study at successive stages of selection with the target of achieving higher number of flower and quantity of large corms (Agayev et al., 2009). Through clonal selection, superior plants of saffron can be identified and selected from among a large number of plants selected from different areas. The selected plants will undergo continuous study and selection for many generations and at the end the selected plants will be assessed for yield with the aim of being released for commercial use. So clonal selection will result in creation of new cultivars of saffron with substantial increase in productivity.

### New concepts in clonal selection

1. Big corm index: Percentage of the weight of big 'corm (>10 g) from the total weight of all corms of the given clone.
2. Multiplication index: Designates how many times the weight of the planted initial corm has increased during four vegetative generations; '
3. Flower creating index: Average corm weight, which was necessary for formation of one flower at a given generation of the clone. "Corm set".
4. Cormset: Indicates the complete set of corms (weight, number and size) dug out from one pit after a certain generation. The "Corm set formula" is developed to show the relationship between the numbers of large, middle-sized and small corms; for example, 8, 3, 7 indicates that there were eight, three and seven large, middle-sized and small corms, respectively.

## RESULTS OF SAFFRON SELECTIONS IN KASHMİR

A survey of fifteen saffron growing villages of Kashmir by SKUAST-K has lead to the identification of ten prominent elite clones through clonal selection with distinct superiority in terms of yield and quality over natural populations .Three conjugate pairs of primers viz., SA-C+SA-T, SA-D+SA-S and SA-E+SA-R produced a consistent banding pattern of 0.5 to 1.0 size in 2% Agarose gel. Perusal of Table 1 revealed that the total number of bands obtained from the conjugate primer pairs from 10 genotypes ranged from 1 to 7 (Table 2). The maximum number of scorable bands (that is, 7) was obtained from SA-D+ SA-S primer pair, whereas, the primer pair SA-C+SA-T and primer pair SA-E+SA-R showed 1 and 2 scorable bands, respectively. The only primer pair, that is, SA-D+SA-S showed maximum number of polymorphic bands (that is, 7). Molecular characterization of ten elite saffron genotypes based on 21 random amplified polymorphic DNA (RAPD) markers revealed a considerable amount of genetic diversity among tested genotypes (Figure 1). Similarity index based on Jaccards coefficient ranged from 0.375 to 0.834 and the maximum similarity coefficient (0.834) was observed between the genotype SMD-45 and SMD-79. The dendrogram based on molecular data divided the tested genotypes in two clusters. However, genotypes including, SD-3, SMD-45, SMD-79 and SMD-68 formed the cluster I and genotypes including, SMD-11, SMD-52, SMD-81, SMD-211, SMD-224 and control formed cluster II at similarity coefficient of 44%, which showed a high level of genetic diversity between two clusters containing different genotypes. A different level of variability was observed among different genotypes within each cluster also. The primer pair, that is, SA-D+SA-S was found to be the best primer pair which showed maximum number of scorable and polymorphic bands (Nehvi et al., 2007a;

**Table 2.** Number of scorable and polymorphic RAPD bands obtained by PCR amplification of DNA of *Crocus sativus* L.

| Primers | Scorable bands | Polymorphic bands | Polymorphism (%) |
|---|---|---|---|
| SA-C + SA-T | 1 | 0 | 0 |
| SA-C + SA-T | 7 | 7 | 100 |
| SA-C + SA-T | 2 | 0 | 0 |
| Total | 10 | 7 | 70 |

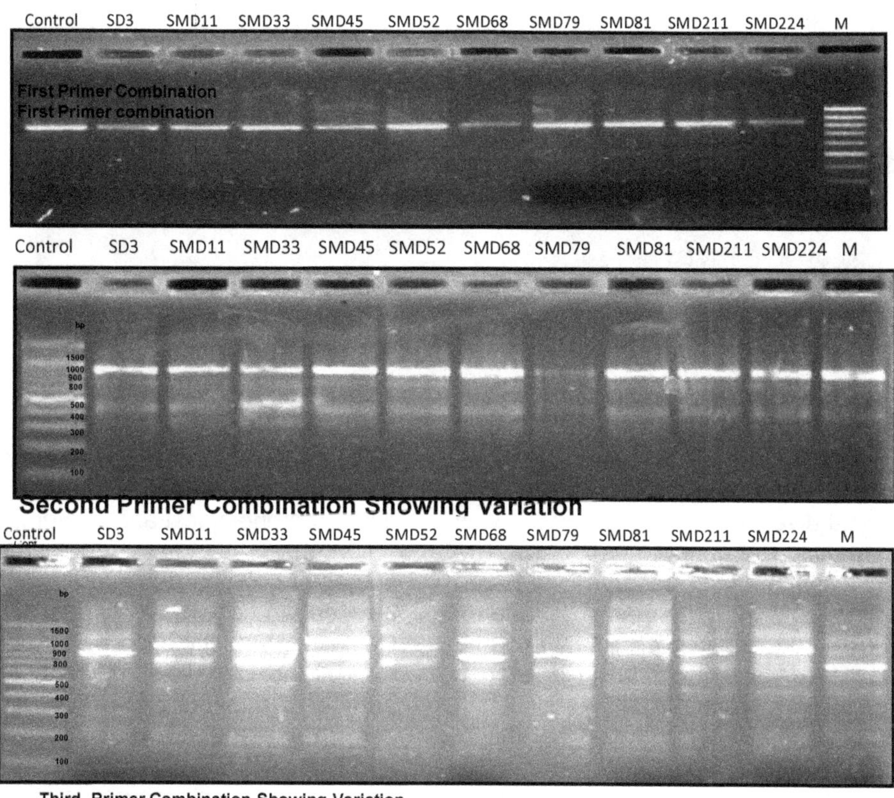


**Figure 1.** Variability at molecular level (Imran et al., 2010).

Imran et al., 2010).

The present investigation was carried out during 2007 on ten elite saffron genotypes viz., SD-3, SMD-11, SMD-31, SMD-45, SMD-52, SMD-68, SMD-79, SM-81, SMD-211, SMD-224 and one sample of natural population being superior in saffron yield and quality. Genetic diversity was studied using polymerase chain reaction (PCR) based amplified polymorphic DNA (RAPD) markers as described by William et al. (1990).

A survey of prominent saffron (229 in number) growing areas of Kashmir comprising 15 villages (Zeevan, Khrew, Wuyan, Ladhoo in Srinagar district; Dusso, Namlabal, Konibal, Chandar, Pampore, Barsu, Lathipora in Pulwama district and Chadora, Chararisharief, Kakawring, Hapatnar in Budgam district) located at an altitude of 1686, 1644, 1597 and 1730 m.a.s.l. The remaining genotypes collected from Zeewan, Khrew, Wuyan, Ladoo, in district Srinagar, Dusso, Namlabal, respectively was carried out during August 2003 and 60 corms samples of uniform weight and size (>10.0 g/3.5 cm) were collected from each location. Data pooled over years revealed that cluster I accommodated 229 genotypes followed by cluster II (2) (Table 1). Konibal, ChadharPampore, Barsoo, Lethipora, Koil in district Pulwama and Chadora, Chariesharief, Kakawring, Hapathanar, Nagm in district Budgam got grouped in cluster I except for genotype SDM-140, SDM-138 from Lethipora (cluster II) SDM-116 from Konibal (cluster III and XI), SDM-61 from Dassu (cluster IV and V); SDM-235 from Ladoo (cluster VI), SDM-45 from Zeewan (cluster VIII) and SDM-224 from Nagam (cluster IX). The pattern of group constellations indicated that the

**Figure 2.** Performance of elite genotypes available at Saffron Research Station, Pampore SKUAST-Kashmir, India.

geographical diversity was not an essential factor to group the genotypes from a particular source. This means that geographical diversity, though important, was not the factor in determining the genetic divergence. The highest intra-cluster distance was observed between genotype SDM-43 and SDM-220 grouped in cluster I. Cluster VIII and X accommodating genotype SDM-45 and SDM-115, recorded maximum inter-cluster distance (445.51) followed by cluster VI and X (416.74), cluster II and cluster XI (416.38) and cluster VIII and cluster XI (415.82). Cluster VIII revealed high cluster mean for saffron yield, average weight of daughter corms and pistil length on account of grouping of high yielding genotype SDM-45, showing saffron yields to the tune of 15.30 mg associated with 17.10% crocin content. Percent contribution of different traits towards divergence revealed strong influence of fresh pistil weight, stigma length and crocin content. Therefore, such characters can be taken as criteria in selection for divergent lines (Makhdoomi et al., 2010).

## Creating new valuable forms induced mutations

Induced mutations play an important role in creating new variations in clonally propagated plants because mutations in such plants may easily get stabilized and can be manipulated. So mutation breeding offers a scope for induction of variability in saffron for its subsequent utilization (Agayev et al., 1975; Nehvi et al., 2010; Muzaferova, 1970). For successful establishment of any crop on commercial scale the availability of adapted cultivars with desired characters is prerequisite, therefore, an attempt was made to create variability for floral, morphological and anatomical traits in saffron using physical and chemical mutagens. Corms at different stages of growth rate were subjected to different doses of physical and chemical mutagens. The study revealed that Colchicine (mostly used as an anti mitotic agent, 0.05%) was found beneficial for enhancing pistil length, leaf number and leaf length with a negative effect on survival corm treatment with radiation dose of 0.2, EMS (0.1%) and Ethidium Bromide (0.2%) showed significant positive effect on survival, increased number of heavier flowers and pistil weight. Mutagens induced morphological and anatomical variants showing increased number of stomata, thicker and broader leaves (Figure 3). Floral variant induced in M1 through 0.2 KRad radiation was not a stable character. Resting bud stage ($1^{st}$ -$15^{th}$ June) has been identified as appropriate stage for inducing variation through physical and chemical mutagens. Plants showing phenotypic variations are being maintained for further observations

## Conclusion

Our results suggest that saffron (*C. sativus* L.) populations are not homogeneous, despite their clonal origin. The application of clonal selection in saffron propagation means that clones with (many) excellent parameters should be selected for, even when these contain any serious defect (or defects). It is clearly not possible to eliminate the identified defects separately from

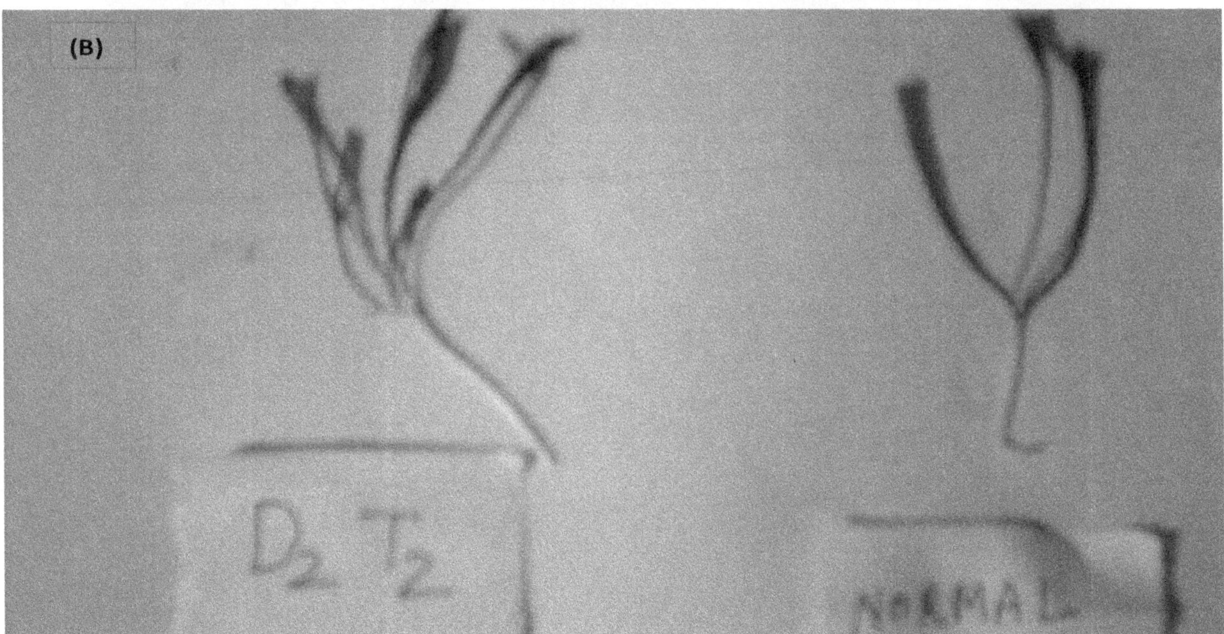


**Figure 3.** Morphological variants induced through (A) Gamma radiation 0.2 kr; (B) Ethidium Bromide 0.2%.

from the rather valuable other attributes due to the sterility sterility of saffron. Therefore, it is essential that those clones designated for selection should necessarily have the entire complex of positive attributes. In pursuit of creating new high-yielding and high quality cultivars of saffron, it is necessary to establish that the existing populations of saffron are rich in terms of genetic and selection potential. We have accomplished this by a method of simple clonal selection, revealing new and extraordinary and superior plants with a complex of positive attributes and creating new cultivars with high practical advantages. Each identified extraordinary or superior plant is potentially a new cultivar; it is only necessary to propagate it and introduce it in practice as a new cultivar.

## REFERENCES

Agayev YM (1993). Some urgent problems in genetics, cytogenetics and breeding of saffron (*Crocus sativus* L.). In:Abstracts book of the second national symposium of saffron and medical plants, Gonabad, pp. 12–13

and breeding ofsaffron. Abstracts of the 2nd Symposium on Saffron and Farming of Medicine Plants. Gonabad, Iran. p. 12.

Agayev YM (1994b). Origin of saffron and its karyotype analyses.Abstracts of the2nd Symposium on Saffron and Farming of Medicine Plants. Gonabad, Iran. p.13.

Agayev YM, Muzaferova RS, Savchenko SP (1975). Results of theexperiments of saffron corm treatments with colchicine solutions. VestnikSelskokhozyaystvennoiNauki. Bull. Agric. Sci. Moscow.10:121-123.

Agayev YM, Shakib AM, Soheilivand S, Fathi M (2007). Breeding of saffron (*Crocus sativus* L): possibilities and problems. ActaHorticulture 739:203–207

Agayev YM, Fernandez JA, Zarife E (2009) Clonal selectionof saffron (*Crocus sativus L.*): the first optimistic experimental results. Euphytica 169: 81-99.

Chichiricco G (1984). Karyotype and meiotic behaviour of the triploid *Crocus sativus* L. Caryologia 37:233-239.

Chichiricco G (1999). Sterility and perspectives for genetic improvement of *Crocus sativus L.* In: Saffron, *Crocus sativus* L. Medicinal and Aromatic Plants – IndustrialProfiles. (Ed. By Negbi, M.). Hardwood Academic Publishers. pp. 127-135.

Dhar AK, Sapru R, Rekha K (1988). Studies on saffron in Kashmir. 1.Variation in natural population and its cytological behavior. Crop Improve. 15:48-52.

Ehsanzadeh P, Yadollahi AA, Maibodi AMM (2004). Productivity, growth and quality attributes of 10 Iranian saffron accessions under climatic conditions of Chahar-MahalBakhtiari, Central Iran. ActaHorticulture 650:183–188

Gowhar Ali, FA Nehvi, Sabeena (2011). Clonal Selection as Tool for Enhancing Productivity in Saffron.Training onAdvances in Saffron Biology, Production and Quality Improvement". SKUAST-Kashmir.

GrilliCaiola M, Di Somma D, Lauretti P (2001) Comparative study on pollen and pistil of *Crocus sativus L.* (Iridaceae) and its allied species. Ann. Bot. Roma. 1: 93-103.

Hassan-Beygy SR, Ghanbarian D, Kianmehr MH, Farahmand M. (2010). Some physical properties of saffron crocus corm. Cercetari Agronomice in Moldova 1: 17-29.

Homes J, Legros M, Jaziri M (1987). *In vitro* multiplication of *Crocus sativus* L.Acta Horticulturae 212:675-676.

Kapinos GE (1965). Biology of development bulbous and cormous plants inApsheron. Azerbaijan Nat. Acad. of Sciences, Baku.

Makhdoomi MI, FA Nehvi, SA Wani (2010). Genetic Divergence in Saffron (*Crocus sativus L.*). Acta Horticulture.850:79-84.

Mathew B (1977). *Crocus sativus* L. and its allies (Iridaceae). Plant Syst. Evol. 128:89-103.

Moraga AR, Rambla JL, Ahrazem O, Granell A, GómezGomez L (2009). Metabolite and target transcript analyses during Crocus sativus stigma development. Phytochemistry 70:1009-1016.

Munshi AM (1992). Genetic variability for important traits in saffron (*Crocus sativus* L.). Crop Res. (Hisar) 5: 326-332.

Munshi AM, Zargar GH (1991). Variation in natural population of saffron(*Crocus sativus* L.) crop in Kashmir and performance of some selected subpopulation. Phyto breed on. 7: 62-67.

Muzaferova RS (1970). Primary results of a study of influences of the mutagenicfactors on changing of saffron. Proceedings of the Institute of Genetics andSelection of the Academy of Sciences of Azerbaijan SSR, Baku, "Elm". p.194-203.

Nehvi FA, Wani SA, Dar SA, Makhdoomi MI, Allie BA, Mir ZA (2007a). Biological interventions for enhancing saffron productivity in Kashmir. Acta Hortic. 739:25–31

Nehvi FA, Wani SA, Dar SA, Makhdoomi MI, Allie BA, Mir ZA (2007b). New emerging trends on production technology of saffron. Acta Horti. 739:375–381.

Nehvi FA, MA Khan, AA Lone, MI Maqhdoomi, SA Wani, V Yousuf, S Yasmin (2010). Effect of Radiation and Chemical Mutagen on Variability in Saffron (*Crocus sativus L.*)Acta. Horti.850:67-73.

Picci V (1987). A summary of experiments on the cultivation of *Crocus sativus L.* in Italy. Atti Convegnosulla Coltivazionedelle Piante Officinali, Ministerodell' Agricoltura e delle Foreste, Italy, pp 119–157

Piqueras A, Han BH, Escribano J, Rubio C, Hellin E, Fernandez JA (1999). Development of cormogenic nodules and microcorms by tissue culture, a new toolfor the multiplication and genetic improvement of saffron. Agronomie. 19:603-610.

Rzakuliyev IM (1959). A study of different specimens of saffron in condition ofApsheron. AGU, UchyonyyeZapiski, Biol. Ser. 5:3-8.

Imran S, FA Nehvi, SA Wani, G Zaffar, MA Khan (2010). Studies in Relation to Molecular Variability in Saffron. Acta Hortic. 850:75-78

Zargar GH (2002). Genetic variation in saffron and importance of quality seed corms. National seminar-cum-workshop, Srinagar.

# 9

# Heterosis studies for earliness, fruit yield and yield attributing traits in bell pepper

**Vinod Kumar Sharma[1], Shailaja Punetha[2] and Brij Bihari Sharma[3]**

[1]Indian Agricultural Research Institute, Pusa Campus Regional Research Station Katrain Dist-Kullu Valley– 175129, India.
[2] Department of Vegetable Science, G. B. Pant University of Agriculture and Technology Pantnagar-263145, India.
[3]Division of Vegetable Science, Indian Agricultural Research Institute, Pusa Campus New Delhi-110012, India.

**The present investigation was envisaged to gather information about the potential and characteristics of the experimental material of bell pepper at the Vegetable Research Block, Department of Vegetable Science, College of Forestry and Hill Agriculture, Gobind Ballabh Pant University of Agriculture and Technology, Hill Campus, Ranichauri (Tehri Garhwal), Uttarakhand (India) during rainy-autumn season, 2009 to 2010. The combined analysis of 6 × 6 diallel revealed significant variation among genotypes in parents, $F_1$ crosses and parents versus $F_1$ crosses for all the studied characters except for pericarp thickness and number of branches for parents, indicating that the used parents in diallel mating design and their crosses were genetically diverse. A wide range of heterosis over better parent and standard check was observed in F1 generation for marketable fruit yield and its attributing traits. The $F_1$ crosses PRC-1 × California Wonder and California Wonder × SSP had revealed the highest significant desirable heterosis both over better parent and standard check (California Wonder) for early maturity. The $F_1$ crosses (Rani Sel-1 × SSP, Rani Sel-1 × Sel-12-2-1, SSP × SP-316 and PRC-1 × California Wonder) showed appreciable heterosis over better parent and standard checks for marketable fruit yield per plant and quality traits in bell pepper. The high heterotic response as observed in these hybrids further supported by the predominant role of non-additive component in the inheritance of the characters studies. These cross combination could be exploited in heterosis breeding programme.**

**Key words:** Heterosis, bell pepper, capsicum, hybrid breeding, horticultural traits, fruit yield.

## INTRODUCTION

Bell pepper (*Capsicum annuum* L. var. *grossum* Sendt.) commonly known as sweet pepper or capsicum, holds a very coveted position as a leading off-season vegetable in Himachal Pradesh, generating cash revenues to the farmers by selling the produce in the neighbouring states and metropolitan cities. In recent years, its demand has increased tremendously with the emergence of pizza industry (Sood et al., 2011). Besides, it also has medicinal properties and hence, recommended for the treatment of dropsy, colic toothache and cholera (Pairce, 1987). Green bell pepper (*C. annuum* L.) is an excellent source of ascorbic acid and a fair source of provitamin A carotenoids (Haytowitz and Matthews, 1984). In addition, peppers are rich in flavonoids (Lee et al., 1995) and other phytochemicals (Duke, 1992). Presently, it is extensively grown in Himachal Pradesh, Uttarakhand, Jammu and

Kashmir, Arunachal Pradesh and Darjeeling and in some parts of West Bengal during summer months and as an autumn crop in Maharashtra, Karnataka, Tamil Nadu and Bihar. Bell pepper transported to distant market in the plains bringing handsome returns to the small and marginal farmers. In spite of its importance, a few varieties are grown commercially. California Wonder an old introduction being grown extensively throughout the country. In recent times, the production scenario of capsicum has changed with the increasing popularity of hybrids cultivated on commercial scale. Hybrids not only led to a boom in the productivity but also helped improving the quality of the produce. Sweet pepper offers much scope of improvement through heterosis breeding method (Sood and Kumar, 2010b). The systematic approach for developing $F_1$ hybrids in any crop depends primarily on magnitude of desirable heterosis. It offers much scope for improvement through heterosis breeding among genetically diverse genotypes, which can be further utilized for the development of desirable recombinants.

As the required goals of increasing productivity in the quickest possible time can be achieved by utilizing heterosis breeding, which is feasible in this crop (Joshi and Singh, 1980); information on the magnitude of heterosis in different cross combination is a basic requisite to assess for identifying crosses that exhibit high amount of exploitable heterosis. The present investigation describes the extent and nature of heterosis in hybrids for earliness, fruit yield and its attributes.

## MATERIALS AND METHODS

The present investigation was envisaged to gather information about the potential and characteristics of the experimental material of bell pepper at the Vegetable Research Block, Department of Vegetable Science, College of Forestry and Hill Agriculture, Gobind Ballabh Pant University of Agriculture and Technology, Hill Campus, Ranichauri (Tehri Garhwal), Uttarakhand during rainy-autumn season, 2009 to 2010.

### Experimental site

#### *Location*

The experimental site is located at about 10 km away from Chamba (on Rishikesh-Gangotri Road) at an altitude about 2000 m above mean sea level. Geographic position of the experimental site lies between 30° 15' N latitude and 78° 30' E longitude under mid hill zone of Uttarakhand, India.

#### *Climate*

The climate is humid temperate. The mean monthly minimum and maximum temperature varied between 6.0 to 16.6 and 17.5 to 23.7°C, respectively during the cropping season. The experimental site experienced average rainfall of 117.84 cm annually, out of which about 61.0% is received during monsoon period. Monsoon arrives in the second fortnight of June and ends in September. The summer is mild and winter is very severe. Light to heavy snow fall is a regular feature of this area during December to February.

#### *Soil*

The soil of the experimental block was silt clay to silt loam in texture, low in available nitrogen (210.0 to 218.5 kg $ha^{-1}$) and phosphorus (11.5 to 13.5 kg $ha^{-1}$) and rich in potash (408.0 to 418.0 kg $ha^{-1}$). The electrical conductivity as measured in 1:2 soil water ratio was 0.5 to 0.7 $dsm^{-1}$. The depth of the soil extends up to 2.0 m.

### Experimental materials

The experimental materials comprised of six genetically diverse bell pepper genotypes (parents) and their fifteen $F_1$ crosses. The main characteristics of the six parental lines are given in Table 1.

### Mating design and material development

The $F_1$ crosses were made manually using a standard procedure of hand emasculation and pollination in a diallel mating design (without reciprocals) during summer- rainy season, 2008 to 2009. The sufficient $F_1$ cross seed was obtained for evaluation in the next summer- rainy season, 2009 to 2010. Selfed seeds of the parents were also obtained during the same season.

#### *Nursery sowing*

The seeds were sown on April 8, 2009 in raised nursery beds under polyhouse at vegetable research block, Hill Campus, Ranichauri. Healthy and disease free seedling of twenty one genotypes (6 parents and 15 $F_1$ crosses) were transplanted on May 25, 2009 (Table 3).

### Experimental design and layout plan

The final trial was laid out on May 25, 2009 comprising of six parents and fifteen $F_1$ crosses were sown in the nursery and transplanted in Randomized Block Design (RBD) with three replications in plot size of 2.0 × 2.0 m. The plants were spaced at 60 cm between row to row and 45 cm plant to plant. The experimental field was prepared by ploughing twice with power tiller up to a depth of 20 cm followed by leveling. The farm yard manure (20 t $ha^{-1}$) was mixed in the soil at the time of field preparation with first ploughing. The chemical fertilizer (70 kg N, 80 kg $P_2O_5$ and 80 kg $K_2O$) were applied as basal dose at the time of field preparation. One third of N, full dose of $P_2O_5$ and $K_2O$ is applied at the time of final field preparation. Remaining two third N was top dressed in two equal amounts and added after 30 and 45 days of transplanting, respectively. The observations were recorded on five randomly selected plants for days to 50% flowering, days to first harvest, number of pickings, fruit weight (g), fruit diameter (cm), fruit length (cm), fruit girth (cm), number of fruit per plant, number of lobs per fruit, pedicel length (cm), pericarp thickness (cm), number of branches per plant, plant height (cm) and marketable fruit yield per plant.

Ascorbic acid content (mg per 100 g fresh weight) from the crushed fruit sample was estimated by the method as described by Ranganna (1986). California Wonder was used as a standard check for estimation of commercial heterosis.

**Table 1.** Distinguished morphological features of bell pepper genotypes.

| S/N | Parents | Source | Fruit color at immature stage | Fruit color at mature stage | No. of lobes | Fruit shape |
|---|---|---|---|---|---|---|
| 1 | PRC-1 | Department of Vegetable Science, Hill Campus, Ranichauri | Dark green | Orange yellow | 3.00 | Rectangular |
| 2 | California Wonder (CW) | IARI, Regional station, Katrain | Medium green | Red | 4.00 | Bell shape |
| 3 | Rani Sel-1 | Department of Vegetable Science, Hill Campus, Ranichauri | Dark green | Red | 3.00 | Bell shape |
| 4 | SSP | Department of Vegetable Science, Hill Campus, Ranichauri | Medium green | Red | 3.00 | Rectangular |
| 5 | Sel-12-2-1 | Department of Vegetable Science, UHF, Solan | Dark green | Red | 3.00 | Triangular |
| 6 | SP-613 | Department of Vegetable Science, UHF, Solan | Medium green | Red | 3.00 | Triangular pointed |

**Statistical analysis**

The mean values of different genotypes for various characters were statistically analyzed using SPAR-1 programme of Doshi and Gupta (1981). SPAR I (developed by Indian Agricultural Statistical Research Institute, New Delhi, India) software was used for statistical analysis. Heterosis was computed by using computer software programme Windowstat 8.0.

## RESULTS

The combined analysis of variance (Table 2) revealed that, significant variation among genotypes in parents, $F_1$ crosses and parents versus $F_1$ crosses for all the studied characters except for pericarp thickness and number of branches for parents, indicating that the parents used in diallel mating design and their crosses were genetically diverse. A wide range of heterosis over better parent and standard check was observed in $F_1$ generation for most of the studied traits. The magnitudes of heterobeltiosis for days to 50% flowering were ranged from -12.86 to 5.83 and -13.33 to 5.58% over better parent and standard check, respectively. Twelve $F_1$ crosses exhibited negative heterosis, of which eleven were found to have significant desirable negative heterosis over better parent for days to 50% flowering. The highest negative heterobeltiosis was recorded in cross combinations namely, California Wonder × SSP (-12.86%), SSP × Sel-12-2-1 (-10.74%) and California Wonder × SP-316 (-9.67). Among 15 $F_1$ crosses, only 11 exhibited significant negative heterosis and two crosses revealed significant positive heterosis over the standard check. The highest desirable significant heterosis was found in $F_1$ crosses PRC-1 × California Wonder (-13.33%) and PRC-1 × SP316 (-13.33%) over standard check, respectively for days to 50% flowering.

The magnitudes of heterosis for days to first harvest in F1 crosses were ranged from -16.14 to 2.15% over better parent. Only eight F1 crosses showed desirable significant heterobeltiosis and remaining crosses exhibited non desirable heterosis. The highest desirable heterobeltiosis was found in the cross combinations PRC-1 × SP316 (-16.1%) followed by SSP × SP316 (-13.00%) and California Wonder × SSP (-11.87%), respectively. Among 15 F1 crosses, only seven exhibited significant negative heterosis over standard check. The $F_1$ crosses California Wonder × SSP and PRC-1 × SSP exhibited highest significant heterosis for days to first harvest in desirable direction to the tune of -11.87% each over standard check. The top three hybrids for days to first harvest over standard check were California Wonder × SSP (-11.87%), PRC-1 × SSP (-11.87%) and PRC-1 × California Wonder (-10.04%) and PRC-1 × SP-316 (-16.14%), SSP × SP-316 (-13.00%) and California Wonder × SSP (-11.87%) for over better parent reported. The estimates of percent heterosis for number of pickings in F1 crosses were ranged from –20.00 to 44.44 and 8.79 to 31.43% over better parent and standard check, respectively. Eleven F1 crosses had exhibited significant positive heterobeltiosis for this trait. The maximum heterosis to the extent of 44.44% was recorded in the cross PRC-1 × Rani Sel-1 followed by Rani Sel-1 × SP-316 (41.66%) over better parent. The highest economic heterosis was recorded in cross Rani Sel-1 × SP-316 to the tune of 31.43% over standard check.

The top three hybrids for number of pickings over standard check were Rani Sel-1 × SP-316 (31.43%), PRC-1 × California Wonder (29.93%) and PRC-1 × Rani Sel-1(24.93%) and PRC-1 × Rani Sel-1 (44.44%), Rani

**Table 2.** Analysis of variance for combined analysis for different horticultural traits in bell pepper.

| SV traits | D. F. | Days to 50% flowering | Days to first harvest | No. of picking | Fruit weight (g) | Fruit diameter (cm) | Fruit length (cm) | Fruit girth (cm) | No. of fruits/ plant | No. of lobs/ fruit | Pedicel length (cm) | Pericarp thickness (cm) | No. of branches /plant | Plant height (cm) | Ascorbic acid content (mg/100 g) | Market-able fruit yield/ plant |
|---|---|---|---|---|---|---|---|---|---|---|---|---|---|---|---|---|
| P | 5 | 12.90** | 69.78** | 2.93** | 530.76** | 0.79** | 5.64** | 3.08** | 23.84** | 1.12** | 0.81** | 0.35 | 0.25 | 206.78** | 378.31** | 52457.47** |
| $F_{1's}$ | 15 | 17.98** | 60.40** | 3.39** | 77.28** | 1.24** | 7.48** | 6.53** | 12.81** | 0.19 | 0.60** | 1.65** | 0.49** | 389.6** | 940.68** | 167673.81** |
| P*Vs* $F_{1's}$ | 1 | 38.13** | 53.15** | 0.16 | 811.76** | 0.05 | 12.38** | 7.23** | 91.13** | 0.65** | 0.84** | 2.68** | 0.13 | 510.56** | 717.91** | 840940.93** |
| Rep | 2 | 0.06 | 2.68 | 3.47** | 5.07 | 0.0011 | 0.62 | 0.81 | 0.51 | 0.09 | 0.00 | 0.17 | 0.88** | 30.6** | 102.70 | 2823.93 |
| Error | 40 | 0.73 | 2.24 | 0.41 | 12.52 | 0.08 | 0.55 | 0.91 | 3.70 | 0.11 | 0.18 | 0.21 | 0.16 | 10.7 | 58.42 | 1957.39 |

** Significant at 1%, * Significant at 5%.

Sel-1 × SP-316 (41.66%) and California Wonder × SSP (27.22%) over better parent. The magnitude of heterosis for fruit weight was ranged from -20.96 to 102.73% over better parent and -31.38 to 49.50% over standard check, respectively. Ten F1 crosses were found with heterobeltiosis effects for fruit weight in F1 crosses over superior parent. The cross combination Rani Sel-1 × Sel-12-2-1 (102.73%) followed by PRC-1 × Sel-12-2-1 (96.94%) had exhibited highest significant heterobeltiosis. Nine F1 crosses out of fifteen were found with significant positive heterosis over standard check. The highest economic heterosis was recorded in cross combination PRC-1 × Rani Sel-1 (49.50%) over standard check. The top three hybrids recorded with highest significant desirable heterosis for fruit weight over better parent were Rani Sel-1 × Sel-12-2-1 (102.73%), PRC-1 × Sel-12-2-1 (97.00%) and SSP × Sel-12-2-1 (83.58%) and PRC-1 × Rani Sel-1 (49.50%), California Wonder × SSP (39.42%) and Rani sel-1 × Sel-12-2-1 (34.49%) over standard check. The magnitude of heterosis for fruit diameter was ranged from -7.82 to 30.64% better parent and -13.59 to 17.81% over standard check, respectively. Seven F1 crosses were found with positive significant heterosis for fruit diameter (cm) over superior parent. The cross combination Rani Sel-1 × Sel-12-2-1 (30.64%) followed by PRC-1 × Sel-12-2-1 (28.58%) had exhibited the highest significant positive heterobeltiosis. Out of fifteen, seven F1 crosses were found with significant positive heterosis over standard check. The highest economic heterosis was recorded in cross combination Sel-12-2-1 × SP-316 (17.81%) over standard check. The top ranking hybrids for fruit diameter were Rani Sel-1 × Sel-12-2-1 (30.64%), PRC-1 × Sel-12-2-1 (28.58%) and SSP × Sel-12-2-1 (22.38%) over better parent and Sel-12-2-1 × SP-316 (17.81%), PRC-1 × Rani Sel-1 (16.12%) and California Wonder × SSP (14.80%) over standard check, respectively (Table 5).

The magnitudes of heterosis for fruit length were ranged from -3.96 to 40.76 and -3.99 to 61.18% over better parent and standard check, respectively. Ten $F_1$ crosses were found with positive significant heterosis for fruit length over superior parent. The cross combination SSP × SP-316 (40.76%) followed by Sel-12-2-1 × SP-316 (36.74%) had exhibited the highest significant positive heterobeltiosis. Among fifteen, eleven F1 crosses were found with significant positive heterosis over standard check. The highest economic heterosis to the extent of 61.18% was recorded in cross combination SSP × SP-316 over standard check. The three top ranking hybrids for fruit length were SSP × SP-316 (40.76%), Sel-12-2-1 × SP-316 (36.74%) and California Wonder × Sel-12-2-1 (34.17%), over better parent and SSP × SP-316 (61.18%), PRC-1 × SP-316 (52.54%), PRC-1 × Rani sel-1 (29.51%) over standard check. The magnitudes of heterosis for fruit girth were ranged from -8.74 to 25.71 and -11.66 to 17.56% over better parent and standard check, respectively. Seven crosses were found with positive significant heterosis for fruit girth over superior parent. The cross combination SSP × Sel-12-2-1 (25.71%) followed by PRC-1 × Sel-12-2-1 (24.28%) had exhibited highest significant positive heterobeltiosis. Out of fifteen, four $F_1$ crosses were found with significant positive heterosis over standard check, the highest being in cross combination Rani Sel-1 × SSP (17.56%) over standard check.

The top three hybrids for fruit girth over better parent were SSP × Sel-12-2-1 (25.71%), PRC-1 × Sel-12-2-1 (24.28%), Rani Sel-1 × Sel-12-2-1 (21.90%), and Rani Sel-1 × SSP (17.56%), PRC-1 × SP-316 (11.245) and SSP × Sel-12-2-1 (9.82%) over standard check, respectively. The estimates of percent heterosis for number of fruit per plant in F1 crosses were ranged from - 10.10 to 156.61 and 43.76 to 155.98% over better parent and standard check, respectively (Table 4). The

**Table 3.** Mean performance of parents and $F_1$ crosses of bell pepper (*Capsicum annuum* L.) genotypes for different horticulture traits.

| S/N | Genotype | Days to 50% flowering | Days to first harvest | Number of picking | Fruit weight (g) | Fruit diameter (cm) | Fruit length (cm) | Fruit girth (cm) |
|---|---|---|---|---|---|---|---|---|
| 1 | PRC-1 | 35.33 | 64.00 | 6.33 | 74.53 | 5.85 | 10.93 | 17.59 |
| 2 | California wonder | 40.00 | 73.00 | 6.66 | 67.33 | 5.53 | 9.41 | 17.14 |
| 3 | Rani Sel.1 | 39.00 | 67.66 | 6.00 | 77.16 | 5.89 | 11.98 | 17.25 |
| 4 | SSP | 39.00 | 70.00 | 8.66 | 79.35 | 5.94 | 10.78 | 17.71 |
| 5 | Se1.12-2-1 | 40.33 | 77.33 | 7.66 | 44.66 | 4.67 | 8.17 | 14.97 |
| 6 | SP-316 | 41.33 | 74.33 | 6.66 | 58.45 | 5.08 | 11.13 | 16.56 |
| 7 | California Wonder × PRC-1 | 42.33 | 71.66 | 7.00 | 71.33 | 5.10 | 11.18 | 16.38 |
| 8 | California Wonder × SSP | 35.00 | 64.33 | 6.33 | 93.87 | 6.36 | 10.32 | 19.06 |
| 9 | California Wonder × Sel-12-2-1 | 37.66 | 75.00 | 6.08 | 58.73 | 4.78 | 10.96 | 15.73 |
| 10 | PRC-1 × California Wonder | 34.66 | 65.66 | 8.66 | 78.33 | 5.40 | 10.47 | 16.86 |
| 11 | PRC-1 × Rani Sel-1 | 35.66 | 67.33 | 8.33 | 100.66 | 6.43 | 12.20 | 19.97 |
| 12 | PRC-1 × SSP | 35.00 | 64.33 | 7.66 | 73.66 | 6.33 | 10.95 | 16.87 |
| 13 | PRC-1 × Sel-12-2-1 | 39.66 | 68.66 | 7.66 | 88.41 | 6.01 | 10.38 | 18.61 |
| 14 | PRC-1 × SP-316 | 34.66 | 62.33 | 5.66 | 65.33 | 4.80 | 14.37 | 17.38 |
| 15 | Rani Sel-1 × SSP | 37.00 | 71.33 | 7.66 | 75.54 | 6.09 | 11.77 | 20.15 |
| 16 | Rani Sel-1 × Sel-12-2-1 | 39.33 | 69.00 | 7.39 | 50.55 | 6.11 | 10.17 | 18.25 |
| 17 | Rani Sel-1 × SP-316 | 38.33 | 72.33 | 8.76 | 60.06 | 4.82 | 11.75 | 16.76 |
| 18 | SSP × Sel-12-2-1 | 36.00 | 70.00 | 7.33 | 82.00 | 5.72 | 10.83 | 18.82 |
| 19 | SSP × SP-316 | 37.33 | 64.66 | 6.66 | 86.61 | 5.58 | 25.18 | 17.46 |
| 20 | (Sel Sel-12-2-1 × SP-316 | 41.66 | 79.00 | 6.10 | 51.62 | 4.55 | 11.17 | 16.94 |
| 21 | California Wonder × SP-316 | 37.33 | 69.66 | 5.33 | 46.20 | 5.36 | 9.04 | 15.14 |
|  | SE± | 0.69 | 1.22 | 0.52 | 2.88 | 0.23 | 0.60 | 0.77 |
|  | CD at 1% | 1.40 | 2.47 | 1.05 | 5.83 | 0.47 | 1.22 | 1.57 |
|  | CD at 5% | 1.88 | 3.31 | 1.41 | 7.81 | 1.27 | 1.64 | 2.10 |

| S/N | Genotype | Number of fruits/ plant | Number of lobs/ fruit | Pedicel length (cm) | Pericarp thickness (cm) | Number of branches/ plant | Plant height (cm) | Ascorbic acid content (mg/100 g) | Marketable fruit yield/ plant (g) |
|---|---|---|---|---|---|---|---|---|---|
| 1 | PRC-1 | 7.62 | 2.90 | 2.84 | 2.93 | 4.00 | 45.90 | 82.43 | 687.59 |
| 2 | California wonder | 6.54 | 4.00 | 3.63 | 3.57 | 3.93 | 55.20 | 87.32 | 490.40 |
| 3 | Rani Sel.1 | 8.32 | 3.20 | 2.90 | 3.01 | 4.60 | 49.00 | 66.68 | 700.84 |
| 4 | SSP | 9.75 | 3.33 | 3.35 | 3.33 | 4.53 | 46.66 | 64.10 | 720.61 |
| 5 | Se1.12-2-1 | 12.85 | 2.66 | 2.13 | 3.04 | 4.20 | 45.20 | 73.46 | 469.51 |
| 6 | SP-316 | 13.39 | 3.05 | 2.73 | 3.78 | 4.00 | 50.26 | 91.52 | 433.95 |
| 7 | California Wonder × PRC-1 | 9.91 | 3.14 | 2.75 | 3.38 | 3.71 | 52.45 | 50.28 | 675.77 |
| 8 | California Wonder × SSP | 12.98 | 3.20 | 3.86 | 4.53 | 4.13 | 47.60 | 93.12 | 1180.13 |
| 9 | California Wonder × Sel-12-2-1 | 11.67 | 2.83 | 2.36 | 2.91 | 3.56 | 53.65 | 71.13 | 557.10 |
| 10 | PRC-1 × California Wonder | 14.55 | 3.10 | 3.06 | 1.97 | 4.23 | 57.60 | 120.18 | 933.05 |
| 11 | PRC-1 × Rani Sel-1 | 14.13 | 3.00 | 3.69 | 4.30 | 4.56 | 52.40 | 85.13 | 1370.06 |
| 12 | PRC-1 × SSP | 13.56 | 3.05 | 3.20 | 4.43 | 4.00 | 50.40 | 84.88 | 865.22 |
| 13 | PRC-1 × Sel-12-2-1 | 9.75 | 3.23 | 3.29 | 3.40 | 3.73 | 48.90 | 88.53 | 995.76 |

**Table 3.** Contd.

| | | | | | | | | | |
|---|---|---|---|---|---|---|---|---|---|
| 14 | PRC-1 × SP-316 | 9.41 | 2.95 | 2.40 | 4.40 | 4.46 | 44.40 | 91.46 | 644.26 |
| 15 | Rani Sel-1 × SSP | 11.35 | 3.11 | 3.40 | 4.00 | 4.66 | 47.20 | 101.69 | 809.92 |
| 16 | Rani Sel-1 × Sel-12-2-1 | 11.79 | 3.26 | 3.63 | 3.55 | 3.96 | 58.90 | 105.89 | 1023.12 |
| 17 | Rani Sel-1 × SP-316 | 10.96 | 3.46 | 2.96 | 3.06 | 3.60 | 51.66 | 46.63 | 581.92 |
| 18 | SSP × Sel-12-2-1 | 14.18 | 2.86 | 3.60 | 3.96 | 4.60 | 45.40 | 73.61 | 830.06 |
| 19 | SSP × SP-316 | 13.58 | 2.80 | 3.43 | 3.96 | 4.60 | 47.00 | 95.63 | 898.38 |
| 20 | (Sel Sel-12-2-1 × SP-316 | 11.56 | 2.36 | 3.20 | 3.38 | 3.60 | 55.60 | 80.13 | 597.22 |
| 21 | California Wonder × SP-316 | 16.76 | 3.13 | 2.96 | 4.76 | 4.20 | 52.40 | 69.57 | 631.53 |
| | SE± | 1.57 | 0.27 | 0.35 | 0.37 | 0.32 | 0.75 | 6.24 | 36.12 |
| | CD at 1% | 3.17 | 0.56 | 0.71 | 0.76 | 0.66 | 2.02 | 12.61 | 73.00 |
| | CD at 5% | 4.24 | 0.75 | 0.95 | 1.02 | 0.89 | 1.51 | 16.87 | 97.67 |

cross combinations California Wonder × SP-316 (156.61%), PRC-1 × California Wonder (121.99%) and California Wonder × SSP (98.57%) had exhibited the significant highest positive heterobeltiosis, respectively. Only ten F1 crosses could reveal significant positive heterobeltiosis over better parent. Out of fifteen, fourteen F1 crosses exhibited significant positive heterosis except PRC-1 × SP-316 over the standard check, respectively. The highest economic heterosis was exhibited by the hybrid California Wonder × SP-316 to the tune of 155.98%. The top three hybrids for number of fruits per plant over better parent were California Wonder × SP-316 (156.61%), PRC-1 × California Wonder (121.99%) and California Wonder × SSP (98.57%) and SP-316 × California Wonder (155.98%), PRC-1 × Rani Sel-1 (155.82%) and California Wonder × SSP (98.27%) over standard check. The perusal of data of plant height revealed that the magnitudes of heterosis were ranged from -3.26 to 30.30 and -2.80 to 6.70% over better parent and standard check, respectively. Only one F1 cross PRC-1 × SP-316 (-3.26%) was found with significant negative heterosis for plant height over superior parent. Among 15 F1 crosses, twelve were found with significant negative heterosis over standard check. The highest desirable economic heterosis was recorded in cross combination Rani Sel-1 × Sel-12-2-1 (6.70%) over standard check for plant height.

The magnitudes of heterosis for number of lobes were ranged from 18.65 to 117.91% over better parent and -46.45 to -21.56% over standard check, respectively. All the fifteen crosses were found with positive significant heterosis for number of lobes per fruit over superior parent. The cross combination Rani sel-1 × Sel-12-2-1 (117.91) followed by PRC-1 × Sel-12-2-1 (112.09) had exhibited highest significant positive heterobeltiosis. None of the crosses was found with significant positive heterosis over standard check. The highest negative heterosis was recorded in cross combination Rani Sel-1 × SP-316 (-21.56%) over standard check. The top three hybrids for number of lobes over better parent were Rani Sel-1 × Sel-12-2-1 (117.91%), PRC-1 × Sel-12-2-1 (112.09%) and SSP × SP-316 (107.02%), respectively. The magnitude of heterosis was ranged from -4.09 to 22.50% and -34.80 to 6.42% over the better parent and standard check, respectively. Only four F1 crosses were found with positive significant heterosis for pedicel length over superior parent. The F1 crosses Rani Sel-1 × Sel-12-2-1 (22.50%) followed by PRC-1 × Sel-12-2-1 (21.25%) had exhibited the highest significant positive heterobeltiosis. Among 15 F1 crosses, none of the cross was found with significant positive heterosis over standard check. The top ranking three hybrids for pedicel length over better parent were Rani Sel-1 × Sel-12-2-1 (22.50%), PRC-1 × Sel-12-2-1 (21.25%) and Sel-12-2-1 × SP-316 (11.25%).

The magnitudes of heterosis for pericarp thickness were ranged from -15.59 to 70.31 and -44.72 to 33.52% over better parent and standard check, respectively. Out of fifteen, only five F1 crosses were found with positive significant heterosis for pericarp thickness over superior parent. The cross combination Rani Sel-1 × Sel-12-2-1 (70.31%) followed by SSP × Sel-12-2-1 (68.75%) had exhibited the highest significant positive heterobeltiosis. As many as four crosses were found with significant positive heterosis over standard check. The highest economic heterosis was recorded in cross combination California Wonder × SP-316 (33.52%) over standard check. The top three hybrids for pericarp thickness over better parent were Rani Sel-1 × Sel-12-2-1 (70.31%), and PRC-1 × Sel-12-2-1 (54.53%), and

**Table 4.** Estimation of Heterosis in $F_1$ crosses of bell pepper for different horticulture traits.

| Cross | P1 × P2 | P1 × P3 | P1 × P4 | P1 × P5 | P1 × P6 | P2 × P3 | P2 × P4 | P2 × P5 | P2 × P6 | P3 × P4 | P3 × P5 | P3 × P6 | P4 × P5 | P4 × P6 | P5 × P6 |
|---|---|---|---|---|---|---|---|---|---|---|---|---|---|---|---|
| 1a | -7.36** | -4.03* | -7.48** | 4.84** | -9.56** | 5.83** | -12.86** | -6.22** | -9.67** | -6.72** | -0.84 | -4.56** | -10.74** | -8.57** | 2.04 |
| 1b | -13.33** | -10.83** | -12.50** | -0.83 | -13.33** | 5.83** | -12.50** | -5.83** | -6.66** | -7.50** | -1.66 | -4.16** | -10.00** | -6.66** | 4.16* |
| 2a | -7.36** | -4.03* | -7.48** | 4.84** | -9.56** | -11.87** | -3.01 | -6.27** | -9.67** | 1.90 | -10.77** | -2.69 | -9.48** | -13.00** | 2.15 |
| 2b | -13.33** | -10.83** | -12.50** | -0.83 | -13.33** | -11.87** | 2.74 | -4.56** | -6.66** | -2.28 | -5.47** | -0.91 | -4.11* | 11.41* | 8.21** |
| 3a | 22.63* | 44.44** | 21.05* | 26.31* | 0.00 | 10.52 | -5.00 | -8.50 | -20.00* | 5.55 | 27.22* | 41.66** | 0.43 | 5.00 | -3.50 |
| 3b | 29.93** | 24.93** | 14.94 | 14.94 | 15.04 | 4.94 | -5.04 | -8.79 | 20.04* | 14.94 | 10.79 | 31.43** | 9.94 | -0.05 | -8.49 |
| 4a | 16.33** | 35.07** | -1.16 | 97.94** | 11.77* | 5.94 | 39.41** | 31.49** | -20.96** | -2.10 | 102.73** | 2.76 | 83.58** | 48.18** | 15.57* |
| 4b | 16.34** | 49.50** | 9.41* | 31.31** | -2.96 | 5.94 | 39.42** | -12.76** | -31.38** | 12.19** | 34.49** | -10.78* | 21.78** | 28.64** | -23.32** |
| 5a | -2.46 | 9.97* | 8.20* | 28.58** | -5.63 | -7.82 | 14.87** | 2.35 | 5.43 | 3.39 | 30.64** | -5.24 | 22.38** | 9.76* | -2.63 |
| 5b | -2.52 | 16.12** | 14.26** | 8.54* | -13.35** | -7.88 | 14.80** | -13.59** | -3.18 | 9.92* | 10.28* | -12.99** | 3.30 | 0.78 | 17.81* |
| 6a | 11.25** | 11.55* | 1.51 | 27.03** | 31.39** | 18.79** | 9.62 | 34.17** | -3.96 | 9.17 | 24.51** | 5.56 | 32.54** | 40.76** | 36.74** |
| 6b | 11.21** | 29.51** | 16.24* | 10.22 | 52.54** | 18.75** | 9.59 | 16.41* | -3.99 | 25.01** | 8.03 | 24.77** | 15.00* | 61.18** | 18.64** |
| 7a | -1.61 | 15.78** | -4.11 | 24.28** | 4.80 | -4.43 | 11.21* | 5.07 | -8.74 | 16.78** | 21.90** | 1.04 | 25.71** | 5.28 | 13.13* |
| 7b | -1.59 | 16.55** | -1.57 | 8.57 | 1.43 | -4.41 | 11.24* | -8.20 | -11.66* | 17.56** | 6.49 | -2.19 | 9.82* | 1.90 | -1.16 |
| 8a | 121.99** | 85.75** | 77.88** | 28.05 | 23.25 | 51.68* | 98.57** | 78.20** | 156.61** | 36.56 | 42.17* | 31.59 | 45.24** | 39.10* | -10.10 |
| 8b | 122.18** | 155.82** | 107.02** | 48.90* | 43.76 | 51.34* | 98.27** | 78.21** | 155.98** | 73.33** | 80.05** | 67.43** | 116.48** | 107.37** | 76.59** |
| 9a | 25.49* | 14.16* | 12.85* | 8.18* | -3.26* | 4.50* | 6.58* | 19.14* | 4.25* | 5.68* | 30.30* | 5.42* | 1.65 | 5.23* | 2.21 |
| 9b | 4.34* | -5.07* | -8.69* | -13.04* | -19.56* | -5.00* | -13.76* | -2.80* | -5.07* | -14.49* | 6.70* | -6.41* | -17.75* | -14.85* | -16.30* |
| 10a | 90.26** | 99.25** | 25.83** | 112.09** | 48.46** | 37.79** | 104.64** | 18.65** | 45.52** | 15.56** | 117.91** | 34.09** | 76.79** | 107.02** | 37.62** |
| 10b | -29.86** | -32.12** | -30.99** | - 26.84** | -33.25** | -28.88** | -27.60** | -35.89** | -29.11** | -29.48** | -26.09** | -21.56** | -35.14** | -36.65** | - 46.45** |
| 11a | 6.89 | 3.44 | 5.17 | 21.25* | 1.72 | 8.39 | -4.09 | 6.25 | 2.73 | -2.60 | 22.50* | 13.66 | 7.50 | *8.19 | *11.25 |
| 11b | -15.51 | 1.83 | -11.84 | -9.18 | -33.88** | -24.24* | 6.42 | -34.80** | -18.27 | -6.33 | 0.09 | -18.27 | -0.82 | -5.41 | 11.84 |
| 12a | 7.85 | 30.01* | 12.54 | 54.53** | -15.59 | -3.28 | 15.20 | 10.93 | 3.48 | 17.24 | 70.31** | 3.48 | 68.75** | 19.76 | 50.00** |
| 12b | -44.72** | 20.54 | 24.18* | -4.66 | 23.24* | -5.22 | 26.98* | -18.39 | 33.52** | 12.04 | -0.46 | -14.09 | 11.11 | 11.11 | -5.13 |
| 13a | -32.72* | 46.70** | 51.13** | 16.02 | 50.00** | 15.34 | 36.00** | -4.37 | 33.39** | 32.89* | 18.05 | 1.88 | 30.19* | 19.00 | 11.16 |
| 13b | 7.71 | 16.20 | 1.78 | -5.08 | 13.65 | -5.42 | 5.17 | -9.24 | 6.87 | 18.74* | 0.76 | -8.39 | 17.04* | 17.04* | -8.39 |
| 14a | -1.11 | 30.89* | 20.36** | 3.04 | -1.85 | -15.98 | 21.81** | -8.21 | -2.36 | 8.59 | 10.99 | -3.14 | 7.69 | 7.69 | -10.27 |
| 14b | 37.61** | -2.51 | -2.79 | 1.37 | 4.73 | -42.42** | 6.63 | -18.54* | -20.33** | 16.44* | 21.25** | -25.99** | -15.70* | 9.51 | -8.24 |
| 15a | 45.81** | 27.67** | 32.42** | 20.51* | 10.95 | -39.00** | 45.27** | -3.17 | -20.33** | 58.63** | 58.84** | -3.06 | 14.84 | 49.18** | 9.07 |
| 15b | 90.25** | 170.37** | 76.42** | 103.05** | 31.37** | 37.79** | 140.64** | 13.59 | 28.77** | 65.15** | 108.62** | 18.66* | 69.25** | 83.19** | 21.78** |

a = percent increase over better parent, b = percent increase over standard check (California Wonder) (1 to 15 horticultural traits).

California Wonder × SP-316 (33.52%), California Wonder × SSP (26.98%) and PRC-1 × SSP (24.18%) over standard check. The estimates of percent heterosis for number of branches per plant ranged from -4.37 to 51.13 and -9.24 to 18.74% over the superior parent and standard check, respectively. Seven F1 crosses had exhibited significant positive heterobeltiosis. The highest heterobeltiosis was recorded in

**Table 5.** Top three parents and cross combinations on the basis of their per se performance and heterotic values.

| Traits | Per se performance | | Heterosis | |
|---|---|---|---|---|
| | Parents | Crosses | BP | SC |
| Days to 50% flowering | PRC-1, Rani Sel-1,<br>SSP | PRC-1 × California Wonder,<br>PRC-1 × SP-316 (34.66),<br>PRC-1 × SSP (35.00) | California Wonder × SSP (-12.86),<br>SSP × Sel-12-2-1 (-10.74),<br>California Wonder × SP-316 (-9.67) | PRC-1 × California Wonder (-13.33),<br>PRC-1 × SP-316 (-13.33)<br>California Wonder × SSP (-12.50) |
| Days to first harvest | PRC-1, Rani Sel-1<br>California Wonder | PRC-1 × SP-316 (62.33),<br>California Wonder × SSP(64.33),<br>PRC-1 × SSP (64.33) | PRC-1 × SP-316 (-16.14),<br>SSP × SP-316 (-13.00),<br>California Wonder × SSP (-11.87) | California Wonder × SSP (-11.87)<br>PRC-1 × SSP (-11.87)<br>PRC-1 × California Wonder (-10.04) |
| No. of pickings | SSP, Rani Sel-1,<br>PRC-1 | Rani Sel-1× SP-316 (8.76),<br>PRC-1 × California Wonder (8.66),<br>PRC-1 × Rani Sel-1(8.33) | PRC-1 × Rani Sel-1 (44.44),<br>Rani Sel-1× SP316 (41.66),<br>Rani Sel-1× Sel-12-2-1 (27.22) | Rani Sel-1 × SP-316 (31.43),<br>PRC-1 × California Wonder (29.93)<br>PRC-1 × Rani Sel-1 (24.93) |
| Fruit weight | SSP, Rani Sel-1,<br>California Wonder | PRC-1 × Rani Sel-1(100.66),<br>California Wonder × SSP(93.87),<br>Rani Sel-1× Sel-12-2-1(90.55) | Rani Sel-1× Sel-12-2-1 (102.73),<br>PRC-1 × Sel-12-2-1 (97.94),<br>SSP × Sel-12-2-1 (83.58) | PRC-1 × Rani Sel-1 (49.50),<br>California Wonder × Rani Sel-1 (39.41),<br>Rani Sel-1× Sel-12-2-1 (34.49) |
| Fruit diameter | SSP, Rani Sel-1,<br>PRC-1 | California Wonder × SSP(6.36),<br>PRC-1 × SSP (6.33),<br>Rani Sel-1× Sel-12-2-1(6.11) | Rani Sel-1× Sel-12-2-1 (30.64),<br>PRC-1 × Sel-12-2-1 (28.58)<br>SSP × Sel-12-2-1 (22.38) | Sel-12-2-1× SP-316 (17.81),<br>PRC-1 × Rani Sel-1 (16.12)<br>California Wonder × SSP (14.80) |
| Fruit length | Rani Sel-1<br>SSP, PRC-1 | SSP × SP-316 (15.18),<br>PRC-1 × SP-316 (14.37),<br>PRC-1 × Rani Sel-1(12.20) | SSP × SP-316 (40.76),<br>Sel-12-2-1× SP-316 (36.74),<br>California Wonder × Sel-12-2-1 (34.17) | SSP × SP-316 (61.18),<br>PRC-1 × SP-316 (52.54),<br>PRC-1 × Rani Sel-1 (29.51) |
| Fruit girth | SSP, Rani Sel-1<br>PRC-1 | Rani Sel-1× SSP (20.15),<br>PRC-1 × Rani Sel-1 (19.97),<br>California Wonder × SSP (19.06) | SSP × Sel-12-2-1 (25.71),<br>PRC-1 × Sel-12-2-1 (24.28),<br>Rani Sel-1× Sel-12-2-1) (21.90) | Rani Sel-1× SSP (17.56),<br>PRC-1 × Rani Sel-1 (16.55)<br>California Wonder × SSP (11.21) |
| No. of fruit/ plant | SP-316,<br>Sel-12-2-1,<br>SSP | California Wonder × SP-316 (16.76)<br>PRC-1 × California Wonder (14.55)<br>SSP × Sel-12-2-1(14.18) | California Wonder × SP-316 (156.61),<br>PRC-1 × California Wonder (121.99)<br>California Wonder × SSP (98.57) | California Wonder × SP-316 (155.98)<br>PRC-1 × Rani Sel-1 (155.82),<br>PRC-1 × California Wonder (122.18) |
| No. of lobs/ fruit | California Wonder<br>SSP, Rani Sel-1 | Rani Sel-1× SP-316 (3.46),<br>Rani Sel-1× Sel-12-2-1 (3.26),<br>PRC-1 × Sel-12-2-1 (3.23) | Rani Sel-1× Sel-12-2-1 (117.91),<br>PRC-1 × Sel-12-2-1 (112.09),<br>SSP × SP-316 (107.02) | Rani Sel-1× SP-316 (-21.56),<br>Rani Sel-1× Sel-12-2-1 (-26.09),<br>PRC-1× Sel-12-2-1 (-26.84) |

**Table 5.** Contd.

| | | | | |
|---|---|---|---|---|
| Pedicel length | CaliforniaWonder, SSP, Rani Sel-1 | California Wonder × SSP (3.86), PRC-1 × Rani Sel-1 (3.69), Rani Sel-1× Sel-12-2-1 (3.63) | Rani Sel-1× Sel-12-2-1 (22.50), PRC-1 × Sel-12-2-1 (21.25), Rani Sel-1 × SP-316 (13.66) | California Wonder × SSP (6.42), Rani Sel-1× Sel-12-2-1 (0.092), SSP × Sel-12-2-1 (-0.826) |
| Pericarp thickness | SP-316, California Wonder, SSP | California Wonder × SP316 (4.76) Rani Sel-1× SSP (4.60), PRC-1 × SP-316 (4.40) | Rani Sel-1× Sel-12-2-1(70.31), SSP × Sel-12-2-1 (68.75), PRC-1 × Sel-12-2-1 (54.53) | California Wonder × SP316 (33.52), California Wonder × SSP (26.98), PRC-1 × SP316 (23.24) |
| No. of branches/plant | Rani Sel-1, SSP, Sel-12-2-1 | Rani Sel-1× SSP (4.66), SSP × Sel-12-2-1 (4.60), SSP × SP-316 (4.60) | PRC-1 × SSP (51.13), PRC-1 × SP-316 (50.00), PRC-1 × Rani Sel-1 (46.70) | Rani Sel-1× SSP (18.74), SP × Sel-12-2-1 (17.04) SSP × SP-316 (17.04) |
| Plant height | Sel-12-2-1, PRC-1, SSP | PRC-1×SSP, PRC-1×SP-16 Rani Sel-1×SSP | PRC-1 × SSP | California Wonder × Sel-12-2-1 PRC-1 × Rani Sel California Wonder × SP-316 |
| Ascorbic acid content | SP-316, California Wonder, PRC-1 | PRC-1 × California Wonder (120.18), Rani Sel-1× Sel-12-2-1 (105.89), Rani Sel-1× SSP (101.69) | PRC-1 × Rani Sel-1 (30.89), California Wonder × SSP (21) PRC-1 × SSP (20.36) | PRC-1 × California Wonder (37.61), Rani Sel-1× Sel-12-2-1 (21.25), California Wonder × SSP (6.63) |
| Marketable fruit yield/plant | SSP, Rani Sel-1, PRC-1 | PRC-1 × Rani Sel-1 (1370.06), California Wonder × SSP (1180.13), Rani Sel-1× Sel-12-2-1 (1023.12) | Rani Sel-1× SP-316 (58.84), Rani Sel-1× Sel-12-2-1 (58.63), SSP × SP-316 (49.18) | PRC-1 × Rani Sel-1 (170.37), California Wonder × SSP (140.64), Rani Sel-1 × Sel-12-2-1 |

PRC-1 × SSP (51.13%) followed by PRC-1 × SP316 (50.00%). On other side, the highest economic heterosis was recorded in cross Rani Sel-1 × SSP to the tune of 18.74% over standard check. The top three hybrids for number of branches per plant over better parent were PRC-1 × SSP (51.13%), PRC-1 × SP-316 (50.00%) and PRC-1 × Rani Sel-1 (46.70%) and Rani Sel-1 × SSP (18.74%), SSP × Sel-12-2-1 (17.04%) and SSP × SP-316 (17.04%) over standard check.

The magnitudes of heterosis for ascorbic acid content were ranged from -15.98 to 30.89 and -25.99 to 37.61% over better parent and standard check, respectively. Only three crosses were found with positive significant heterosis for ascorbic acid content in F1 generation over superior parent. The cross combination PRC-1 × Rani Sel-1 (30.89%) followed by California Wonder × SSP (21.81%) had exhibited the highest significant positive heterobeltiosis. Three crosses out of fifteen were found with significant positive heterosis over standard check. The highest economic heterosis to the extent of 37.61% was recorded in cross combination PRC-1 × California Wonder over standard check. The top three hybrids for ascorbic acid content over better parent were PRC-1 × Rani Sel-1 (30.89%), California Wonder × SSP (21.81%) and PRC-1 × SSP (20.36%) and PRC-1 × California Wonder (37.61%), Rani Sel-1 × Sel-12-2-1 (21.25%) and Rani Sel-1 × SSP (16.44%) over Standard check.

The magnitudes of heterosis for marketable fruit yield per plant were ranged from -39.00 to 58.84% and -13.59 to 170.37% over better parent and standard check, respectively. Eight crosses were observed with positive significant heterosis for marketable fruit yield per plant in F1 generation over superior parent. The cross combination Rani Sel-1 × Sel-12-2-1 followed by Rani Sel-1 × SSP had manifested the highest significant positive heterobeltiosis to the tune of 58.84 and 58.63%, respectively. Out of fifteen F1 crosses, fourteen were found with significant positive heterosis over standard check. The highest economic heterosis was recorded in cross combination PRC-1 x Rani

Sel-1 to the tune of 170.37% over standard check followed by California Wonder × SSP (140.64%) and Rani Sel-1 × Sel-12-2-1 (108.62%), respectively.

The top ranking three hybrids for marketable yield over better parent were Rani Sel-1 × Sel-12-2-1 (58.84%), Rani Sel-1 × SSP (58.63%) and SSP × SP-316 (49.18%) and PRC-1 × Rani Sel-1 (170.37%), California Wonder × SSP (140.64%) and Rani Sel-1 × Sel-12-2-1 (108.62%) over standard check.

## DISCUSSION

Heterosis is the increase of size, yield and vigor through cross-breeding rather than interbreeding; without increases, there is no heterosis. Heterosis has a slightly more extensive coverage than hybrid vigor; that is, though all hybrid vigor is heterosis, not all heterosis can be with equal propriety termed hybrid vigor. Heterosis breeding is a potential method to achieve improvement in production and productivity of bell pepper that otherwise cannot be achieved through existing traditional methods. Negative heterosis for days to 50% flowering is considered desirable for earliness. In the present study, twelve $F_1$ crosses exhibited significant negative heterosis for days to 50% flowering. The cross combination California Wonder × SSP (-12.76%) followed by SSP × Sel-12-2-1 (-10.72%) had revealed the highest heterobeltiosis for this trait. Sujiprihati et al. (2007) had also been reported desirable negative heterosis for days to 50% flowering. Heterosis for early fruit yield was manifested through negative desirable heterosis for days to first harvest. The $F_1$ crosses California Wonder × SSP and PRC-1 × SSP exhibited highest significant heterosis in desirable direction to the tune of - 11.87% each over standard check. The highest economic heterosis was recorded for number of fruit pickings in cross Rani Sel-1 × SP-316 to the tune of 31.43% over standard. Similar findings had also been reported by Prasad et al. (2003). Such cross could be utilized for hybrid breeding for production of fruits over a long time by increasing picking. The highest economic heterosis for fruit weight was recorded in cross combination PRC-1 × Rani Sel-1 (49.50%) over standard check.

Heterobeltiotic effects for fruit weight in bell pepper were earlier reported by Gomide et al. (2008). The top ranking hybrids for fruit diameter were Rani Sel-1 × Sel-12-2-1 (30.64%), PRC-1 × Sel-12-2-1 (28.58%) and SSP × Sel-12-2-1 (22.38%) over better parent and Sel-12-2-1 × SP-316 (17.81%), PRC-1 × Rani Sel-1 (16.12%) and California Wonder × SSP (14.80%) over standard check. Significant heterosis for fruit diameter was also reported earlier by Ahmed et al. (2003), Rajesh and Gulshan (2001) and Prasad et al. (2003). Significant positive highest magnitude of heterosis for fruit girth was recorded in SSP × Sel-12-2-1 (25.71%), PRC-1 × Sel-12-2-1 (24.28%), Rani Sel-1 × Sel-12-2-1 (21.90%) over the better parent. The maximum useful heterosis was exhibited by the cross combination SSP × Sel-12-2-1 to the tune of -8.74 to 25.71% over standard check. High value of heterosis favours the development of hybrid, which was in conformity with the findings of Bhagyalakshmi et al. (1991). The highest heterosis to the tune of 155.98% over standard parent was exhibited by the hybrid California Wonder × SP-316. Significant heterosis for number of fruit per plant had also been observed earlier by Joshi (1986). Among the traits, fruit per plant exhibited maximum heterosis over standard check 'Aishwarya' followed by marketable fruit per plant, harvest duration and fruit yield per plant by Sood and Kumar (2010b). Shrestha et al. (2011) were found hybrid of 5AVS7 × SP32 exhibited the highest heterosis for fruit number (104.0%) and yield (141.2%) per plant. Hybrids of 5AVS7 × SP45, 5AVS7 × SP32 and 5AVS8 × SP48 had highest positive standard heterosis on fruit yield per plant over Special, Fiesta and President.

The results, thus suggested that heterosis breeding may be utilized to exploit the non-additive components followed by selection in segregating generations. The parents SP-316 and Sel-12-2-1 could be utilized in future breeding programme to develop hybrids/pure lines having more number of fruits per plant. The $F_1$ cross Rani Sel-1 × Sel-12-2-1 (22.50%) followed by PRC-1 × Sel-12-2-1 (21.25%) had exhibited the highest significant positive heterobeltiosis. Significant heterosis for pedicel length had also been reported earlier by Ahmed et al. (2003) and Gomide et al. (2008). The $F_1$ crosses PRC-1 × SP-316 (-3.26%) had exhibited the highest significant positive heterobeltiosis for plant height. Similar finding were also reported earlier by Ahmed et al. (2003) and Rajesh and Gulshan (2001). Significant heterosis was found over better parent in the cross Sel-1 × Sel-12-2-1 (117.91%), PRC-1 × Sel-12-2-1 (112.09%) and SSP × SP-316 (107.02%). Similar results were also obtained earlier by Sujiprihati et al. (2007). The highest economic heterosis for number of branches was recorded in cross Rani Sel-1 × SSP to the tune of 18.74% over standard check. The significant heterosis for number of branches had over better parent and standard check also been reported earlier by Joshi (1986) and Sujiprihati et al. (2007). The highest useful heterosis for the trait had exhibited by PRC-1 × California Wonder to the tune of 37.61% over standard check. Similar findings for ascorbic acid were also reported by Vandana et al. (2002) and Gomide et al. (2008).

The highest economic heterosis for fruit yield per plant was recorded in cross combination PRC-1 × Rani Sel-1 to the tune of 170.37% over standard check followed by California Wonder × SSP (140.64%) and Rani Sel-1 × Sel-12-2-1 (108.62%), respectively. Significant desirable heterosis for fruit yield per plant was also reported earlier by Zecevic (1997), Ahmed et al. (2003), Milerue and Nikornpun (2006) and Sujiprihati et al. (2007). On the basis of heterosis and per se performance, PRC-1 × Rani

Sel-1 was the best cross-combination followed by California Wonder × SSP and Rani Sel-1 × Sel-12-2-1 for fruit yield per plant (Table 5). These cross combinations also had high heterosis for most of the yield contributing traits. These hybrids offer high scope for the exploitation of heterosis for improving horticultural traits. These cross-combinations could be utilized as hybrid breeding and can also be released as hybrids after further field testing.

## REFERENCES


Ahmed N, Hurra M, Wani SA Khan SH (2003). Gene action and combining ability for fruit yield and its component characters in sweet pepper. Capsicum Eggplant Newslett. 22:55-58.

Bhagyalakshmi PV, Shankar CR, Subramanyam D, Babu VG (1991). Heterosis and combining ability studies in chillies. Indian J. Gen. Pl, Breed. 51(4):420-423.

Doshi SP, Gupta KC (1981). Development of two programme packages in Fortran IV for analysis of diallel set data. A report. Indian Agricultural statistical Research Institute (ICAR), New Delhi.

Duke JA (1992). Biologically Active Phytochemicals and Their Activities. CRC Press, Boca Raton, FL.

Gomide ML, Maluf WR, Gomes LAA (2008). Heterosis and combining capacity of sweet pepper lines (*Capsicum annuum L.*). Ciencia-e-Agrotechnologi 27(5):1007-1015.

Haytowitz DB, Matthews RH (1984). Composition of Foods: Vegetables and Vegetable Products Raw, Processed, Prepared. Agric. Handbook Number 8-11. U.S. Dept. Agric., Washington, DC.

Joshi S, Singh B (1980). A note on hybrid vigour in sweet pepper (*Capsicum annuum* L.). Haryana J. Hort. Sci. 9:90-92.

Joshi S (1986). Results of heterosis breeding on sweet pepper (*Capsicum annuum L.*). Capsicum Newslett. 5:33-34.

Lee Y, Howard, LR, Villalon B (1995). Flavonoid and ascorbic acid content and antioxidant activity of fresh pepper (*Capsicum annuum*) cultivars. IFT Abstract. 55:79

Milerue N, Nikorpun M (2000). Studies on heterosis of chili (*Capsicum annuum* L.). Kasetsart J. Nat. Sci. 34:190-196.

Pairce LC (1987). Vegetable: Characterstics, Production and Marketing. John Wiley and Sons, New York. P. 325.

Prasad NBC, Madhavi RK, Sadashiva AT (2003). Heterosis studies in chilli (*Capsicum annuum* L.). Indian J. Hort. 60:69-74.

Rajesh K, Gulshan L (2001). Expression of heterosis in hot pepper (*Capsicum annuum* L.). Capsicum and eggplant Newslett. 20:38-41.

Ranganna S (1986). Hand book of analysis and quality control for fruit and vegetable products. Tata Mc Graw-Hill, Publishing Company Ltd., New Delhi. pp. 105-106.

Shrestha SL, Binod PL, Won HK (2011). Heterosis and Heterobeltiosis Studies in Sweet Pepper (*Capsicum annuum* L.). Hort. Environ. Biotechnol. 52(3):278-283.

Sood S, Naveen K (2010a). Heterosis for fruit yield and related horticultural traits in bell pepper. Int J Vegetable Sci. 16(4):361-373.

Sood S, Kumar N (2010b).Heterotic expression for fruit yield and yield components in intervarietal hybrids of sweet pepper (*Capsicum annuum* L. var. grossum Sendt.). SABRAO J. Breed. Gene. 42(2):106-116.

Sood S, Naveen K, Chandel KS, Parveen S (2011). Determination of genetic variation for morphological and yield traits in bell pepper (*Capsicum annuum* var. *grossum*). Indian J. Agric. Sci. 81(7):590-594.

Sujiprihati S, Yunianti R, Syukur M, Undang (2007). Pendugaan nilai heterosis dan daya gabung beberapa komponen hasil pada persilangan dialel penuh enam genotipe cabai (*Capsicum annuum* L.). Bul. Agron. 35:28-35.

Vandana P, Ahmed Z, Kumar N (2002). Heterosis and combining ability in diallel crosses of sweet pepper (*Capsicum annuum* L). Veg. Sci. 29(1):66-67.

Zecevic B (1997). Heterosis effect on some cultivar hybrids of pepper (*Capsicum annuum* L.). Review of research work at Faculty of Agriculture (Yugoslavia). 42(1):169-181-64.

# 10

# Leaf senescence and physiological characters in different adzuki bean (*Vigna angularis*) cultivars (lines)

**Hui Song[1], Xiaoli Gao[1], Baili Feng[1], Huiping Dai[2], Panpan Zhang[1], Jinfeng Gao[1], Pengke Wang[1] and Yan Chai[1]**

[1]State Key Laboratory of Crop Stress Biology on Drought Regions, Northwest A&F University, Yangling, Shaanxi 712100, P. R. China.
[2]College of Life Science, Northwest A&F University, Yangling, Shaanxi 712100, China.

**The aim of the study is to examine the relation of leaf senescence and its physiological characters at different flower internodes from flowering to maturing, to explore the aging mechanisms of adzuki bean leaf, to find out the intrinsic yield-forming mechanisms and to provide a theoretical basis for high-yield breeding and production of adzuki beans. A field experiment was conducted, the high-yielding (2000-75 and JiHong 9218) and the low-yield varieties (HongBao1 and WanXuan1), all adopted in the summer planting ecological region of China, were grown in 2008 and 2009, and their leaf physiological characters, such as chlorophyll contents, net photosynthetic rates (Pn), superoxide dismutase (SOD), catalase (CAT), peroxidase (POD) activities and malondialdehyde (MDA) contents were determined. The results indicated that the chlorophyll contents, $P_n$, SOD and CAT activities gradually decreased from 15 days after the varieties flowered to maturing, but their POD activities and MDA contents gradually increased when leaves senescence started. Leaf senescence initiated from the low internodes and gradually moved toward to the upper internodes after the plants flowered. Compared with low-yielding varieties, high-yielding varieties maintained higher contents of chlorophyll contents, $P_n$ and SOD, CAT in the late stages, and thus resulting significantly higher grain yields. The overall data indicated that yield is positively correlated with leaf chlorophyll and $P_n$, as well as SOD and CAT activities and negatively associated with POD activities and MDA accumulations at the late growth stage. Therefore, effective inhibition of leaf senescence or prolonging the functional period of leaves at the late growing stage plays an important role in raising yield.**

**Key words:** Adzuki bean, leaf senescence, chlorophyll content, seed yield.

## INTRODUCTION

Adzuki bean (*Vigna angularis* (Willd.) Ohwi & H. Ohashi), one of the major food legumes of China, has a short growth period and is tolerant to drought, poor soil fertility and salinity (Lin et al., 2002). In traditional Chinese medication, adzuki is commonly used for many purposes including diuretics and antidotes, and symptom alleviations of dropsy and beriberi (Itoh and Furuichi, 2009). Recent studies indicate that adzuki bean is widely planted in the northeast and northern parts of China. Strong interests in planting adzuki bean have aroused in China because of its high quality.

Leaf senescence, which occurs at the final leaf development stage, is a critical process for plants because the process exerts influence on their fitness and leaf nutrient relocation into their seeds. Leaf senescence involves a coordinated process under the control of the highly regulated genetic programs at cellular, tissue, organ, and organism levels (Nooden, 1988). Theoretically, delayed leaf senescence after flowering and increased photosynthesis at the seed filling stage can improve seed dry matter accumulation and then resulting in yield increase (Hayati et al., 1995). The onset of leaf senescence is typically characterized as an orderly procedure which involves catabolism of chlorophyll, proteins, lipids and nucleic acids along with nutrient remobilization to developing grains, and ended with plant death (Breusegem and Dat, 2006). According to the free radical theory, leaf senescence is that if reactive oxygen species are metabolically imbalanced, their reactive oxygen species accumulation substantially increases the product (MDA) content of lipid peroxidation, and the damage caused by oxygen free radicals exerts direct influence on the progress of aging. Superoxide dismutase (SOD), catalase (CAT) and peroxidase (POD), which play their roles coherently, can eliminate excessive reactive oxygen species, maintain their balance and protect membrane structure, delay the onset of aging (Kukavica and Velijovic, 2004). In the recent years, researches on leaf senescence, which mainly focus on crops, such as wheat (Zhang et al., 2006; Feng et al., 2009), maize (Pommela et al., 2006; Efeoğlu et al., 2009), soybean (Kaschuk et al., 2010), mung bean (Batish et al., 2006; Gao et al., 2007), indicate that one of the main reasons for leaf senescence is that plants produce reactive oxygen species, which can lead to leaf damage and even leaf death.

However, there are few research reports on morphological and physiological changes of adzuki bean. The study investigated the dynamic changes of the chlorophyll contents, $P_n$ and antioxidative enzymes (SOD, CAT, and POD) activities and MDA contents in the leaves of four adzuki bean varieties (two high-yield versus two low-yield varieties) from flowering to ripening. The objectives of the study were to reveal the leaf senescence mechanism of adzuki bean and the relationship between the metabolic properties and yield of the bean, and to provide theoretical guidance for effectively improving field managements for the adzuki bean.

## MATERIALS AND METHODS

### Experimental design

The study was conducted in No.1 Agricultural Experiment Station of the Northwest A&F University, Yangling, Shaanxi, China. Located in the Loess Plateau (108°E and 34°N). The station is covered by a sub-humid warm temperate climate of which the average annual rainfall is 660 mm (mainly distributing from July to September), and the elevation above sea level is 520 m. Two high-yield cultivars (2000-75 and JiHong 9218), and two low- yield cultivars (HongBao1 and WanXuan1) of adzuki bean were planted in 2008 and 2009. The P-K fertilizers were separately applied at 225 and 30 kg $hm^{-2}$ days before the sowing date of June 12. The four adzuki bean cultivars were cultivars participating in the State Regional Bean cultivar Test of China and their yield components are presented in Table 1. The areas of the plots were 2 m × 5 m and in each plot there were six rows of bean planted of which the spaces were 0.4 m. The field management practices for the study were the same as required by the State Regional Adzuki Bean Test of China. The design of the study was a Randomized complete block design with three replications.

Leaf sampling was done within 8:00 to 9:00 am. Every 7 days at the full-blooming stage and sample Leaves were chosen from the 5th, 6th, 7th, and 8th flowering nodes of the stems (the 5th leaf grew on the first flowing node). The middle leaflets of the sample ternate leaves were viewed as the reference for screening leaf-growing nodes for leaf sampling. After collected, the samples were placed in a cooler box, brought to lab, cleaned with water, and blotted with filter paper.

### Measurement of chlorophyll content

The sample leaf chlorophyll were extracted by the 80% acetone-soaking extraction (Heath and Packer, 1968) and measured with UVIKON810 spectrophotometer (Kontron Instruments, Zurich, Switzerland).

### Measurement of net photosynthetic rate

The photosynthetic parameters were measured with a portable LI-6400 photosynthesis system (LI-COR Inc., USA), and while the system was used to measure the parameters, its leaf temperatures and flow rate were set separately at 27°C and 500 $\mu molm^{-2} s^{-1}$, and its photosynthetically active radiation provided with a red-blue light source was set at 1,000 $\mu molm^{-2} s^{-1}$.

### Antioxidative enzyme extractions and activity assays

0.500 g frozen leaves were homogenized in 50 mM potassium phosphate buffer (pH 7.8) containing 1 mM ethylene diamine tetraacetic acid, 3 mM 2-mercaptoethanol, and 2% (w/v) polyvinylpyrrolidone in a chilled mortar and pestle. The homogenate was centrifuged at 20,000 g for 20 min at 4°C and the supernatant was used for enzyme assays.

Superoxide dismutase (SOD) activity was assayed by the nitroblue tetrazolium method (Dhindsa et al., 1981). The 3 ml reaction mixture contained 50 mM Na–phosphate buffer (pH 7.3), 13 mM methionine, 75 mM NBT, 0.1 mM EDTA, 4 mM riboflavin, and 0.2 ml of enzyme extract. This reaction was started by the addition of riboflavin, and the glass test tubes were shaken and placed under fluorescent lamps (160 $\mu mol\ m^{-2}\ s^{-1}$). The reaction proceeded for 5 min and was then stopped by switching off the light. Absorbance was measured at 560 nm. Blanks or controls were run in the same manner but without illumination or enzyme, respectively. One unit of SOD was defined as the amount of enzyme that produced 50% inhibition of NBT reduction under assay conditions.

Activities of peroxidase (POD) and catalase (CAT) were determined by modified the method of Chance and Maehly (1955). Samples were homogenized with acetone on ice. The POD reaction solution (3 ml) contained 50 mM phosphate buffer (pH 7.8), 25 mM guaiacol, 200 mM $H_2O_2$, and 0.5 ml of enzyme extract. Changes in absorbance of the reaction solution at 470 nm were determined every 30 s. One unit of POD activity was defined as an absorbance change of 0.01 unit·$min^{-1}$. The CAT reaction solution (3 ml) contained 50 mM phosphate buffer (pH 7.0), 200 Mm $H_2O_2$, and

**Table 1.** Yield components of different Adzuki bean varieties.

| Variety | Plants /ha | Pods /plant | Seeds/pod | 1000-grain weight (g) | Seeds yield (kg/ha) |
|---|---|---|---|---|---|
| 2000-75 | 100000 | 28.52 | 6.33 | 159.3 | 1579.3 |
| JiHong 9218 | 100000 | 27.91 | 6.54 | 158.4 | 1558.0 |
| HongBao 1 | 100000 | 24.82 | 6.12 | 152.1 | 1337.7 |
| WanXuan 1 | 100000 | 22.31 | 6.08 | 146.5 | 1316.9 |

Each point represents the mean of six biological replicates.

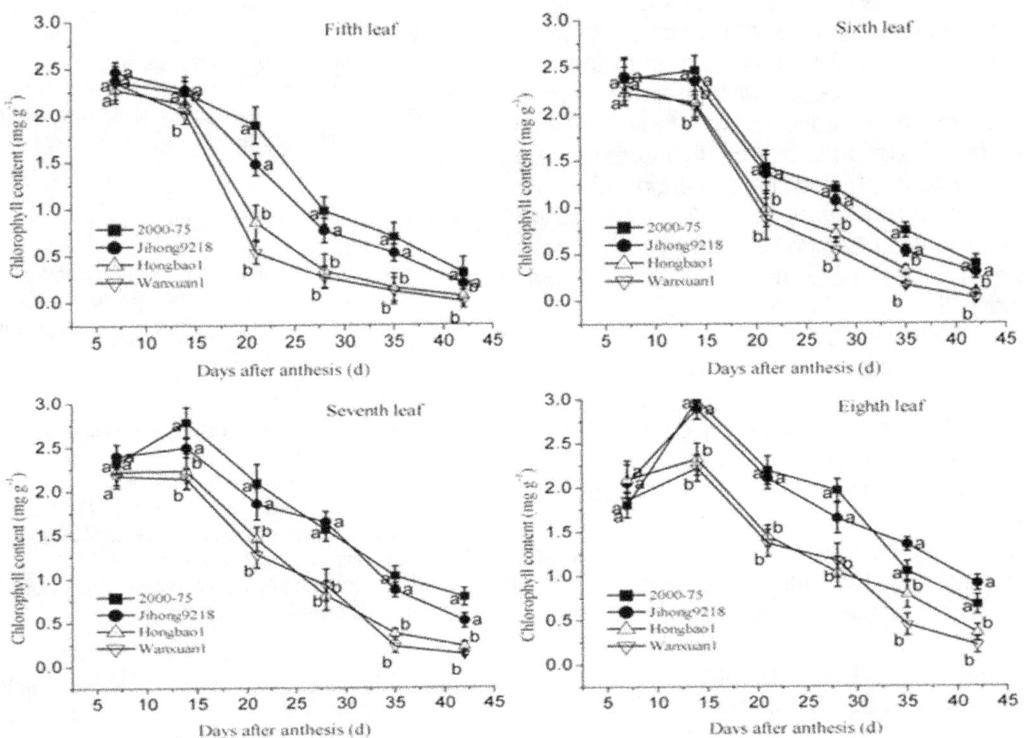


**Figure 1.** Trend of chlorophyll content in the leaves of four adzuki bean varieties at different flowering nodes. Data were the average ± SD in two years (2008 and 2009); a and b, the significant levels in net photosynthetic rate at the same day after anthesis between four varieties and relative ab controls at $p < 0.05$.

50 ml of enzyme extract. The reaction was initiated by adding the extract. Changes in absorbance of the reaction solution at 240 nm were read every 30 s. One unit of CAT activity was defined as an absorbance change of 0.01 units·min$^{-1}$.

### Measurement of MDA content

The MDA contents were determined by modified TBA method (Heath and Packer, 1968). Generally, the crude enzyme extracts will present colors while when TBA reacts with MDA. The MDA contents were calculated depending on the absorbance subtraction at the wavelengths of 532 and 600 nm.

### Statistical analysis

The data were processed using SAS software with the figures accomplishing by Origin 8. The difference significances among the various agronomic traits of the cultivars were tested by analyses of variance (ANOVA) for the various agronomic variables using SAS (SAS Institute, 2003) PROC MIXED procedure followed by Tukey multiple comparison tests.

## RESULTS

### Leaf chlorophyll contents

Figure 1 shows that the leaf Chl contents of the different cultivars tended to decline at the flowering and fruiting stages, and the leaf senescence of the cultivars initiated at their lower nodes and gradually moved up to the upper nodes after their flowering. The leaf Chl contents of 2000-75 and JiHong 9218 were higher than those of HongBao1 and WanXuan1. When HongBao1 and WanXuan1

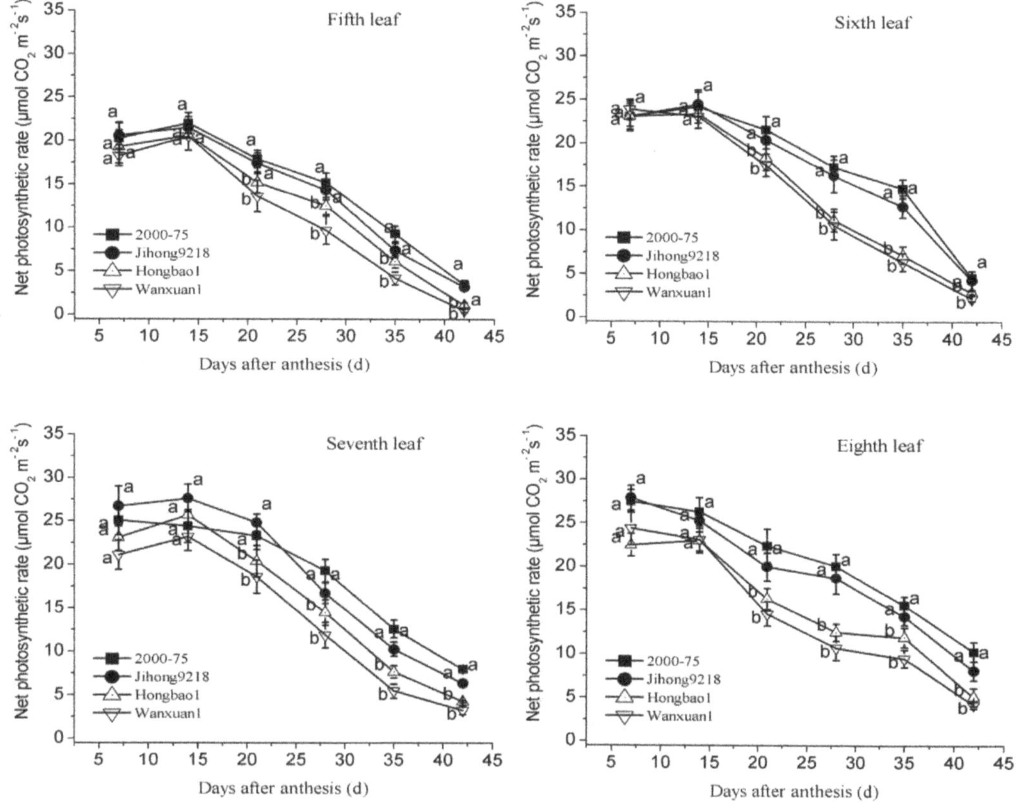


**Figure 2.** Trend of net photosynthetic rate in the leaves of four adzuki bean varieties at different flowering nodes.Data were the average ± SD in two years (2008 and 2009); a and b were the significant levels in net photosynthetic rate at the same day after anthesis between four varieties and relative a,b controls at $p < 0.05$.

approached to their maturity, their leaves became nearly dry and their leaf Chl content was hardly detectable.

## Leaf net photosynthetic rates

The weighted average leaf $P_n$ of the high-yield and low-yield cultivars at the different flowering nodes at the flowering and fruiting stages are shown in Figure 2. The $P_n$ tended to decline at the flowering and fruiting stages. A the different flowering nodes, Te leaf photosynthetic capacities of the high-yield and low-yield cultivars decreased from the eighth leaf through seventh and sixth leaves to the fifth leaf. Compared with the two low-yield cultivars, 2000-75 and JiHong 9218 had considerably higher net leaf photosynthetic rate, indicating that they had higher photosynthetic efficiencies.

## SOD and CAT activities

Figure 3 shows the leaf SOD activities of the adzuki bean cultivars after their anthesis. In general, the cultivars tended to present increasing SOD activities from their lower nodes to their upper nodes and initiate leaf senescence at their lower nodes, and their leaf SOD activities and leaf senescence initiations appeared significantly different, which was similar to what was shown on the CAT activities in Figure 4. High-yield2000-75 and JiHong 9218 showed higher leaf SOD and CAT activities than HongBao1 and WanXuan1. This indicated that the high-yield cultivars had stronger physiological capacities than the low-yield cultivars.

## POD activities

Figure 5 shows that the POD activities of the different adzuki bean cultivars generally tended to increase at the flowering and fruiting stages. Comparison of the POD activity increments at the different flowering nodes showed that the POD activity increment of the fifth leaves appeared the largest, the POD activity increment of the eighth leaves appeared the lowest, and the POD activity increments of the seventh and sixth leaves stood between the former two. After their anthesis, the POD activities of the different cultivars generally tended to increase but differed significantly. High-yield 2000-75 and JiHong 9218 showed lower leaf POD activities than low-yield HongBao1 and WanXuan1.

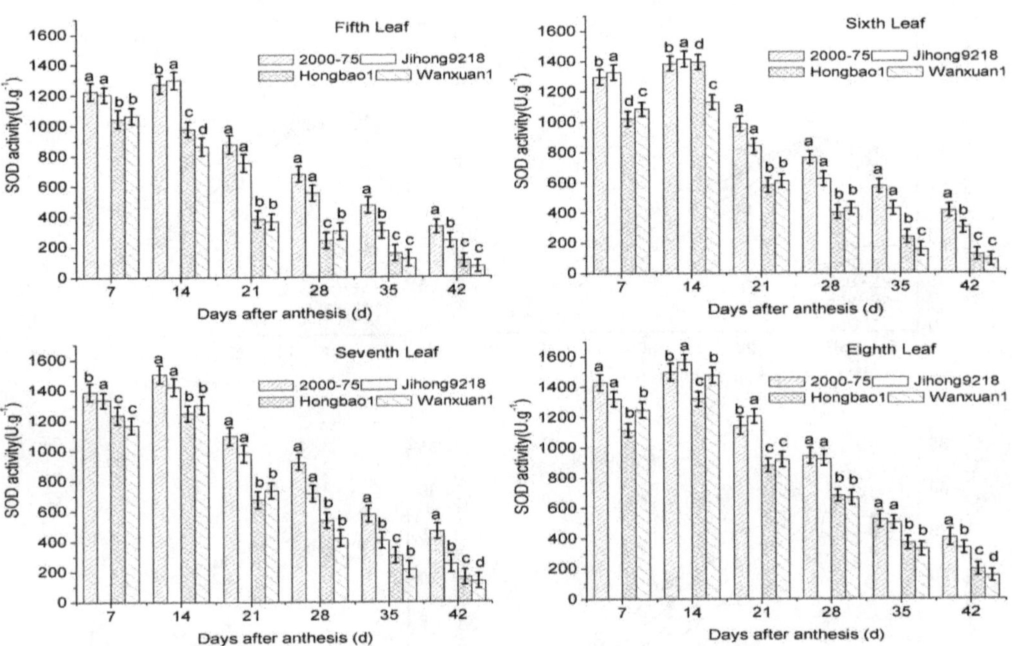


**Figure 3.** Trend of SOD activity in the leaves of four adzuki bean varieties at different flowering nodes. Date presented as mean of two years (2008 and 2009) ±SD. a, b and c represents significant difference in SOD activity at the same day after anthesis from control at $p < 0.05$.

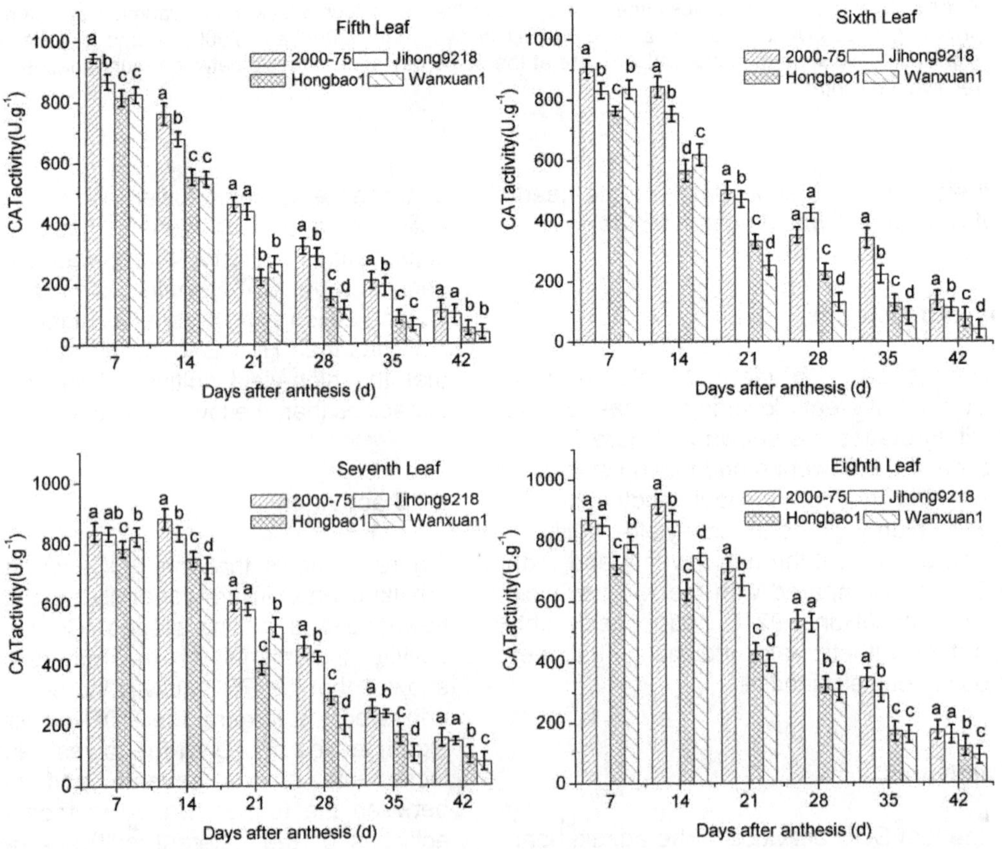


**Figure 4.** Trend of CAT activity in the leaves of four adzuki bean varieties at different flowering nodes. Date presented as mean of two years (2008 and 2009) ±SD. a, b and c represents significant difference in CAT activity at the same day after anthesis from control at $p < 0.05$.

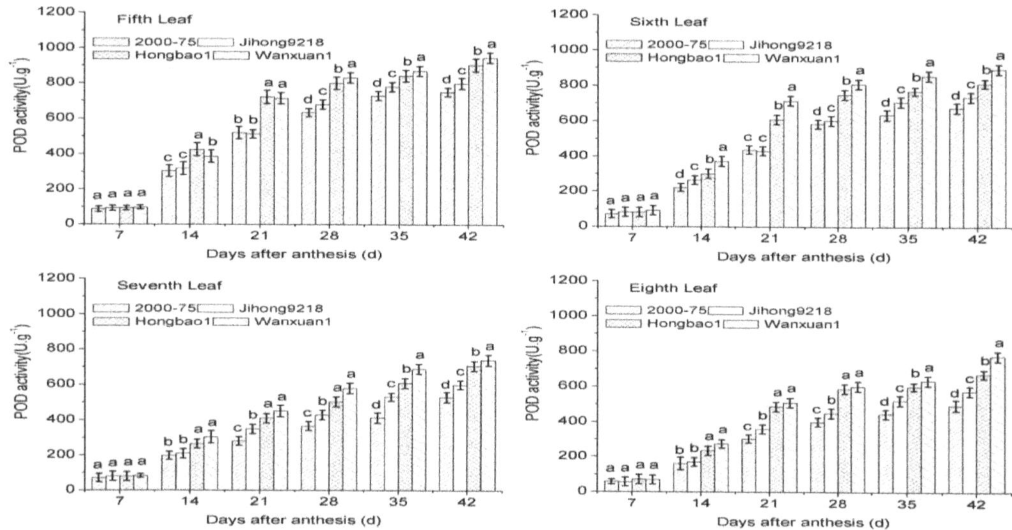


**Figure 5.** Trend of POD activity in the leaves of four adzuki bean varieties at different flowering nodes. Date presented as mean of two years (2008 and 2009) ±SD. a, b and c represents significant difference in POD activity at the same day after anthesis from control at $p < 0.05$.

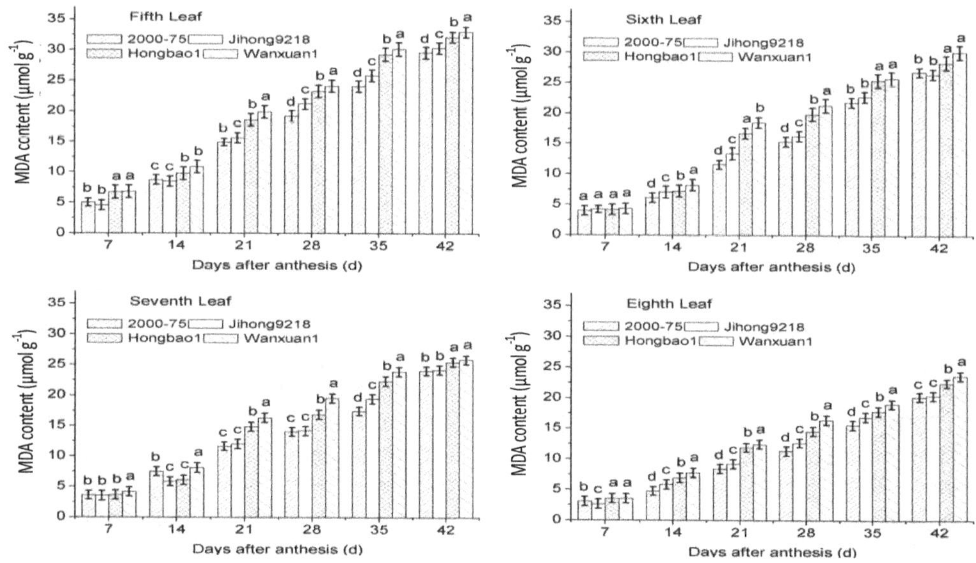


**Figure 6.** Trend of MDA content in the leaves of four adzuki bean varieties at different flowering nodes. Date presented as mean of two years ±SD (2008 and 2009). a, b and c represents significant difference in MDA content at the same day after anthesis from control at $p < 0.05$.

## MDA contents

Figure 6 shows the weighted average MDA contents of the high-yield cultivars and low-yield cultivars at the different flowering nodes. The MDA contents of the different cultivars generally tended to rise as the cultivars approached to their maturities. It can be seen from Figure 6 that under the simulated conditions, the MDA contents of the cultivars increased from the lower nodes to the upper nodes after they flowered. The MDA contents of 2000-75 and JiHong9218 were considerably lower than those of the low-yield cultivars on most of the sampling dates.

## Yield components

Table 1 presents four yield components of the cultivars,

**Table 2.** Correlation of seeds yield and physiological indicators of adzuki bean at the late growth stage.

| Indicator | Seeds yield | Chlorophyll content | $P_n$ | SOD | CAT | POD |
|---|---|---|---|---|---|---|
| Chl | 0.932** | | | | | |
| $P_n$ | 0.966** | 0.957** | | | | |
| SOD | 0.971** | 0.916** | 0.908** | | | |
| CAT | 0.860* | 0.893* | 0.975** | 0.834* | | |
| POD | -0.824* | -0.870* | 0.902* | -0.970** | -0.931** | |
| MDA | -0.863* | -0.955** | -0.842* | -0.892* | -0.950** | 0.872* |

Each point represents the mean of three biological replicates. Significance levels: *$p < 0.05$.

plant/ha, pod number per plant, seed number pod$^{-1}$, and 1000-seed weight. Compared to the yields of HongBao1 and WanXuan1, the yield of 2000-75 and JiHong9218 was significantly higher.

## Correlations between the yields and physiological parameters

The correlations between the Physiological parameters of the functional leaves at the late growth stage and the yield per plant were analyzed (Table 2). The $P_n$ activities of SOD and the yield per plant appeared significantly and positively correlated from the jointing stage to the maturing stage. In addition, appeared a negative correlation between the POD activities (r = -0.824) and MDA contents (r = -0.863) and the yield per plant.

## DISCUSSION

Some researchers have indicated that the functioning durations of crop leaves and particularly the life spans of them after anthesis are crucially important for crop yield (Masulaux et al., 2001; Dai et al., 2011a). The study showed that the chlorophyll contents of the leaves of the cultivars peaked approximately 15 days after the varieties flowered, and the leaf Chl contents of 2000-75 and JiHong 9218 were higher than those of HongBao1 and WanXuan1. When HongBao1 and WanXuan1 approached to their maturity, their leaves became nearly dry and their leaf Chl content was hardly detectable. Therefore, it was important to seed yield formation of adzuki bean to effectively control or slow down the process of leaf chlorophyll degradation.

Photosynthesis is the primary process for crops to form their dry matters and grain yields. However, the photosynthetic carbon cycle will be compromised because the imbalance among vigorous metabolism, nutrition stress and energy metabolism causes free radicals to form (McDonald et al., 1969; Zelitch, 1982; Evans et al., 1984; Jiang et al., 1988).This study showed that the net photosynthetic rates of the cultivars tended to decline from flowering to ripening, and the net photosynthetic rates of the high-yield varieties. When the adzuki bean cultivars nearly reached their complete maturity, the leaves of the low-yield varieties (HongBao1 and WanXuan1) became completely dried up without showing any photosynthetic capacities. It follows that that the high-yield varieties had higher capacity of providing more carbohydrates to fill up their seeds.

Active oxygen species cause deterioration of membrane lipids, leading to increased leakage of solutes from membranes (Dai et al., 2011a). Lipid peroxidation which leads to impairment of membrane function is the system most easily ascribed to oxidative damage and also most frequently measured (Dai et al., 2011b). Malondialdehyde (MDA), a decomposition product of polyunsaturated fatty acids, has been utilized as a biomarker for lipid peroxidation that may occur in the presence of reactive oxygen species (Dai et al., 2011a, b). The leaf SOD and CAT activities of the cultivars first increased, then peaked and finally declined from the flowering stage to the maturing stage, and their leaf MDA contents substantially increased during the same period of time. Comparatively, the SOD and CAT activities of the low-yield varieties 2000-75 and JiHong9218 was high and presented lower decrements, and the MDA contents of them was lower and increased slower. Consequently, it needs further studies to understand how the international action mechanisms of different cultivars operate to minimize oxygen damage under variable environmental conditions.

POD has a dual function on protective enzyme systems and the damage factor of plant senescence (Zhang and Krikham, 1994; Srivallia and Khanna, 2001). The study found that the POD activities increased sharply when the leaf senescence started. Which was completely the same as what happened in Mung bean (Gao et al., 2008).

## ACKNOWLEDGEMENTS

The study was supported by the Special Funds (200903007) for Scientific Research of Agriculture Ministry public-interest industry (agriculture), and Cyrus Tang Specific Plant Genetics and Breeding program of the Northwest A&F University (No. 50).

## ABBREVIATIONS

**CAT,** Catalase; **Chl,** chlorophyll content; **MDA,** malondialdehyde; **POD,** peroxidase; $P_n$, net photosynthetic rate; **SOD,** superoxide dismutase.

### REFERENCES


Batish DR, Singh HP, Setia N, Kaur S, Kohli RK (2006). 2-Benzoxazolinone (BOA) induced oxidative stress, lipid peroxidation and changes in some antioxidant enzyme activities in mung bean (*Phaseolus aureus*). Plant Physiol. Biochem. 44:819-827.

Breusegem FV, Dat JF (2006). Reactive oxygen species in plant cell death. Plant Physiol. 141:384-390.

Chance M, Maehly AC (1955) Assay of catalases and peroxidases. Methods Enzymol. 2:764-775

Dai HP, Jia GL, Lu C, Wei AZ, Feng B L, Zhang SQ (2011a). Studies of synergism between root system and leaves senescence in Broomcorn millet (*Panicum miliaceum* L.). J. Food Agric. Environ. 9:132-135.

Dai HP, Zhang PP, Lu C, Jia GL, Song H (2011b). Leaf senescence and reactive oxygen species metabolism of broomcorn millet (*Panicum miliaceum* L.) under drought condition. A .J.C.S. 5:1655-1660.

Dhindsa RS, Plumb-Dhindsa P, Throne TA (1981). Leaf senescence correlated with increased levels of membrane permeability and lipid peroxidation and decreased levels of superoxide dismutaseand catalase. J. Exp. Bot. 32:93-101

Efeoğlu B, Ekmekçi Y, Çiçek N (2009). Physiological responses of three maize cultivars to drought stress and recovery. S. Afr. J. Bot. 75:34–42.

Evans LT (1984). Morphological and physiological changes among rice varieties used in the Philippines over the seventy years. Field Crops Res. 8:105-125.

Feng BL, Yu H, Hu YG, Gao XL, Gao JF, Gao DL, Zhang SW (2009). The physiological characteristics of the low canopy temperature wheat (*Triticum aestivum L.*) genotypes under simulated drought condition. Acta Physiol. Plant 31:1229-1235.

Gao XL, Gao JF, Feng BL, Chai Y, Jia ZK (2007). Photosynthetic performance in the leaf of different mung bean Genotypes. Acta Agric. Sinica 33:1154-1161.

Gao XL, Sun JM Gao JF, Feng BL, Chai Y, Jia ZK (2008). Leaf senescence and reactive oxygen metabolism in different genotypes of mung bean. Sci. Agric. Sinica 41:2873-2880.

Hayati R, Egli DB, Crafts-Brandner SJ (1995). Carbon and nitrogen supply during seed filling and leaf senescence in soybean. Crop Sci. 35:1063-1069.

Heath RL, Packer L (1968). Photoperoxidation in isolated chloroplasts: 1. Kinetics and stoichiometry of fatty acid peroxidation. Arch. Biochem. Biophys. 125:189-198.

Itoh T, Furuichi Y (2009). Lowering serum cholesterol level by feeding a 40% ethanol-eluted fraction from HP-20 resin treated with hot water extract of adzuki beans to rats fed a high-fat cholesterol diet. Nutrition 25:318-321.

Jiang GZ (1988). Physiological and ecological characteristics of high yielding varieties in rice plants I. Yield and dry matter production. Crop Sci. 57:132-138.

Kaschuk G, Hungria M, Leffelaar PA, Giller KE, Kuyper TW (2010). Differences in photosynthetic behaviour and leaf senescence of soybean dependent on $N_{-2}$ fixation or nitrate supply. Plant Biol. 12:60-69.

Kukavica B, Veljovic JS (2004). Senescence related changes in the antioxidant status of ginkgo and birch leaves during autumn yellowing. Physiol. Plants 122:321-327.

Lin RF, Chai Y, Liao Q, Sun SX (2002).Minor Grain Crops in China. Agricultural Science and Technology Press, Beijing, China. pp. 192-209.

Masulaux C, Quilleré I, Gallais A, Hirel B (2001). The challenge of remobilization in plant nitrogen economy. A survey of physio-agronomic and molecular approaches. Ann. Appl. Biol. 138:8-81.

McDonald DJ, Stansel JW, Gilmore EC (1969). Photosynthesis studies, Rice. J. 74:55.

Nooden LD (1988). whole plant senescence. In: Nooden LD. and Leopold AC (eds) Senescence and Aging in plants. Academic press, San Diego. pp. 391-439.

Pommela B, Gallais A, Coque M, Quilleré I, Hirel B, Prioul JL, Andrieu B, Floriot M (2006). Carbon and nitrogen allocation and grain filling in three maize hybrids differing in leaf senescence. Eur. J. Agron. 24:203-211.

Srivallia B, Khanna CR (2001). Induction of new is forms of superoxide dismutase and catalase enzymes in the leaf of wheat during monocarpic senescence. Biochem. Biophys. Res. Commun. 288:1037-1042.

Zelitch I (1982).The close relationship between net photosynthesis and crop yield. Biol. Sci. 32:796-802.

Zhang CJ, Chen GX, Gao XX, Chu CJ (2006). Photosynthetic decline in flag leaves of two field-grown spring wheat cultivars with different senescence properties maize. S. Afr. J. Bot. 72:15-23.

Zhang JX, Kirkham MB (1994). Drought stress induced changes in activities of superoxide dismutase, catalase, and peroxidase in wheat species. Plant Cell Physiol. 35:785-791.

# Modelling greenhouse air temperature using evolutionary algorithms in auto regressive models

**R. Guzmán-Cruz[1]*, E. Olvera-González[2], I. L. López-Cruz[3] and R. Montoya-Zamora[1]**

[1]División de Investigación y Posgrado, Facultad de Ingeniería, Universidad Autónoma de Querétaro, Cerro de las Campanas s/n, Col. Las Campanas, C.P. 76010 Querétaro, Qro., México.
[2]Universidad Autónoma de Zacatecas, Jardín Juárez 147, Centro Histórico, Zacatecas, Zacatecas, México.
[3]Posgrado en Ingeniería Agrícola y Uso Integral del Agua, Universidad Autónoma de Chapingo, C.P. 056230 Chapingo, Méx., México.

**This paper presents comparison of genetic algorithms (GAs) and evolutionary programming (EP) to estimate parameters of a linear auto regressive model with external input (ARX) and an auto regressive moving average model with external input structures (ARMAX) that predict the behavior of air temperature within a greenhouse. Data groups were used to estimate and validate models and these data groups were 20:80, 33.33:66.67, 50:50, 66.67:33.33 and 80:20%. The objective was to determine which evolutionary algorithm generates parameter values that give the best prediction of the environment in a greenhouse located in the central region of Mexico. Simulation and analysis of the ARX and ARMAX model's performance show that these models under-estimate measurements. Furthermore, the estimations of the inside temperature have a better fit when the parameter identification of an ARX structure is calculated by means of GAs, so that, there is a better fit of the simulated data to measured data when the 20% of the data are used to estimate and 80% of the data are used to validate the model.**

**Key words:** Auto regressive moving average model with external input structures (ARMAX) model, auto regressive model with external input (ARX) model, genetic algorithms, evolutionary programming, parameter identification.

## INTRODUCTION

Nowadays, the competitiveness in vegetable market has been growing, so products with better quality are necessary. However, the high-tech farming in Mexico is limited, so, crop production in greenhouses is a technique that has been applied in this country recently, because growing crops in controlled environments offers many advantages to the farmer such as; a better crop quality, low water and fertilizer consumption and greater yields in any period of the year. Crop growth within greenhouse depends mainly on climatic conditions in which the crop is developing. However, climate inside greenhouses is a very complex system which depends highly on external climatic conditions and greenhouse design. Therefore, for adequate development of the crop, it is necessary to implement control strategies taking into account variables such as: air temperature, air humidity content and $CO_2$ concentration. These control systems are based on mathematical models that predict the climate conditions within a greenhouse, taking into account, external climate

*Corresponding author. E-mail: ros.guzman@gmail.com.

conditions (Castañeda-Miranda et al., 2006).

A tool for describing the environment variables of greenhouses is a mathematical model. In the literature, many models based on physical laws have been presented (Bot, 1983; Deltour et al., 1985; de Zwart, 1996; Wang and Boulard, 2000; Tap, 2000; Castañeda-Miranda et al., 2007; Ruiz-Palacios and Cotrino-Badillo, 2010), and in these models a detailed description of climate conditions inside the greenhouse is shown. However, some of these models are of high order and with many parameters which are difficult to adjust due to the non-linear behavior of the greenhouse climate model. Other kinds of model are empirical or black box models, these are based on observations between inputs and outputs but they do not give an explanation of the underlying mechanisms, such as the auto regressive models with external input (ARX), auto regressive moving average models with external input (ARMAX) (Uchida-Frausto et al., 2003; Boaventura-Cunha et al., 1996; Lopez-Cruz et al., 2007) and neuronal networks (NN) (Seginer et al., 1994; Uchida-Frausto and Pieters, 2004; Pantil et al., 2008; Rahimikhoob, 2010). These models have the advantage of being generated quickly, as they are obtained experimentally when establishing the input-output relationships of the system through parameters identification techniques (Ljung, 1999).

Uchida-Frausto et al. (2003) investigated an ARX and ARMAX that can be used to describe the air temperature of a greenhouse. They found out that these models can describe the behavior of the greenhouse during most parts of the year, except for periods of ventilation, due to the fact that the behavior of the greenhouse becomes highly nonlinear when the strategies of control are imposed. They also observed that ARX models have a better performance than the ARMAX models. López-Cruz et al. (2007) presented a procedure to obtain a dynamic linear model (ARX) to predict the behavior of air temperature inside a greenhouse in Chapingo, Mexico. The goodness of fit between the temperatures simulated and observed, and the residual analysis, indicated than ARX second order or superior models predict adequately the behavior of the temperature inside the greenhouse. Pantil et al. (2008) accomplished a study that included an ARX model, an ARMAX model and an auto regressive model with neural networks. The models worked better when the fit was made out to fix temperature series. Although, ARX models worked better than ARMAX models, however, NNARX model results are closer to the measurements.

Generally, the model structure of both ARX and ARMAX models is determined by classical local search algorithms, such as the least squares estimation method or nonlinear optimization algorithms (Ljung, 1999), which can converge to local minimum. Such problems are referred to as multimodal (Eiben and Smith, 2003).

However, in recent years, global optimization methods have been increasingly used to solve these kind of problems (Michalewicz, 1994) due to the advantages of obtaining possible global optimal solutions. Accordingly, there is a need to develop algorithms based on this methodology applied to the greenhouse temperature. According to Eiben and Smith (2003), evolutionary algorithms (EAs) could be an excellent alternative to provide an answer to the challenge of deploying automated solution methods for more complex problems and more rapidly. EAs are stochastic search methods that include evolution strategies (ES), evolutionary programming (EP) and genetic algorithms (GAs) (Michalewicz, 1994).

Therefore, the objective of the present work was to compare results obtained by means of linear auto regressive models for predicting the behavior of air temperature inside a greenhouse when the estimation of coefficients in both ARX and ARMAX models is done using two EA techniques, such as Gas and EP in order to identify the most appropriate technique to estimate these parameters that allow better description of the temperature within a greenhouse.

Recently, some researchers have applied global methods like evolutionary techniques, and in particular genetic algorithms, to perform calibration processes (Coelho et al., 2005; Wang et al., 2005; Blasco et al., 2007; Nannen and Eiben, 2007). For instance, Herrero et al. (2007a) carried out a non-linear modelling for a climate model of a greenhouse in which they estimated the feasible parameter set (FPS) when the identification error was bound simultaneously by several norms. For the optimization task a special evolutionary algorithm was presented which characterized the FPS by means of a discrete set of models that were well distributed along the FPS. Later, Herrero et al. (2007b) used a multi-parameter fit of a non-linear climate model for temperature and humidity of a greenhouse where roses are grown.

## MATERIALS AND METHODS

Measurements of greenhouse climate were obtained from a greenhouse located at Amazcala facilities which are part of the Biotronic Laboratory of the Universidad Autónoma de Querétaro (UAQ), Querétaro, México, during the period of June 2005 to September 2005. The greenhouse has double slanted roof, covered with single plastic layer and the ground wss covered with a white canvas. It is 108 m long, oriented North-South direction and consists of 12 spans, each 52 m wide and 5.9 m high (4.2 m to the gutter). The greenhouse is equipped with side ventilation on all four walls. Furthermore, it has a climate monitoring system developed by UAQ researchers. The monitoring system generates a historical database of both outside and inside climate variables (Figure 2) which were measured at a sampling interval of 5 min, that is, 26500 samples during all time period.

For modeling purposes, the outside and inside climate variables measured were air temperature, relative humidity, solar radiation and wind velocity (Figure 3). The climate variables were measured with sensors of Global Water Instrumentation, Inc. installed inside and outside the greenhouse. Inside the greenhouse, pair of resistive temperature detectors (RTDs) which have an accuracy of ±0.1°C were installed in locations that did not receive direct sun-light. In addition, there were two capacitive sensors to measure the relative humidity (RH) which have an accuracy rate of ±2% RH. These were located in the centre of two spans of the greenhouse at a height of 2.20 m. Outside the greenhouse, similar sensors were used for air temperature and humidity. An anemometer to measure the wind velocity (m $s^{-1}$) which has an accuracy of 0.1 m $s^{-1}$ over the range of 5 to 24 m $s^{-1}$ was also used and there was a pyranometer (LI-COR Inc) to measure the global radiation (W $m^{-2}$) with an error less than 5% located at a distance of 6 m from the greenhouse and a height of 5 m.

Air temperature inside a closed environment can be modeled by ARX and by ARMAX considering that inputs and outputs are measured by sensors. For a system with one input and one output (SISO), the model is given by Equation (1) (Ljung, 1999; Aguado and Martínez, 2003; Ljung, 2005):

$$y(t)+a_1y(t-1)+...+a_{n_a}y(t-n_a)= b_1u(t-n_k)+...+b_{n_b}u(t-n_k-n_b+1)+e(t)+c_1e(t-1)+...+c_{n_c}e(t-n_c) \quad (1)$$

where *y(t)* is the output of the ARX and ARMAX models for t = 1, t-1,... t-$n_a$; u(t) is the input for t = t-$n_k$, t-$n_k$ -1,..., t-$n_k$-$n_b$ +1; $n_a$ is the number of samples passed in the time of the output; $n_k$ is the delay time of the input u(t), e(t) is the white noise associated with the input and t is discrete time.

To evaluate the temperature inside a closed environmental using ARX and ARMAX models, more input variables are required, so, the models have multiple inputs and one output (MISO). The structures ARX and ARMAX for MISO systems are defined by Equation (2) and (3), respectively:

$$A(q)y(t)=B(q)u(t-n_k)+e(t) \quad (2)$$

$$A(q)y(t)=B(q)u(t-n_k)+C(q)e(t) \quad (3)$$

Where A(q) and B(q) are matrices and C(q) is a vector, all defined by Equation (4) to (6):

*A(q)*:

$$1+a_1q^{-1}+...+a_{n_a}q^{-n_a} \quad (4)$$

*B(q)*:

$$1+b_1q^{-1}+...+b_{n_b}q^{-n_b} \quad (5)$$

*C(q)*:

$$1+c_1q^{-1}+...+c_{n_c}q^{-n_c} \quad (6)$$

and the operator $q^{-1}$ is the backward shift operator:

$$q^{-1}u(t)=u(t-1) \quad (7)$$

For a system in which the number of inputs is given by $n_y$ and the number of outputs by $n_u$, *A(q)* and *B(q)* are $n_y$ by $n_y$ and $n_u$ by $n_u$ matrices, respectively, whose elements are polynomials in the shift operator $q^{-m}$ (with m any natural number), the entries $a_{ij}(q)$ and $b_{ij}(q)$ of the matrices *A(q)* and *B(q)*, respectively, can then be written as:

$$a_{ij}(q)=\delta_{ij}+a_{1_{ij}}z^{-1}+\cdots+a_{n_{a_{ij}}}z^{-n_{a_{ij}}} \quad (8)$$

and

$$b_{ij}(q)=b_{1_{ij}}z^{-n_{k_{ij}}}+\cdots+b_{n_{b_{ij}}}z^{-n_{k_{ij}}-n_{b_{ij}}+1} \quad (9)$$

Where $\delta_{ij}$ represents the Kronecker symbol.

From the above, it is clear that the ARX structure for a given system can be defined by means of the number of poles $n_a$, the number of zeros $n_b$*-1* and the number of time that delays $n_k$. The definition of the ARMAX structure additionally requires the order of the measured error $n_c$ to be known. The matrices A(q), B(q) and C(q) are determined by means of off-line parameter identification methods (Uchida-Frausto et al., 2003).

Types of selected models depend on subgroups of data used for estimating and validating the ARX and ARMAX models. The data subgroups of 20:80, 33.33:66.67, 50:50, 66.67:33.33 and 80:20% were evaluated to estimate and validate the models. On one hand, the parameters of the matrix A(q), B(q) and the vector C(q) are typically estimated by offline methods using the System Identification Toolbox available in Matlab software (Ljung, 2005). Using this Matlab Toolbox, it is possible to evaluate many models; this is the reason for its use as adjustment criteria to select the model that presents more accurate values. On the other hand, these parameters are estimated using GAs and EP. At last, the results are compared to determine which one of gives the best fit. To achieve a good fit of a greenhouse climatic model, it is necessary to estimate the suitable values for the coefficients implicated in the autoregressive model. That is, for each structure of ARX and ARMAX models, we need to obtain the coefficients $a_1,...,a_{n_a}$, $b_1,...,b_{n_b}$ and $c_1,...,c_{n_c}$ (for ARMAX) and the order of the model given by the values, the parameters $n_a$, $n_b$, $n_c$ (for ARMAX), $n_k$, based on the information provided by the inputs and outputs in order to get the best fit between the measured values and the estimated values by the model.

To determine the coefficients of ARX and ARMAX models that better fit the simulated to the measured data, a method to use is to minimize the sum of square errors (J):

$$J(p)=\sum_{i=1}^{N}(\bar{y}(t_i,p)-y(t_i))^2 \qquad p^*=\arg\min J(p) \quad (10)$$

Where $\bar{y}(t_i,p)$ is the simulated output, *y* in time $t_i$, $y(t_i)$ is the measurement *y* in time $t_i$, *N* is the number real measurement (time), *p* is the set of parameters (coefficients of the model) and *p** are the parameters that reduced *J(p)* to the minimum.

In the current work, the minimization of Equation (10) is a non-linear multivariable optimization problem that can be solved by using evolutionary algorithms, such as: GAs and EP since they are global optimization methods. The structure of any EA is the same (Eiben and Smith, 2003) as is shown in Figure 1. Differences among evolutionary techniques consist of the kind of selection, mutation and crossover operators applied to find the optimum value of the parameters for the optimization function. In this case, the kind of selection, the crossover and the mutation used for each EA is presented in Table 1 and Figure 1. Some general statistics are used to analyze the obtained results with the percent of data group considered to validate the model, such as: the correlation coefficient *r*, the percentage standard error of the prediction

**Table 1.** Comparative table of the techniques of evolutionary algorithms.

| | Genetic algorithms | Evolutionary programming |
|---|---|---|
| Representation | Real-valued | Real-valued |
| Parent selection | Deterministic: by mean tournament | Deterministic (each parent creates one offspring) |
| Recombination | 2-point crossover | None |
| Mutation | Non uniform | Gaussian perturbation |
| Survival selection | Generational | Determinations: $(\mu+\mu)$ |

**Table 2.** Structures of best ARX and ARMAX models obtained using the percentages of the data for estimation and validation of the models.

| Data subgroup (%) | ARX | | | ARMAX | | | |
|---|---|---|---|---|---|---|---|
| | na | nb | nk | na | nb | nc | nk |
| 20:80 | 1 | 3 1 8 10 | 1 1 1 1 | 1 | 2 2 2 2 | 1 | 1 1 1 1 |
| 33.33:66.67 | 1 | 10 9 10 10 | 1 1 1 1 | 1 | 2 2 2 2 | 1 | 1 1 1 1 |
| 50:50 | 1 | 9 1 10 10 | 1 1 1 1 | 1 | 2 2 2 2 | 1 | 1 1 1 1 |
| 66.67:33.33 | 1 | 10 1 9 1 | 1 1 1 1 | 1 | 2 2 2 2 | 1 | 1 1 1 1 |
| 80:20 | 1 | 10 10 1 1 | 1 1 1 1 | 1 | 2 2 2 2 | 1 | 1 1 1 1 |

**Table 3.** Statistical results of greenhouse air temperature when the parameter identification for ARX and ARMAX model is typical and using EA with the data group of 20:80%.

| Method | ARX | | | | ARMAX | | | |
|---|---|---|---|---|---|---|---|---|
| | r | E | %SEP | AVR | r | E | %SEP | AVR |
| Typical | 0.2345 | -71.9648 | 377.0408 | 72.9648 | 0.5286 | -3.0448 | 88.7727 | 4.0448 |
| GAs | 0.9296 | 0.8335 | 18.0094 | 0.1665 | 0.7468 | 0.5537 | 29.4889 | 0.4463 |
| EP | 0.9184 | 0.8428 | 17.5015 | 0.1572 | 0.914 | 0.8317 | 18.1064 | 0.1683 |

(*%SEP*), the efficiency coefficient (*E*) and the average relative variance (*ARV*). These estimators are used to determine the way in which the model can explain the total variance of the data (Ríos-Moreno et al., 2006). Having a perfect relation, *r* and *E* should be near 1 and the values of *%SEP* and *ARV* near 0.

## RESULTS AND DISCUSSION

Table 2 shows the structures obtained when evaluating the subgroups of data to estimate and to validate the ARX and ARMAX models. In the case of ARX model, an evaluation was made of 100,000 models, given that the maximum value for $n_a$ (one output variable) and for $n_b$ (four input variables) was 10 and for $n_k$ one, therefore, the resulting number of combinations (structures of models) is $10^5$ (Lopez-Cruz et al., 2007). In the case of ARMAX model, for selection of the most appropriate structure, the maximum number of poles $n_a$ and the maximum time delay $n_k$ were set at 4. The maximum number of zeros was set at 5 ($n_b$ was set at 2) and the order of the measured error $n_c$ was set at 1 (Patil et al., 2008).

From Tables 3 to 7 the statistical results obtained when evaluating the data subgroups to validate the ARX and ARMAX models are shown. In the same way, from Figures 4 to 8 the behavior of the ARX and ARMAX models with structure showed in Table 2 and their data groups corresponding to the validation are shown.

Looking at Figures 4 and 5 it can be observed that there was a better fit when the parameters of an ARX structure were found with EA and with the data group 20:80% and 33.33:66.67%.

In addition, it is shown that ARX and ARMAX models were under-estimated to measured data when the 20 or 33.33% of the data were only used. The worst case was obtained when estimations were done with the typical ARX model and the 20% of the data. Correlation *r* takes values between 0.23 (typical ARX) and 0.92 (identification with GAs), in fact, the efficiency (E = 0.83), percentage standard error of the prediction (%SEP = 18)

**Table 4.** Statistical results of greenhouse air temperature when the parameter identification for ARX and ARMAX model is typical and using EA with the data group of 33.33:66.67%.

| Method | ARX | | | | ARMAX | | | |
|---|---|---|---|---|---|---|---|---|
| | r | E | %SEP | AVR | r | E | %SEP | AVR |
| Typical | 0.509 | -4.7948 | 106.2551 | 1.3544 | 0.7323 | -0.3544 | 51.3698 | 5.7948 |
| GAs | 0.8867 | 0.6691 | 25.3921 | 0.3309 | 0.9159 | 0.8246 | 18.4881 | 0.1754 |
| EP | 0.8965 | -0.1286 | 46.8923 | 1.1286 | 0.916 | 0.7582 | 21.7068 | 0.2418 |

**Table 5.** Statistical results of greenhouse air temperature when the parameter identification for ARX and ARMAX model is typical and using EA with the data group of 50:50%.

| Method | ARX | | | | ARMAX | | | |
|---|---|---|---|---|---|---|---|---|
| | r | E | %SEP | AVR | r | E | %SEP | AVR |
| Typical | 0.8241 | 0.8241 | 31.5013 | 0.5093 | 0.8847 | 0.6507 | 26.0884 | 0.3493 |
| GAs | 0.9186 | 0.842 | 17.5443 | 0.158 | 0.9176 | 0.8418 | 17.5572 | 0.1582 |
| EP | 0.9301 | 0.8267 | 18.377 | 0.1733 | 0.9182 | 0.5107 | 30.8753 | 0.4893 |

**Table 6.** Statistical results of greenhouse air temperature when the parameter identification for ARX and ARMAX model is typical and using EA with the data group of 66.67:33.33%.

| Method | ARX | | | | ARMAX | | | |
|---|---|---|---|---|---|---|---|---|
| | r | E | %SEP | AVR | r | E | %SEP | AVR |
| Typical | 0.908 | 0.8205 | 18.6999 | 0.1795 | 0.9289 | 0.8531 | 16.9187 | 0.1469 |
| GAs | 0.9231 | 0.8467 | 17.2832 | 0.1533 | 0.9142 | 0.8358 | 17.8877 | 0.1642 |
| EP | 0.9198 | 0.8442 | 17.42 | 0.1558 | 0.9117 | 0.6816 | 24.9064 | 0.3184 |

**Table 7.** Statistical results of greenhouse air temperature when the parameter identification for ARX and ARMAX model is typical and using EA with the data group of 80:20%.

| Method | ARX | | | | ARMAX | | | |
|---|---|---|---|---|---|---|---|---|
| | r | E | %SEP | AVR | r | E | %SEP | AVR |
| Typical | 0.9236 | 0.852 | 16.9825 | 0.148 | 0.9371 | 0.8678 | 16.0484 | 0.1322 |
| GAs | 0.9274 | 0.8593 | 16.5543 | 0.1407 | 0.9117 | 0.7594 | 21.6507 | 0.2406 |
| EP | 0.8632 | 0.6706 | 25.334 | 0.3294 | 0.499 | 0.2487 | 38.2599 | 0.7513 |

and the average relative variance (ARV = 0.16) got the best values when the parameter identification was performed with GAs (Table 3).There was improvement in the estimations done with the typical ARX and ARMAX models when 50 or 66.67% of the data were used for these estimations. In Table 6, a better fit can be observed when ARMAX structure is used with the data group 66.67:33.33%. In this case, correlation *r* takes its maximum value (0.92) when parameter identification is carried out by classical method, furthermore good results occur when parameter identification is performed with GAs, that is, E got values of 0.83 to 0.84, %SEP was about 17 and ARV was about 0.17. Finally, the best results were obtained with ARMAX structure for the data group 80:20%. These were followed by the parameter identification with GAs.

Similarly, typical ARX and ARMAX models were underestimated to measure data when the data groups: 20:80, 33.33:66.67 and 50:50% were selected to estimate and validate the models. However, there were a better fit when the parameter identification was developed using GAs and EP. In general, every method applied to the data

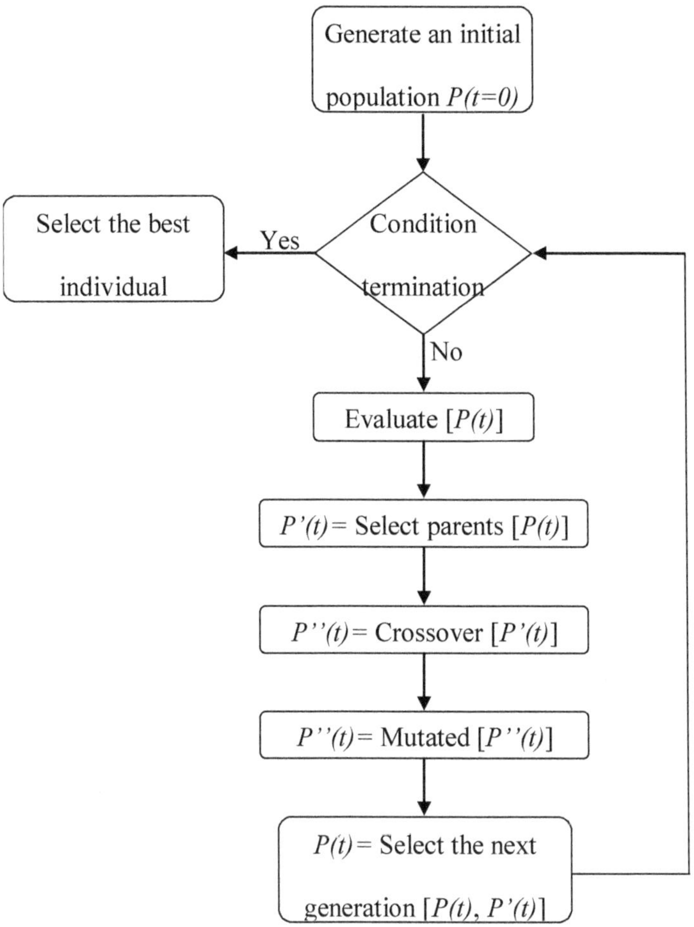


**Figure 1.** Diagram of the general scheme of an EA.

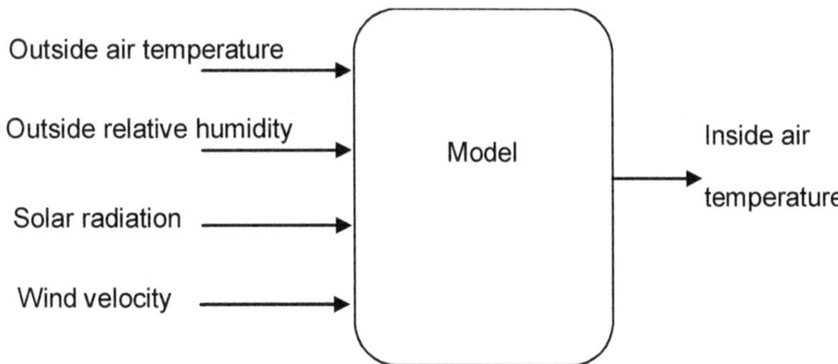


**Figure 2.** Input and output variables.

Groups; 66.67:33.33 and 80:20% had an acceptable behavior.

Results for ARX and ARMAX structures (Tables 3 to 7) within the five data groups using GAs and EP were obtained from 20 runs in each case and the selection was made considering the parameters that minimized the error between measured and estimated data. In general, better results were obtained with the parameter values estimated by EA than with those calculated with the least-squares method.

Some models present a more complex structure than others; from the point of view of the Control Theory, the

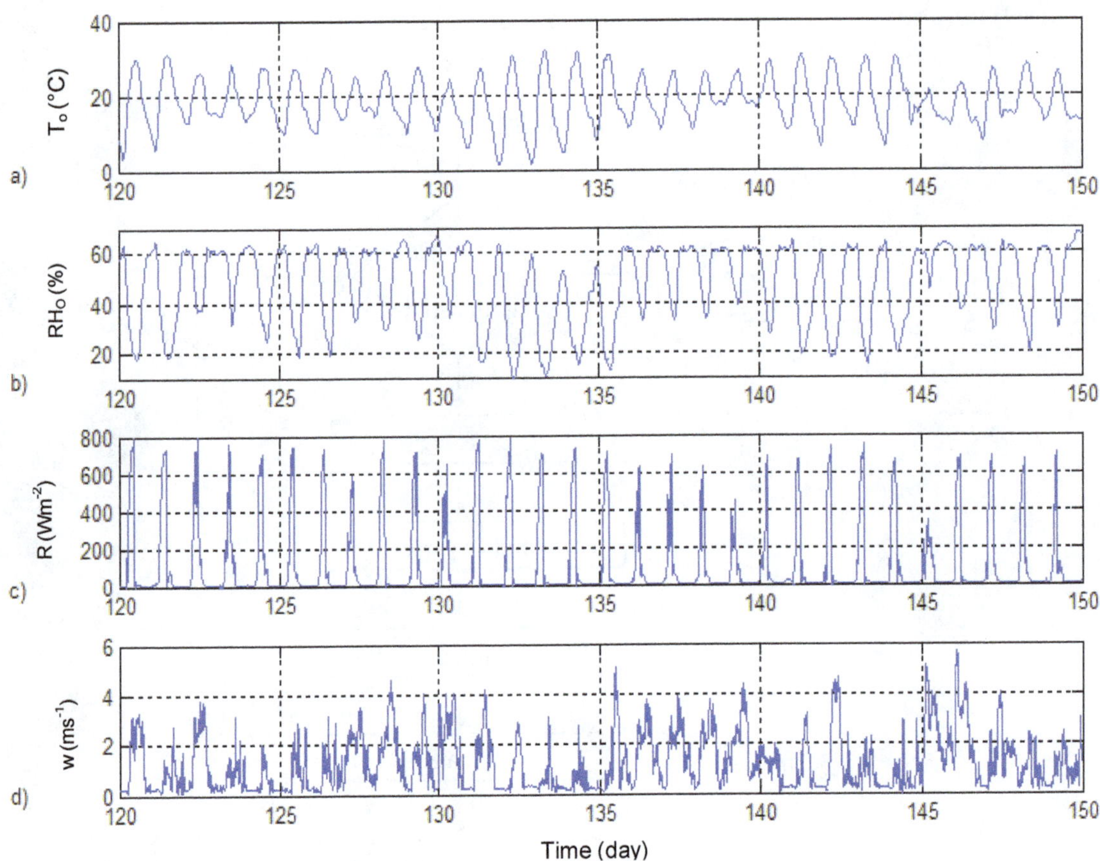


**Figure 3.** Input variables in a period of thirty days: (a) outside air temperature, (b) outside relative humidity, (c) solar radiation and (d) wind velocity.

preferred model is the one with the simplest structure with respect to the values of $n_a$, $n_b$ and number of parameters. Thus, the design process of a control system is facilitated (Lopez-Cruz et al., 2007). In addition, if a model can be obtained using the 20% of data and validated with the other 80%, it is more reliable than if 80% is used for identification and only 20% for validation. For this reason, it is important to select the simplest model with the most acceptable behavior and the least computational cost.

As it has been mentioned, different data subgroups were considered, this to determine the structures of the ARX and ARMAX models and to compare them to know which of them has the best fit. The relevance of using different data subgroups consists of identifying a mathematical model that the behavior of the greenhouse temperature can be describe as much as possible, that is, it can consider the number of data as big as possible for validation of the model. The aforementioned is achieved when evolutionary techniques were applied, in particular GAs. According to the results, from the 100% of available data, when the parameter identification of an ARX structure is performed by means of GAs there is a better fit of the simulated data to measured data from June to September when just the 20% of the data are used to estimate and 80% of the data are used to validate the model. This is more reliable than if 80% of data is used for identification and 20% for validation. In this case, the mathematical model explains the behaviour of a big validation data set. Otherwise, as it was demonstrated by Uchida-Frausto et al. (2003), it will be necessary to recalibrate the model periodically or use recursive algorithms to determine new structures when there are new scenarios (Ljung, 1999; Aguado and Martínez, 2003). However, implementation of the GA for the identification of parameters of the mathematical model allows delaying recalibration so that the model can be used by prolonged periods.

In general, the fit of measured data increases while the prediction error decreases when the estimation of the coefficients of both ARX and ARMAX models was carried out by GAs. This fact occurs because the search space is larger than with classical local search algorithms, such as the least squares estimation method or nonlinear optimization algorithms, which can converge to local minima.

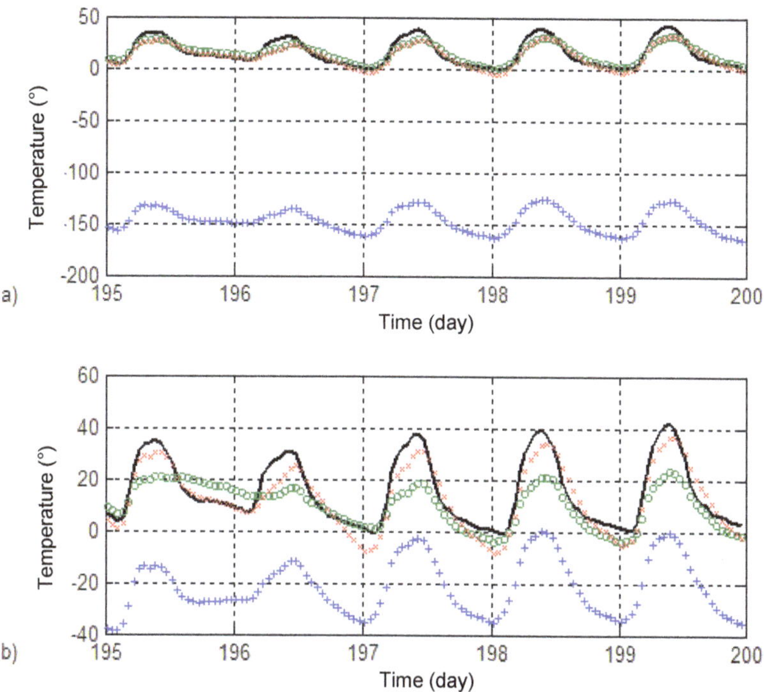


**Figure 4.** Measured and simulated data after parameter identification for: (a) typical ARX structure, GAs and EP method and (b) typical ARMAX structure, GAs and EP method for data group of 20:80%; ---, real data; +++, typical ARX and ARMAX; ooo, GAs; xxx, EP.

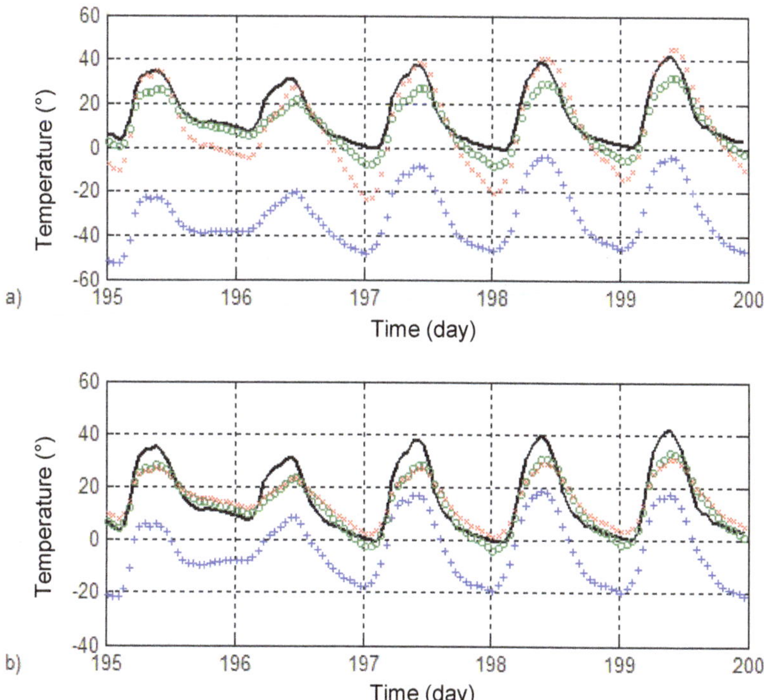


**Figure 5.** Measured and simulated data after parameter identification for: (a) typical ARX structure, GAs and EP method and (b) typical ARMAX structure, GAs and EP method for data group of 33.33:66.67%; ---, real data; +++, typical ARX and ARMAX; ooo, GAs; xxx, EP.

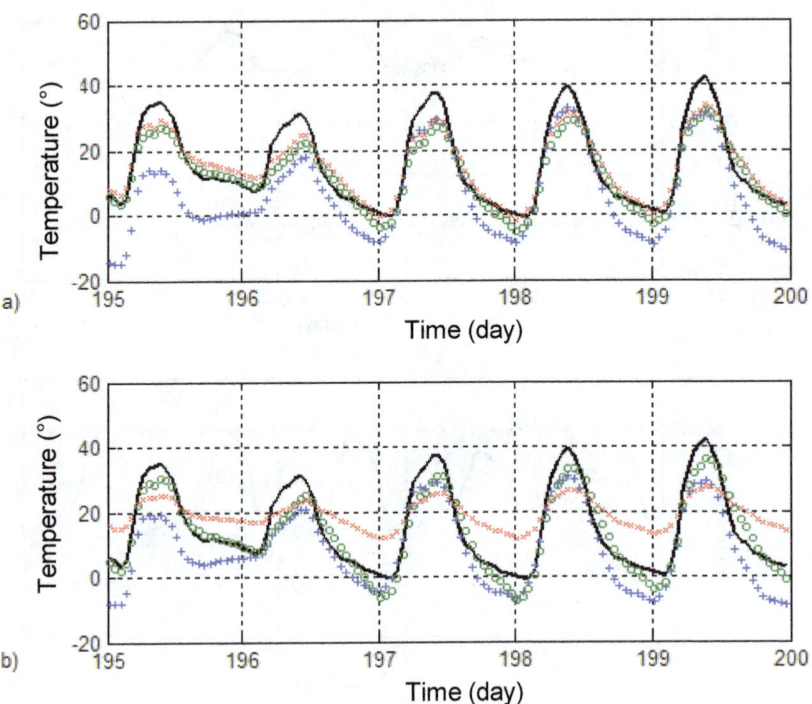


**Figure 6.** Measured and simulated data after parameter identification for: a) typical ARX structure, GAs and EP method and b) typical ARMAX structure, GAs and EP method for data group of 50:50%; ---, real data; +++, typical ARX and ARMAX; ooo, GAs; xxx, EP.

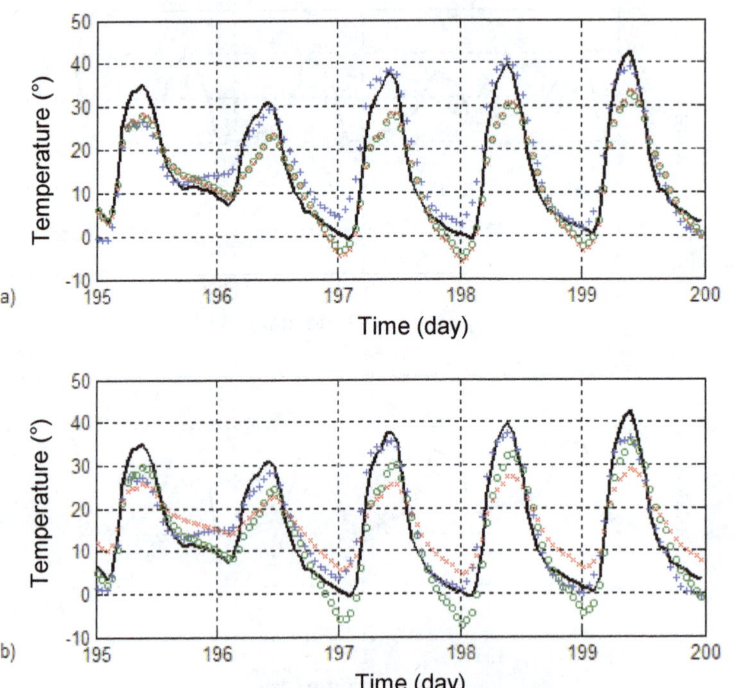


**Figure 7.** Measured and simulated data after parameter identification for: (a) typical ARX structure, GAs and EP method and (b) typical ARMAX structure, GAs and EP method for data group of 66.67:33.33%; ---, real data; +++, typical ARX and ARMAX; ooo, GAs; xxx, EP.

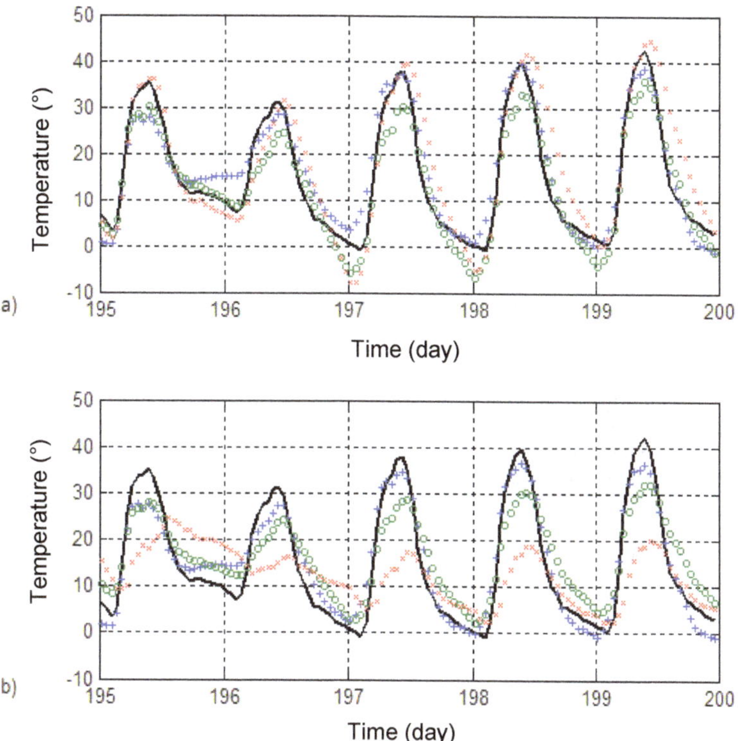


**Figure 8.** Measured and simulated data after parameter identification for: a) typical ARX structure, GAs and EP method and b) typical ARMAX structure, GAs and EP method for data group of 80:20%; ---, real data; +++, typical ARX and ARMAX; ooo, GAs; xxx, EP.

## Conclusions

In the present work, ARX and ARMAX models that describe the air temperature within a greenhouse located in the central region of Mexico were used but these models showed under-estimation of measured data. In addition, global search methods as evolutionary algorithms (GAs and EP) were used to carry out parameter identification given ARX and ARMAX structures; the number of identified parameters varied from 10 to 40 for the case of ARX structure using the five data. In general, there is a better fit for the air temperature within the greenhouse when the parameter identification of an ARX structure with the data group 20:80% was performed by means of GAs.

Overall results obtained with GAs show that this method is more effective and efficient than the EP method to find parameters for auto regressive models to predict air temperature inside a greenhouse. For the model obtained using 20% of data, there is enough identification and validation with the other 80% of them is which is more reliable than if 80:20% is used. Moreover, these results were obtained with the: ARX structure with $n_a$ = 1, $n_b$ = [3 1 8 10], and $n_k$ = [1 1 1 1] and the data group 20:80%.

## ACKNOWLEDGMENT

This research was partially supported by CONACyT.

### REFERENCES

Aguado BA, Martínez M (2003). Automática y Robótica: Identificación y Control Adaptativo. Prentice Hall Ed. Madrid.

Blasco X, Martínez M, Herrero JM, Ramos C, Sanchis J (2007). Model based predictive control of greenhouse climate for reducing energy and water consumption. Comput. Electron. Agric. 55:49-70.

Boaventura-Cunha J, Ruano AEB, Couto C (1996). Identification of greenhouse climate dynamic models. Sixth International Conference on computers in agriculture. Cancún, México. pp. 161-171.

Bot GPA (1983). Greenhouse climate: form physical processes to a dynamic model. PhD thesis, Wageningen Agricultural University, The Netherlands.

Castañeda-Miranda R, Ventura-Ramos E, Peniche-Vera RR, Herrera-Ruiz G (2006). Fuzzy Greenhouse Climate Control System based on a Field Programmable Gate Array. Biosyst. Eng. 94(2):165-177.

Castañeda-Miranda R, Ventura-Ramos E, Peniche-Vera RR, Herrera-Ruiz G (2007). Analysis and simulation of a greenhouse physical model under weather conditions of the central region of Mexico. Agrociencia 41(003):317-335.

Coelho JP, de Moura Oliveira PB, Bonaventura Cunha J (2005). Greenhouse air temperature predictive control using the particle swarm optimisation algorithm. Comput. Electron. Agric. 49:330-344.

de Zwart HF (1996). Analyzing energy-saving options in greenhouse cultivation using a simulation model. PhD thesis, Wageningen

Agricultural University, The Netherlands.
Deltour J, de Halleux D, Nijskens J, Coutisse S, Nisen A (1985). Dynamic modelling of heat and mass transfer in greenhouses. Acta Horticulturae 174:119-126.
Eiben AE, Smith JE (2003). Introduction to evolutionary computing. (Natural computing series). Springer-Verlag, Berlin, Germany.
Herrero JM, Blasco X, Martínez M, Ramos C, Sanchis J (2007a). Robust identification of non-linear greenhouse model using evolutionary algorithms. Control Eng. Pract. 16:515–530.
Herrero JM, Blasco X, Martínez M, Ramos C, Sanchis J (2007b). Non-linear robust identification of a greenhouse model using multi-objetive evolutionary algorithms. Biosyst. Eng. 98:335–346
Ljung L (1999). System Identification, Theory for the user. Prentice Hall Ed.
Ljung L (2005). System Identification Toolbox for use with MATLAB. The Matworks Inc.
López-Cruz IL, Rojano-Aguilar A, Ojeda-Bustamante W, Salazar-Moreno R (2007). ARX models for predicting greenhouse air temperatura: a methodology. Agrociencia 41:181-192.
Nannen V, Eiben AE (2007). Efficient relevance estimation and value calibration of evolutionary algorithm parameters. IEEE Congress on Evolutionary Computation, IEEE. pp. 103–110.
Michalewicz Z (1994). Evolutionary computation techniques for nonlinear programming problems. Int. Trans. Oper. Res. 1(2):223–240.
Patil SL, Tantau HJ, Salokhe VM (2008). Modelling of tropical greenhouse temperature by auto regressive and neural network models. Biosyst. Eng. 99:423-431.
Rahimikhoob A (2010). Estimation of evapotranspiration based on only air temperature data using artificial neural networks for a subtropical climate in Iran. Theor. Appl. Climatol. 101:83-91.
Ríos-Moreno GJ, Trejo-Perea M, Castañeda-Miranda R, Hernández-Guzmán VM, Herrera-Ruiz G (2006). Modelling temperature in intelligent buildings by means of autoregressive models. Autom. Constr. 16(5):713-722.
Ruiz-Palacios FO, Cotrino-Badilo CE (2010). Identifying a greenhouse climate model by using subspace methods. Ingeniería e Investigación 30(2):157-167.
Seginer I, Boular T, Baley BJ (1994). Neural network models of the greenhouse climate. J. Agric. Eng. Res. 59:203-216.
Tap F (2000). Economics-based optimal control of greenhouse tomato crop production. PhD Thesis. Wageningen Agricultural University, Wageningen, The Netherlands.
Uchida-Frausto H, Pieters JG, Deltour JM (2003). Modelling greenhouse temperature by means of auto regressive models. Biosyst. Eng. 84(2):147-157.
Uchida-Frausto H, Pieters JG (2004). Modelling greenhouse temperature using system identification by means of auto neural networks. Neurocomput 56:423:428.
Wang S, Boulard T (2000). Predicting the microclimate in a naturally ventilated plastic house in a Mediterranean climate. J. Agric. Eng. Res. 75:27-38.
Wang W, Zmeureanu RY, Rivard H (2005). Applying multi-objective genetic algorithms in green building design optimization. Build. Environ. 40:1512-1525.

# 12

# Growth assessment of endangered *Aframomum sceptrum* (Braun) under different planting regime for sustainable utilization

**Lawal I. O.[1], Ige P. O.[1], Awosan E. A.[1], Borokini T. I.[2] and Amao O. A.[1]**

[1]Forestry Research Institute of Nigeria (FRIN), Ibadan, Oyo State, Nigeria.
[2]National Centre for Genetic Resources and Biotechnology (NACGRAB), Nigeria.

**The study investigated the growth assessment and germination study of endangered *Aframomum sceptrum* (Braun) under different planting regime for sustainable utilization. The result of various planting regime showed that shoot height was highest in seeds planted 5 days after extraction (35.36 cm) closely followed by 3 days after extraction (31.36 cm) and the least value was found in 14 days after extraction. In the same vain, base diameter was highest in seeds planted 5 days after extraction, while the least was recorded in seeds sown after 3 days of extraction. Number of leaves was highest for seeds planted 3 days after extraction (11 cm) and the least value was recorded for the ones planted 14 days after extraction (8 leaves). Differences in all the growth parameters assessed were significant at $p < 0.05$. There were positive effects in growth performance of this specie in the entire planting regime. Therefore, this study shows a positive domestication and multiplication of this species.**

**Key words:** *Aframomum sceptrum*, planting regime, sustainable utilization, dormancy, germination.

## INTRODUCTION

*Aframomum sceptrum* (Braun) is an herbaceous, perennial and aromatic species classified in the monocotyledonous family of Zingiberaceae, native to Ethiopia where it is called Korarima. In Nigeria, it is commonly found in derived savanna area. In Benue State, the Idoma people called it Ugbenya. The plant consists of an underground rhizome, a pseudostem, and several broad leaves and resembles *Elettaria* species morphologically (Eyob et al., 2008). Mature plant can reach a height of 1 to 2 m. It sets seed after 3 to 5 years of planting depending on the planting materials used and it continue to bear seeds for a number of decades. It occurs as a cultivated crop only in Ethiopia. The seed of *A. sceptrum* is mainly used as sources of spices in traditional Ethiopian dishes and in Nigeria. It is a source of income for growers as its seeds fetches high prices in local and export markets. *A. sceptrum* parts are used in traditional medicine for humans and cattle. Also, it is an important plant for soil conservation as the rhizomes and leaves spread on the ground covering and protecting the soil from erosion in hilly areas (Eyob et al., 2008). Recent attempts in Ethiopia to encourage farmers to cultivate the *A. sceptrum* plant have not been successful due to several production constraints. In Nigeria, it is neglected and underutilized. Among all production constraints, farmers emphasized that lack of improved varieties with improved agronomic practices like propagation techniques had contributed to decrease in production (Eyob et al., 2009). The slow seed germination and growth of the subsequent seedlings were concerns of *A. sceptrum* growers. The germination of *A. sceptrum* seeds faces certain problems. There might be some kind of

**Table 1.** Analysis of variance of planting regime and weeks on height.

| SV | DF | SS | MS | F |
|---|---|---|---|---|
| Planting regime | 4 | 411.64 | 102.90906 | 194.17** |
| Weeks | 2 | 557.52 | 278.76161 | 525.96** |
| Error | 38 | 20.32 | 0.53 | |
| Total | 44 | 989.48 | | |

** = Significant at p < 0.05.

dormancy, possibly associated with the hard seed coat. Dormancy as a result of impermeable seed coat was reported from seeds of *Elettaria* species (Sulikeri and Kololgi, 1977). Low food reserve in the seed endosperm might be a reason for the very slow growth of the seedlings. Enhancement of korarima seed germination is important in propagation and breeding program as well as for testing and using germplasms (Bhattacharya and Khuspe, 2001). Although, *A. sceptrum* is mainly propagated by vegetative method using 1 year old rhizomes, the need for bulk of rhizome as planting materials and slow multiplication rate of the rhizomes became another critical problem (Polat, 1997). Also, the destructive harvestings and malhandling of the rhizomes for vegetative propagation seems not to be feasible because there is always the possibility of losing the mother plant during this process. Despite the fact that *A. sceptrum* is a useful crop with a high potential as income source and other purposes, only limited efforts have been made to improve this crop using traditional and modern biotechnological approaches (Echeverrigaray et al., 2003). To achieve such an improvement, proper agronomic and tissue culture procedures, which assure successful and efficient propagation, need to be developed. To date, only two tissue culture studies have been reported (Tefera and Wannakrairoj, 2004), but have no reports on agronomic practices such as seed germination procedures. Therefore, the overall goal of the present study was to investigate the effects of different seed treatment methods, to evaluate *in vitro* performances of different cultivars and *in vitro* growth response of ex-plant sources of *A. sceptrum*.

## MATERIALS AND METHODS

Matured seeds of *A. sceptrum* were collected from Akinsola village in Ido-Local Government Area of Oyo State between the months of September to October, 2009. The plant was collected by Lawal Ibraheem, and was further identified by Mr P. O Daramola of Forestry Research Institute of Nigeria, while the herbarium specimen was kept in Forest Herbarium Ibadan. The seeds were extracted from its outer layer coat. Washed and sterilized river sand was filled into perforated plastic trays and later transferred to black polythene pot for growth evaluation/assessment and to assess best sowing or planting regime. These were investigated under mist propagated chamber. The 20 seedlings of *A. sceptrum* were sown according to their planting regime at both germination and seedling stages. Watering was carried out once daily. The study set up consisted of freshly sown immediately after extraction, sown after 3, 5, 7 and 14 days, respectively.

### Growth variables

Plant height (cm), base diameter and number of leaves were respectively assessed for 3 weeks. Analysis of variance was carried out to ascertain the relationship between the germination and growth pattern and the treatment used (planting regime) for the species over time (weeks).

## RESULTS AND DISCUSSION

### Effect of planting regime on plant height

The result of the plant height assessed was significantly influenced by the planting regime and time (weeks) at $P < 0.05$ (Table 1). This could be due to the juvenility of the seedlings, which are in their initial period of active growth and have high levels of auxins to enhance rooting as reported by Oni et al. (2005). Plant height improved with increased period of planting. It was observed that seeds sown after 5 days of extraction (45.32 cm) performed better than others (Figure 1). The least of 21.02 cm was obtained from seeds planted after 14 days of extraction. This revealed that the seed dormancies are broken after 5 days of extraction and the height growth is also improved

### Effect of planting regime on number of leaves

The significant effects of planting regime and time of planting (weeks) reflected on leaves number (Table 2). Mean leaf numbers did not follow similar trends for the various planting regime with the seedlings having the maximum leaf numbers (14 leaves) in seeds sown after 3 days of extraction at the 3$^{rd}$ week (Figure 2). This was followed by seeds freshly sown (10 leaves) and the least number of leaves was observed in seeds sown after 14 days (8 leaves). This revealed that the delay in sowing of the seed affect the leaves production of the plant. This needs to be noted, if the purpose of production is meant for leaves harvesting.

### Effect of planting regime on base diameter

Base diameter is a vital parameter needed in growth and

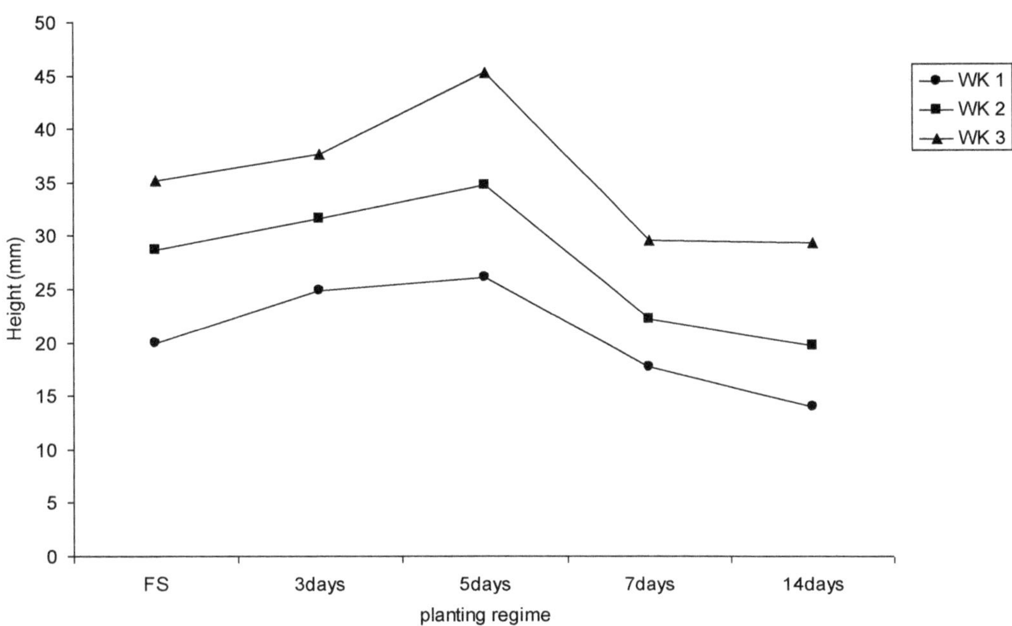


**Figure 1.** Chart of the plant height after 3 weeks of assessment. FS, Freshly sown; DF, Degree of freedom; SV, source of variation; SS, sum of squares; MS, mean of sum of squares; F, F-values.

**Table 2.** Analysis of variance of planting regime and weeks on number of leaves.

| SV | DF | SS | MS | F |
|---|---|---|---|---|
| Planting regime | 4 | 20.10 | 5.02 | 35.86** |
| Weeks | 2 | 17.52 | 8.76 | 62.57** |
| Error | 38 | 5.17 | 0.14 | |
| Total | 44 | 42.79 | | |

** = Significant at $p < 0.05$.

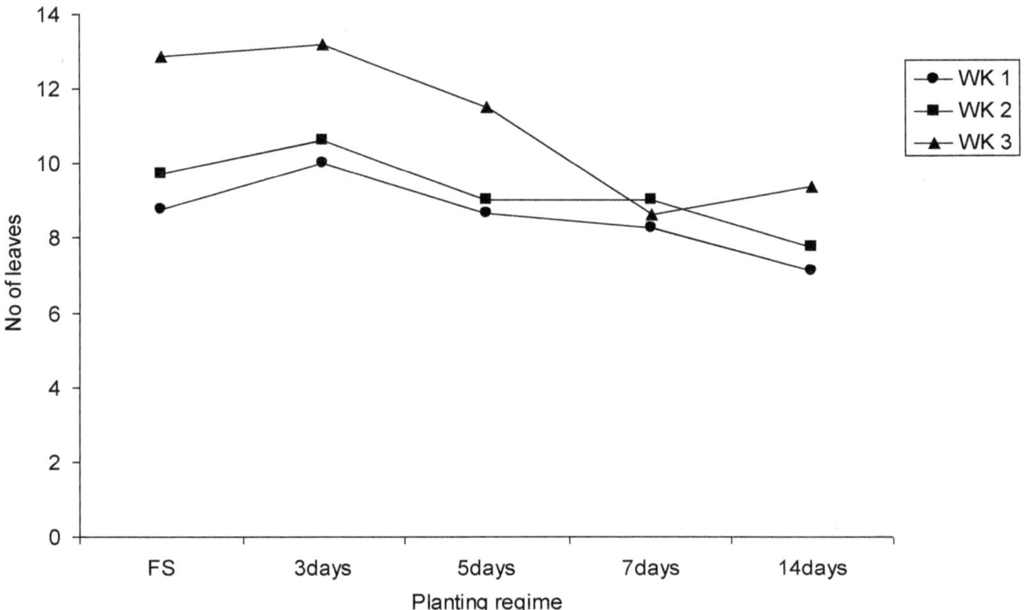


**Figure 2.** Chart of number of leaves produced after 3 weeks of assessment. **FS,** Freshly sown.

**Table 3.** Analysis of variance of planting regime and weeks on base diameter.

| SV | DF | SS | MS | F |
|---|---|---|---|---|
| Planting regime | 4 | 1.36 | 0.34075 | 17.00** |
| Weeks | 2 | 0.11 | 0.06 | 3.00** |
| Error | 38 | 0.75 | 0.02 | |
| Total | 44 | 2.22 | | |

** = Significant at p < 0.05.

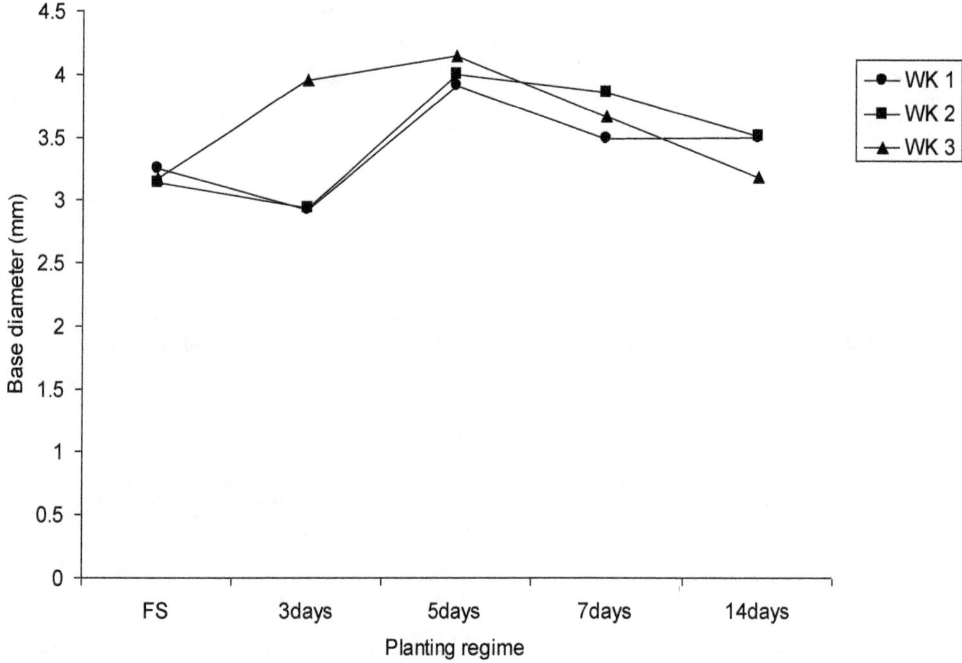


**Figure 3.** Chart of the plant base diameter after 3 weeks of assessment. FS, Freshly sown.

yield assessment of a plant. Volume of tree/plant is a function of its base diameter and height (Avery and Burkhart, 2002). The effect of the planting regime and weeks on the base diameter was also significant (Table 3). It was also observed that the base diameter was highest for seeds left for 5 days after extraction before sowing and the least was observed in seeds sown after 3 days of extraction (Figure 3).

## Conclusion

From this study, optimum growth of *A. sceptrum* can be obtained in the seeds sown after 5 days which shows that the seed does not required mechanical or chemical method breaking its dormancy. Planting regime indicated a contributory effect on early seedlings growth revealed a positive effect in the growth performance of *A. sceptrum* in the entire study. This study has therefore showed a possible domestication and multiplication of this species to avoid its extinction and to make it available for sustainable utilization in any capacity.

## REFERENCES

Avery TE, Burkhart HE (2002). Forest Measurements, 5th ed. McGraw-Hill Higher Education, New York, USA. P. 456.

Bhattacharya J, Khuspe SS (2001). *In-vitro* and *in-vivo* germination of papaya (*Carica papaya* L.) seeds. Sci. Hort. 91:39-49.

Echeverrigaray S, Fracaro F, Santos ACA, Paroul N, Wasum R, AttiSerafini L (2003). Essential oil composition of south Brazilian populations of Cunila galioides and its relation with the geographical distribution. Biochem. Syst. Ecol. 31:467-475.

Eyob S, Appelgren M, Rohloff J, Tsegaye A, Messele G (2008). Traditional medicinal uses and essential oil composition of leaves and rhizomes of korarima (*Aframomum corrorima* (Braun) P.C.M. Jansen) from Southern Ethiopia. S. Afr. J. Bot. 74:181-185.

Eyob S, Tsegaye A, Appelgren A (2009). Analysis of korarima (*Aframomum corrorima* (Braun) P.C.M. Jansen) indigenous production practices and farm based biodiversity in southern Ethiopia. Genet. Resour. Crop Evol. 56:573-585.

Oni PI, Uzokwe N, Adeyanju, BA (2005). Evaluation of different pretreatment techniques and potting media on seedlings emergence and early growth in *Pterocarpus osun*. Proceedings of the 30th Annual Conference of the Forestry Association of Nigeria in Kaduna, Kaduna State, Nigeria. P. 579.

Polat AA (1997). Determination of germination rate coefficients of loquat seeds and their embryos stratified in various media for different durations. Turk. J. Agric. For. 21:219-224.

Sulikeri GS, Kololgi SD (1977). Seed viability in Cardamom (*Elettaria cardamomum* Maton). Curr. Res. 6: 163-164.

Tefera W, Wannakrairoj S (2004). Micropropagation of korarima (*Aframomium corrorima* (Braun) Jansen). Sci. Asia 30:1-7.

# Evaluation of carnation (*Dianthus caryophyllus L.*) varieties under naturally ventilated greenhouse in mid hills of Kumaon Himalaya

**Ajay Kumar Singh, D. K. Singh, Balraj Singh, Shailja Punetha and Deepak Rai**

Research Station and KVK, Lohaghat (GBPUA&T, Pantnagar) Uttarakhand, 262 524, India.

**Among eight Carnation varieties viz. Diana, Aurturo, White Dona, Pink Dona, Soto, Red King, Tuareg and Dona., evaluated under naturally ventilated greenhouse and subjected to uniform treatment and cultural package of practices, the variety Red King was found best with respect to number of branches (8.0), number of flowers/plant (35.6), fresh weight of flower (8.38 g), dry weight of flower (2.66 g), flower diameter (7.83 cm) and vase life an important post harvest quality parameter was observed to be the superior in variety Red King (29.3 days) followed by cv Tuareg (24 days) and Pink Dona (21.3 days). On the basis of present experiment it was concluded that cultivar Red King were found to be promising with respect to yield as well as flower quality parameters and found suitable for commercial cultivation under naturally ventilated green house in mid hills of Kumaon Himalayas.**

**Key words:** Carnation, greenhouse, varieties, vase life.

## INTRODUCTION

Carnation (*Dianthus caryophyllus* L.), a member of family Caryophyllaceae is one of the leading cut flower crops in the world florist trade and ranks with in top ten cut flower of the world. It is half hardy perennial with branching stems and timid joints, leaves are linear, glaucous, in opposite or decussate pairs. Each stem forms terminal flowers which is bisexual or occasionally unisexual. The hybrids have remarkable long flowering period which produces blooms continuously in mild weather.

The crop grown in open field is exposed to aberration of environmental conditions and attack by different pest and diseases, resulting in poor quality flowers. A carnation variety varies with region, season, genotypes and growing environment. In India, depending upon the regions, there is a wide difference in temperature, light intensity and humidity which not only affect the yield and quality of the flowers but also limit their availability for a particular period of a year. To produce quality flowers, carnation need to be grown under cover, that is, in greenhouse which provides the plants with the optimum condition of light, temperature, humidity and carbon dioxide etc for proper growth and to achieve maximum yield of best quality flowers (Bhalla et al., 2006). Though, there are different types of the greenhouses, naturally ventilated polyhouses are preferred in mild climate in which temperature is reduced by ventilation (Ryagi et al., 2007).

The mid hills of Uttarakhand provides an ideal condition for growing variety of flowers and quality carnation cut flowers can be grown under naturally ventilated green houses. Considering the importance of the flower crop, an experiment was conducted during 2010-12 to evaluate eight carnation varieties for their yield and quality attributes under naturally ventilated green house condition at Research Station and KVK, Lohaghat, (GBPUA&T, Pantnagar), Champawat, Uttarakhand which

**Table 1.** Effect of growing condition on flower yield attributes of eight carnation varieties under naturally ventilated greenhouse.

| Varieties | No. of branches/plants | No. of flowers/plants | Stem length (cm) | No of leaf pairs/stem | Fresh weight of flower (g) | Dry weight of flower (g) | Flower diameter (cm) | Vase life (days) |
|---|---|---|---|---|---|---|---|---|
| Diana | 3.66 | 27.00 | 66.30 | 9.00 | 5.70 | 1.19 | 7.30 | 20.30 |
| Aurturo | 5.00 | 23.30 | 69.30 | 9.00 | 6.70 | 1.55 | 7.00 | 19.30 |
| White Dona | 5.60 | 23.60 | 56.60 | 8.00 | 4.00 | 1.01 | 7.00 | 18.30 |
| Pink Dona | 7.30 | 29.30 | 53.40 | 10.30 | 5.70 | 1.01 | 6.80 | 21.30 |
| Soto | 5.60 | 24.00 | 62.30 | 9.30 | 5.30 | 1.22 | 6.80 | 17.60 |
| Red King | 8.00 | 35.60 | 75.00 | 9.60 | 8.30 | 2.66 | 7.80 | 29.30 |
| Tuareg | 5.33 | 22.00 | 68.30 | 11.00 | 6.80 | 1.63 | 5.70 | 24.00 |
| Dona | 6.60 | 28.00 | 65.50 | 10.60 | 6.60 | 1.63 | 7.50 | 20.00 |
| SEm | 0.53 | 5.90 | 4.71 | 0.47 | 0.54 | 0.152 | 0.27 | 0.68 |
| CD at 5% | 1.61 | 10.54 | 14.29 | 1.45 | 1.65 | 0.46 | 0.83 | 2.08 |

is situated at an latitude of 29 60'; Longitude of 80 1' and altitude of 1700 to 1800 m from MSL in North West Himalaya of India.

#### MATERIALS AND METHODS

The eight varieties of carnations viz. Diana, Aurturo, White Dona, Pink Dona, Soto, Red King, Tuareg and Dona were evaluated for their yield and quality attributes. The raised beds of 80 cm width with 50 cm path in between were made under greenhouse. The planting of rooted cuttings of eight varieties were done in four rows on a bed with spacing of 15 × 15 cm. Experiment was designed in randomized block design with eight varieties as treatments and three replication of each treatment. Uniformly two pinching has been done in all treatments with first pinching all main shoots were pinched retaining six pair of leaves from base at 21 to 25 days after planting and second pinching with half pinching of the laterals produced from single pinching has been done to all the varieties. Uniform package of practices were followed throughout the cropping season to grow a successful crop. Flowers are harvested at paint brush stage retaining 2 to 3 nodes at the base of the stalk.

The data were recorded for flower yield attributes like number of branches per plants, number of flower per plant, stem length (cm), number of leaf pairs per stem and flower quality attributes like flower diameter (cm), flower fresh weight (gm), flower dry weight (gm) and flower vase life (days) using the standard method. The collected data were subjected to statistical analysis as suggested by Cochran and Cox (1992).

### RESULTS AND DISCUSSION

Among eight varieties of carnations, significantly maximum number of branches per plants was recorded with variety Red King (8.0) which was at par with variety Soto (7.3). Significantly higher cut flower stem length was recorded with variety Red King (75 cm) as compared to all other verities (Table 1). The similar results with respect to varieties were reported by Bhautkar (1994), Mahesh (1996) and Patil (2001). This may be due to fact that removal of apical portion by pinching which neutralizes the effect of apical dominance. Similar variations with pinched variety were reported by Patel and Arora (1983), Khanna et al (1986) in carnation. Among the eight varieties, the significantly more number of flowers per plant were recorded in Red King (35.6) and minimum flowers per plant were observed with Variety Tuareg (22). Similar flower yield variations in carnation varieties were reported by Gill and Arora (1988), Naveen et al. (1999) and Patil (2001). Ubukata (1999) reported that single and half pinching increased the number of flowers in three carnation varieties which may be due to fact that there was increase in the temperature under greenhouse from February month onwards and in single and half pinching affects in increase of the laterals which might be due to high temperature (Figures 1 and 2). Similarly, Ramesh et al. (2002) reported that pinching twice increase the number of the flowers per plants compared to single pinching. The increase in the number of flowers per plants and yield of flower might be attributed to the development of large number of auxillary shoots as a result of cessation of the terminal growth by pinching (Narayana and Jayanthi, 1991). These results are in conformity with Yassin and Pappiah (1990) and Singh and Baboo (2003) in Chrysanthemum.

Variety Redking has significantly higher flower fresh weight and dry weight followed by variety Turang (Table 1). Similar results were also observed by Ateeque et al. (1994), El-Shafei (1977), Singh and Sangama (2003), where they observed that increase in leaves dry matter may be related to the increase in the number of leaves. Flower quality with respect to flower diameter and flower vase life was significantly higher in variety Red King as compared to all varieties grown under similar growing condition in greenhouse. Maximum flower diameter (7.83 cm) and flower vase life (29.33 days) were observed with Red King. Variation in vase life could be attributed to the variation in ability to produce ethylene and sensitivity to it among the different varieties. The similar variation in flower diameter and vase life of the different varieties of

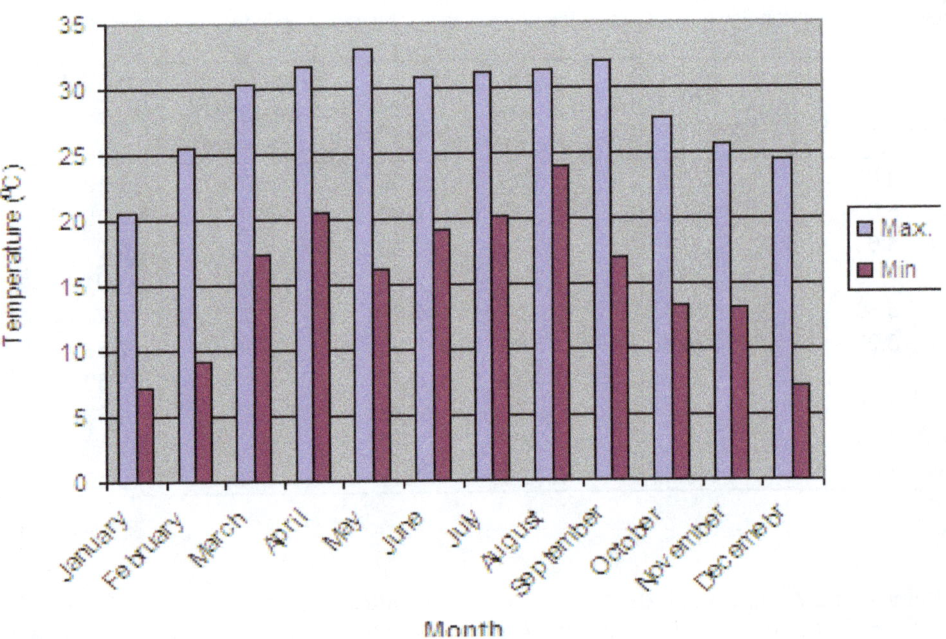


**Figure 1.** Average monthly temperature under greenhouse.

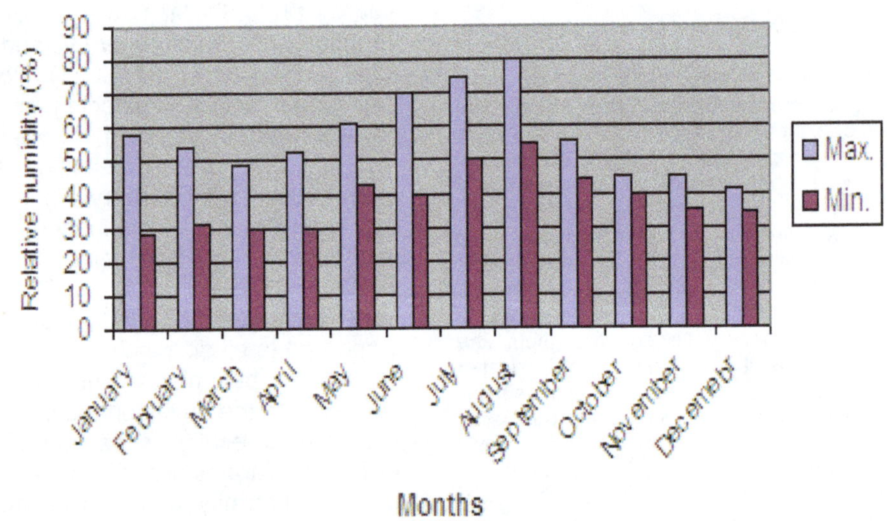


**Figure 2.** Average monthly relative humidity under greenhouse.

the carnation was reported by Reddy et al. (2004), Ramesh et al. (2002), Singh and Sangama (2003), Naveen et al. (1999), Sathisha (1997) and Gill and Arora (1988).

## ACKNOWLEDGEMENT

The authors are thankful to the NAIP project - Protected cultivation of High value vegetables and cut flowers - a value chain approach running at Research and KVK, Lohaghat for the financial assistance in carrying out the experiment.

## REFERENCES

Ateeque M, Malewar U, More M (1994). Influence of phosphorous and boron on yield and chemical composition of sunflower. Soils Fert. Abstr. 57:10373.

Bhalla R, Sharma S, Dhiman SR, Ritu J (2006). Effect of biofertilizers and bistimulant on growth and flowering in standard carnation (*Dianthus caryophyllus* Linn.). J. Ornam. Hortic. 9(4):282-285.

Bhautkar R (1994). Carnation cultivation in green house under

Mahabaleshwar condition. J. Maharastra Agric. Uni. 19:292-293.
Cochran WG, Cox MC (1992). Experimental design. John Wiley Sons, Inc, New York. pp. 106-117.
El-Shafei SA (1977). The effect of spacing and fertilization levels on the growth and flowering of carnation. Archi. Fer. Gartenbeu 25:347-54.
Gill APS, Arora JS (1988). Performance of sim carnation under subtropical climate condition of Punjab. Indian J. Hortic. 45:292-93.
Khanna K, Arora JS, Singh J (1986). Effect of spacing and pinching on growth and flower production of carnation. Indian J. Hortic. 43:148-52.
Mahesh K (1996). Variability studies in carnation (*Dianthus caryophyillus* L.) M.Sc (Agri.) Thesis, University of Agricultural Sciences, Bangalore. India.
Narayana GJV, Jayanthi R (1991). Effect of cycocel and Meleic Hydrazide on growth and flowering of African marigold (*Tagetes erecta*). Prog. Hortic. 23(1-4):114-18.
Naveen KP, Singh B, Siddhu SS, Voleti SR (1999). Effect of growing environment on carnation flowering. J. Ornamental Hortic. 2:139-40.
Patil RT (2001). Evaluation of standard carnation (*Dianthus caryophyillus* L.) cultivars under protected cultivation M.Sc (Agric.) Thesis, University of Agricultural Sciences, Dharwad, India.
Patel KS, Arora JS (1983). Effect of pinching, sources and doses of N on growth and flower production of carnation (Dianthus caryophyllus) Cv. Marguerite white. Indian J. Hortic. 40: 92-97.
Ramesh K, Singh K, Reddy BS (2002). Effect of planting time, Photoperiod, GA3 and pinching on carnation. J. Ornam. Hortic. 5:20-23.
Reddy BS, Patil JRT, Kulkarni BS (2004). studies on vegetative growth, flower yield and quality of standard carnation (*Dianthus caryophyillus* L.) under low cost polyhouse condition. J. Ornam. Hortic. 7(3-4):217-220.
Ryagi VY, Mantur SM, Reddy BS (2007). Effect of pinching on growth yield and quality of flower of carnation varieties grown under polyhouse. Karnataka J. Agric. Sci. 20(4):816-818.
Sathisha S (1997). Evaluation of Carnation (*Dianthus caryophyillus* L.) cultivar under low cost green house. M.Sc (Agri.) Thesis. University of Agricultural Sciences, Bangalore, India.
Singh KD, Sangama (2003). Evaluation of post harvest quality of the some cultivars of carnations flowers grown in greenhouse. J. Ornamental Hortic. 6(3):274-276.
Singh MK, Baboo R (2003). Response of Nitrogen, potassium and pinching levels on growth and flowering of Chrysanthemum cv. Jayanthi. J. Ornamental Hortic. 6(4):390-93.
Ubukata M (1999). Evaluation of one half pinch method of spray carnation cultivation in Hokkaido. Bull. Hokkaido Prefectural Agric. Exp. Station. 77:39.
Yassin G Md, Pappiah CM (1990). Effect of pinching and manuring on growth and flowering of chrysanthemum cv. MDU-1. South Indian Hortic. 38(4):232-233.

# 14

# Out breeding for yield and horticultural attributes in indigenous eggplant germplasm

**S. Ramesh Kumar[1] , T. Arumugam[2], V. Premalakshmi[2], C. R. Anandakumar[3] and D. S. Rajavel[4]**

[1]Department of Horticulture, Vanavarayar Institute of Agriculture, Tamil Nadu Agricultural University, Manakkadavu, Pollachi-642 103, Tamil Nadu, India.
[2]Department of Horticulture, Agricultural College and Research Institute, Tamil Nadu Agricultural University, Madurai-625 104, Tamil Nadu, India.
[3]Centre for Plant Breeding and Genetics, Tamil Nadu Agricultural University, Coimbatore-641003, Tamil Nadu, India.
[4]Department of Crop Protection, Agricultural College and Research Institute, Killikulam, Tamil Nadu Agricultural University, Tuticorin- 628 252, Tamil Nadu, India.

**Heterotic effects and genetic components of variation for qualitative and quantitative characters were estimated in eggplant (*Solanum melongena* L.). Forty hybrids generated by crossing four testers (males) with ten lines (females) were studied along with parents for studying heterosis and gene action for calyx length, fruit pedicel length, shoot borer infestation, fruit borer infestation, little leaf incidence, ascorbic acid content, total phenols content, number of fruit per plant and fruit yield per plant during rainy season of 2010-2011. The crosses obtained by L × T method possessed variation in terms of growth, yield and quality traits. Mean fruit yield per plant ranged from 2.85 to 1.04 kg. Among the 40 hybrids, the ones obtained from the cross 'Alagarkovil Local' × 'Annamalai' ($L_4 \times T_1$), 'Palamedu Local' × 'Punjab Sadabahar' ($L_5 \times T_3$), 'Palamedu Local' × 'EP 65' ($L_5 \times T_4$) and 'Keerikai Local' × 'KKM 1' ($L_7 \times T_2$) were suitable for heterosis breeding. Average performance of parents indicated that lines 'Alavayal local' ($L_1$), 'Sedapatty local' (Green) ($L_2$) and the tester 'Annamalai' ($T_1$) were good parents for further breeding to exploit high yield and low pest and disease incidences. Performance of these hybrids needs to be further evaluated in multiple locations or on farm trial prior to commercial use.**

**Key words:** Hybrid vigour, brinjal germplasm, selection, yield attributes.

## INTRODUCTION

Eggplant (*Solanum melongena*) belongs to the family Solanaceae. Brinjal grows throughout the tropical and sub-tropical regions of the world. It is also widely grown and common vegetable crop in India. It is locally known as 'Kathirikkai' or Aubergine' or Badanekkai' and is popular among the rural people. India is the major producer of brinjal in the world and it is grown in an area of 0.61 million ha with an estimated annual production of 13.37 million tons with a productivity of 17.3 tons/ ha. In Tamilnadu the production was 8.5 lakh tonnes from 0.75 lakh ha of area (Anonymous, 2010). Immature fruits are used as vegetable and extensively used in various culinary preparations. Nutritive value of brinjal is well compared with tomato (Choudhary, 1976). They are also known to have alkaloid solanine in roots and leaves. Some medicinal uses of brinjal include treatment of diabetes, asthma, cholera, bronchitis and dysuria. Fruits and leaves are administered to lower blood cholesterol levels. Fruits are rich source of minerals like Ca, Mg, P and fatty acids (Dhankhar and Singh, 1984).

There are number of local cultivars with wide range of variability in size, shape and color of fruits available in India and for this we can easily fulfill the gap by developing high yielding hybrid variety (Prabakaran, 2010).

The productivity of local genotypes ranged from 10t/ha to 18t/ha. Though a fairly common crop, to-date there is only limited work has been done for evolving hybrids/hybrid derivatives of high yield potential and better quality in Tamil Nadu (India) by using local germplasm.

Furthermore, very limited attempt has been made for genetic improvement of available indigenous types in this crop. Heterosis or hybrid vigor can play a vital role in increasing the yield quality of eggplant. Identification of potential parents in brinjal on the basis of progeny performance requires a large number of crosses, which is very laborious. L × T is a mating design whereby the selected parents are crossed in a certain order to predict the combining ability of the parents and elucidate the nature of gene action involved in the inheritance of the traits (Abhinav and Nandan, 2010).

The phenomenon of heterosis of F1 hybrids can also reflect specific combining ability (SCA) and general combining ability (GCA) of parental lines. The combining ability works as the basic tool for improved production of crops in the form of F1 hybrids (Dhillion, 1975).

Heterotic studies can also provide the basis for exploitation of valuable hybrid combinations and their commercial utilization in future breeding programs (Chowdhury et al., 2010). The combining ability and heterosis work as the principal methods for screening of germplasm and determination of the ability of the genotypes to be included or not in a breeding program on the basis of their GCA, SCA, reciprocal, and heterotic effects. Recently, it has been divulged that the utilization of hybrid vigor is most effective for the improvement of different characters and the combining ability is the fundamental tool for enhancing the productivity/yield of different crops in the form of F1 hybrids (Pachiyappan et al., 2012). Considering the above idea in mind the present investigation was undertaken with the objectives (i) to determine the magnitude of heterosis in the hybrids (ii) to study the gene action of different traits and (iii) development of hybrids for higher fruit yield and quality using indigenous genotypes.

## MATERIALS AND METHODS

### The experimental site

The experiment was conducted during *kharif* (rainy season) 2010-2011 at College Orchard, Agricultural College and Research Institute, Madurai, Tamil Nadu, India which is situated at 9°5 latitude and 78°5 longitude and at an elevation of 147 m above MSL.

### Experimental materials

Genetic improvement in eggplant germplasm has been initiated in Department of Horticulture, Agricultural College and Research Institute, Madurai, Tamil Nadu, India during 2009-2011 using indigenous germplasm for increasing the overall production of eggplant and release of hybrids for commercial use. Around 33 eggplant germplasm were collected from in and around the Tamil Nadu state having wider range of diversity and variability and evaluated under field condition for yield, quality and other desirable traits. Among them, 10 germplasm were selected based on color (consumer preference), shape, size and yield and used as female parents for crossing programme. Ten lines (females) were crossed with four testers (males) through Line × Tester mating design to derive the 40 $F_1$ hybrids. The details of different parents used in the present study were narrated in Table 1.

### Selfing

All the fourteen parents were maintained in a homozygous condition by continuous selfing. Hundred percent selfing was achieved by bagging the flowers with white butter paper covers, a day prior to anthesis. When the fruits have attained marble stage, the covers were removed and the fruits were tagged. Fully ripe labelled fruits were collected individually and seeds were extracted by fermentation method. Seed moisture content was reduced to eight per cent by shade drying and the seeds were stored in butter paper covers for further use.

### Crossing technique

The fourteen parents were used for crossing to get 40 hybrids. A separate crossing block was maintained for the production of hybrid seeds.

In the female parent, healthy, long or medium styled flower buds, which were likely to open on the next day, were selected for emasculation. Emasculation was carried out between 3.00 and 5.00 pm and bagged with butter paper covers. Similarly, in the male parents, a few selected flower buds for collection of pollen grains were bagged without emasculation to avoid contamination by foreign pollen. Pollen from the bagged flowers of the pollen parents were collected between 7.00 and 8.00 am in the next day and dusted on to the stigma of the emasculated flowers of the respective ovule parents. The flowers were bagged with butter paper covers and then labelled. The covers were removed after ensuring proper fruit set. The crossed fruits were harvested at full ripe stage and seeds were extracted by fermentation method. The seeds were shade dried to eight percent moisture content and stored in butter paper covers for further breeding programme.

### Experimental design

Forty hybrids along with 14 parents were raised in a Randomized Complete Block Design with three replications.

### Nursery and cultivation aspects

Thirty days-old seedlings raised in the nursery beds were transplanted on the ridges adopting a spacing of 60 x 60 cm. Thirty plants were maintained for each hybrid and parent in each replication. Recommended cultural practices were followed uniformly to all the hybrids and parents as per the Tamilnadu Agricultural University Crop Production Guide (2005). Observations were recorded in five randomly selected plants in each replication.

**Table 1.** Brief description of the parental genotypes used in experiments.

| S/N | Name of the types | Flower colour | Flower bearing | Fruit bearing | Fruit shape | Fruit colour | Calyx fleshness | Calyx type | Source | Symbol |
|---|---|---|---|---|---|---|---|---|---|---|
| **Lines** | | | | | | | | | | |
| 1. | Alavayal Local | Dark purple | Cluster | Cluster | Round | Light purple | Fleshy | Persistent | Alavayal, Madurai D.t ,Tamil Nadu | $L_1$ |
| 2. | Sedapatty Local (Green) | Purplish white | Cluster | Cluster | Oval | Purplish green | Fleshy | Persistent | Sedapatty, Madurai D.t, Tamil Nadu | $L_2$ |
| 3. | Kariapatty Local | Purplish white | Cluster | Cluster | Round | Green striped | Fleshy | Persistent | Kariapatty, Virdhunagar D.t, Tamil Nadu | $L_3$ |
| 4. | Alagarkovil Local | Light purple | Cluster | Cluster | Round | Green striped | Fleshy | Persistent | Alagarkovil, Madurai D.t, Tamil Nadu | $L_4$ |
| 5. | Palamedu Local | Purplish white | Cluster | Cluster | Round | Light blue | Non fleshy | Persistent | Palamedu, Madurai D.t, Tamil Nadu | $L_5$ |
| 6. | Melur Local | Purplish white | Cluster | Cluster | Round | Purple | Fleshy | Persistent | Melur, Madurai D.t, Tamil Nadu | $L_6$ |
| 7. | Keerikai Local | Purplish white | Cluster | Cluster | Oval | Purplish green | Non fleshy | Persistent | Sempatty, Dindigul D.t, Tamil Nadu | $L_7$ |
| 8. | Nilakottai Local | White | Cluster | Cluster | Oblong | Green striped | Non fleshy | Persistent | Nilakottai, Dindigul D.t, Tamil Nadu | $L_8$ |
| 9. | Singampunari Local | Light purple | Cluster | Cluster | Round | Purplish green | Fleshy | Persistent | Singampunari, Sivagangai D.t, Tamil Nadu | $L_9$ |
| 10. | Sedapatty Local (Blue) | Purplish white | Cluster | Cluster | Round | Purple striped | Fleshy | Persistent | Sedapatty, Madurai D.t, Tamil Nadu | $L_{10}$ |
| **Testers** | | | | | | | | | | |
| 1. | Annamalai | Purple | Cluster | Cluster | Long | Purple | Non fleshy | Non persistent | Vegetable Research Station, Palur, Tamil Nadu | $T_1$ |
| 2. | KKM 1 | Purple | Cluster | Cluster | Egg shaped | White | Fleshy | Persistent | Agricultural College and Research Institute, Tuticorin, Tamil Nadu | $T_2$ |
| 3. | Punjab Sadabahar | Purple | Cluster | Cluster | Long | Purple | Non fleshy | Non persistent | Tamil Nadu Agricultural University, Coimbatore | $T_3$ |
| 4. | EP 65 | Purple | Cluster | Cluster | Oval | Dark purple | Non fleshy | Non persistent | Vegetable Research Station, Palur, Tamil Nadu | $T_4$ |

**Data collected**

The data recorded for nine biometrical traits *viz*., calyx length, fruit pedicel length, shoot borer infestation, fruit borer infestation, little leaf incidence, ascorbic acid content, total phenols content, number of fruits per plant and fruit yield per plant in 14 parents and forty hybrids were used for estimating heterosis.

***Calyx length (cm)***

The calyx length of five randomly selected fruits at vegetable maturity in the third harvest was recorded and the mean is expressed in centimeter (cm).

***Fruit pedicle length (cm)***

The pedicle length of five randomly selected fruits at vegetable maturity in the third harvest was recorded and the mean is expressed in centimeter (cm).

***Shoot borer infestation (%)***

The number of shoots affected by borer and total number of shoots per plant were recorded and the percent of shoot borer infestation was worked out.

***Fruit borer infestation (%)***

The numbers of fruits affected by borer and total number of fruits harvested were recorded and the percent of fruit borer infestation was worked out.

***Little leaf incidence (%)***

The number of plants affected by little leaf and total number plants available was recorded and the percent of little leaf incidence was worked out.

***Number of fruits per plant***

Fruits at vegetable maturity were harvested and counted at each harvest and the cumulative number of fruits per plant is expressed.

***Fruit yield per plant (kg)***

The weight of fruits in each plant was recorded at each harvest and the total weight of fruits over all the harvests were recorded as yield per plant and expressed in kilogram (kg).

**Table 2.** Estimates of variance (L×T) components, rainy season 2010-2011.

| Character | GCA variance | SCA variance | $\delta^2$ A | $\delta^2$ D | Ratio of $\delta^2$ A: $\delta^2$ D |
|---|---|---|---|---|---|
| Plant height (cm) | 67.02 | 98.15 | 2.91 | 98.15 | 0.02 |
| Days to first flowering | 2.06 | 14.97 | 0.20 | 14.97 | 0.01 |
| Number of branches per plant | 4.52 | 17.34 | 0.26 | 17.34 | 0.01 |
| Fruit length (cm) | -0.01 | 2.15 | 0.05 | 2.15 | 0.02 |
| Fruit pedicel length (cm) | 0.07 | 0.41 | 0.03 | 0.41 | 0.07 |
| Fruit circumference (cm) | 1.17 | 4.40 | 0.06 | 4.40 | 0.01 |
| Calyx length (cm) | 0.15 | 0.38 | 0.007 | 0.38 | 0.01 |
| Number of fruits per plant | 11.80 | 62.29 | 0.75 | 62.29 | 0.01 |
| Average fruit weight (g) | 19.12 | 51.88 | 0.83 | 51.88 | 0.01 |
| Shoot borer infestation (%) | 5.48 | 19.66 | 0.25 | 19.66 | 0.01 |
| Fruit borer infestation (%) | 1.58 | 18.65 | 0.03 | 18.65 | 1.60 |
| Little leaf incidence (%) | 6.61 | 26.49 | 0.32 | 26.49 | 0.01 |
| Ascorbic acid content (mg/100g) | 1.45 | 6.84 | 0.09 | 6.84 | 0.01 |
| Total phenols content (mg/100g) | 47.28 | 159.27 | 1.67 | 159.27 | 0.01 |
| Fruit yield per plant (kg) | 0.02 | 0.20 | 0.02 | 0.20 | 0.10 |

GCA-General combining ability; SCA-specific combining ability; $\delta^2$ A-additive variance; $\delta^2$ D-dominance variance.

***Quality traits***

**Ascorbic acid (mg/100 g):** Ascorbic acid content was estimated by volumetric method as suggested by AOAC (2001).

$$\text{Ascorbic acid} = \frac{0.5\text{ mg}}{V_1\text{ ml}} \times \frac{V_2\text{ ml}}{5\text{ ml}} \times \frac{100\text{ ml}}{\text{weight of the sample}} \times 100$$

**Total phenols (mg/100 g):** Folin ciocalteau reagent method was followed for estimating the total phenols (Bray and Thrope, 1954).

***Selection of hybrids***

The selections were made in the $F_1$ hybrids based on fruit shape, colour, size and fruit yield per plant. The superior hybrids were selected and selfed. The seeds were collected from the selfed fruits and stored for further breeding programme.

**Data analysis**

Line x tester analysis was carried out to test parents and hybrids with respect to their general and specific combining ability, respectively. The line x tester analysis of combining ability gives useful information regarding the choice of parents and elucidates the nature and magnitude of various types of gene action for the expression of yield and yield attributing characters. The data on the hybrids and parents were subjected to L × T analysis. The assumption of null hypothesis was tested for differences among the genotypes as detailed by Panse and Sukhatme (1967). The general combining ability effects of the parents and specific combining ability effects of the crosses were worked out as suggested by Kempthorne (1957).

The magnitude of heterosis in hybrids was expressed as percentage of increase or decrease of a character over mid parent ($d_i$), better parent ($d_{ii}$) and standard hybrid ($d_{iii}$) and was estimated following the formula of Fonseca and Patterson (1968). The significance of magnitude of the relative heterosis, heterobeltiosis and standard heterosis was tested at error degrees of freedom by the formula as suggested by Turner (1953). It is estimated as follows:

$$\text{Standard heterosis } (d_{iii}) = \frac{\overline{F1} - \overline{SV}}{\overline{SV}} \times 100$$

## RESULTS

### Line × tester analysis

Line × tester analysis was carried out to detect the gene action of different traits. The significance level and mean square value of all the investigated traits are shown in Table 2. The total variance is further partitioned into several components, like variance due to lines, testers and their interactions. We observed a significant genetic variation among lines and testers for all traits. Heterosis is the superiority of an F1 hybrid produced through crossing of two genetically different individuals over the mean of its parents or the better parent. Standard heterosis was estimated for fruit pedicel length, calyx length, shoot borer infestation, fruit borer infestation, number of fruits per plant, ascorbic acid content, total phenols content and fruit yield per plant. Heterosis was obvious in different crosses for all the characters and the magnitude varied significantly for traits and crosses. The hybrid vigor can help to increase the yield by several times than open pollinated variety.

### Performance of germplasm and hybrids

In any statistical analysis of data, average performance is

the true realized mean of the recorded data and this is a direct estimate based on the observation and not on assumption. Among parents, line $L_8$ produced the longest fruit pedicel followed by $L_2$. Testor $T_2$ had the shortest fruit pedicel. The parents $L_1$, $L_2$, $L_4$, $L_8$ and $T_4$ exceeded the mean value. The longest calyx was for parent $L_7$ followed by $L_6$ and the shortest was for $L_8$. Among testers, $T_3$ had the longest calyx. Nine parents had significant values for this character. The fewest fruit per plant was for $L_5$ and the most was for $T_1$ followed by $L_9$ among parents. The parents $L_2$, $L_3$, $L_6$, $L_9$, $T_1$, $T_2$ and $T_3$ exceeded the mean value. Parents $T_2$ and $L_8$ had the highest and lowest borer infestation, respectively. The females $L_1$, $L_2$, $L_5$, $L_6$, $L_7$, $L_9$, and $L_{10}$ and males $T_3$ and $T_4$ had higher values over the mean. Fruit borer infestation was highest in $T_2$ and lowest in $T_4$ among testers. Seven parents had significant little leaf incidence values for this trait. Among parents, the highest ascorbic acid content was for $T_1$ and the lowest was for $L_{10}$. The parents $L_1$, $L_3$, $L_4$, $L_6$, $L_8$, $T_1$, $T_3$ and $T_4$ had high values for ascorbic acid content. Six parents had significant mean values for total phenol. The line $L_5$ had the highest yield per plan followed by $L_2$, while the lowest yield was in $L_9$. Six of 14 parents had higher fruit yield per plant over the grand mean. The hybrid $L_3 \times T_4$ produced the longest fruit pedicel length followed by $L_6 \times T_1$. The hybrid $L_7 \times T_4$ had the shortest fruit pedicel length. Of 40 hybrids, 13 had significantly higher values than the grand mean. The longest calyx was in the hybrid $L_3 \times T_4$ followed by $L_5 \times T_3$. Thirty one hybrids had significant values for this trait. Of 40 hybrids, 16 crosses had higher mean values than the grand mean of fruit per plant. Among hybrids, the $L_3 \times T_3$ had the least borer infestation, followed by $L_1 \times T_3$. Twenty-four hybrids had significant mean values for this trait. The hybrid $L_8 \times T_1$ had the lowest and $L_8 \times T_3$ the highest borer infestation. Twenty-eight hybrids had lower borer infestation than the mean. Little leaf incidence was highest in the hybrid $L_{10} \times T_2$. The hybrid $L_5 \times T_3$ had the minimum little leaf incidence. The hybrid, $L_{10} \times T_2$ had the highest little leaf incidence. Twenty-three hybrids had lower values than the grand mean. The hybrid $L_1 \times T_3$ had the highest, followed by $L_1 \times T_1$; $L_{10} \times T_1$ (9.63 mg 100 $g^{-1}$) had the lowest ascorbic acid content. Among the 40 hybrids, 17 had higher values than the grand mean. Hybrid $L_7 \times T_2$ had the highest value and $L_2 \times T_4$ the lowest value for this trait. Nineteen hybrids exceeded the mean value. The highest fruit yield was for $L_7 \times T_2$ followed by $L_1 \times T_1$, $L_4 \times T_1$ and the lowest was in $L_3 \times T_4$ for this trait. Fourteen hybrids had values than the grand mean.

For fruit pedicel length standard heterosis over the standard variety. In 26 hybrids there was positive significant heterosis over the standard variety. Ten hybrids did not have heterosis and 3 ($L_7 \times T_3$, $L_7 \times T_4$ and $L_8 \times T_3$) had significant negative heterosis. For fruit circumference standard heterosis varied. Thirty-one hybrids had significant, positive, heterosis over the standard variety. Four hybrids had no heterosis. For calyx length standard heterosis was negative, and significant, in all hybrids except $L_5 \times T_3$ which had no heterosis over the standard variety. For numbers of fruit per plant significant heterosis over the standard variety occurred in 33 hybrids; 9 had positive and 23 had negative heterosis. For shoot borer infestation, fifteen hybrids had significant negative heterosis. For fruit borer infestation, the highest significant negative standard heterosis was for the hybrid $L_8 \times T_1$. Seventeen hybrids had significant negative heterosis and 7 had significant positive heterosis over the standard variety. For little leaf incidence, negative heterosis is a beneficial for this trait. Of 40 hybrids, 2, 7 and 16 hybrids had significant negative heterosis over mid-, and better parent and the standard check, respectively. For ascorbic acid content positive standard heterosis was highest for $L_1 \times T_3$. Seven hybrids had significant and positive $d_{iii}$. For total phenol content of 40 hybrids, 19 had positive and significant $d_{iii}$ values with the highest in $L_7 \times T_2$. For fruit yield per plant, there was an appreciable amount of heterosis in $F_1$s over the standard check. Expression of superiority over the standard check occurred in 7 crosses. More hybrids had economic heterosis in crosses involving the tester $T_2$.

## DISCUSSION

Knowledge of the relative importance of additive and non-additive gene action is essential to breeders for development of efficient hybridization. Panse (1942) stated that if additive genetic variance is greater, the chance of fixing superior genotypes in early segregating generation would be greater; if dominant and epistatic interactions are predominant, selection should be postponed to later generations and appropriate breeding should be adopted to obtain useful genotypes. The analysis of combining ability estimates (Table 2) indicated that non-additive gene action was operating for all characters studied because variance due to GCA and SCA were significant. Variance due to SCA was higher in magnitude than GCA for all traits. This supports the predominance of non-additive gene effects on governing expression of most characters.

The parents had significant differences for all characters. Variance due to lines were significant for all traits indicating existence of enormous amount of genetic variability for growth and yield attributes among the lines (females). Similarly, testers (males) had significant differences for all traits. The interaction between lines x testers was also significant for yield, quality and other traits studied (Table 3).

The hybrids chosen for heterosis breeding based on significant mean value, SCA effects and standard heterosis are shown in the Tables 4a, b and c. Mean performance and heterosis are important parameters to assess potential of $F_1$ hybrids. Among these, mean

**Table 3.** Analysis of variance for parents and hybrids with respect to 9 characters, rainy season 2010-2011.

| Source | df | FPL | CL | NF/P | SBI | FBI | LLI | ACC | TPC | FY/P |
|---|---|---|---|---|---|---|---|---|---|---|
| Hybrids | 39 | 1.4005* | 1.0970* | 161.9557* | 54.4746* | 63.3535* | 69.4393* | 12.9752* | 533.5026* | 0.6288* |
| Lines | 9 | 2.1125* | 2.5695* | 197.8137* | 104.8571* | 80.0248* | 128.8896* | 24.3787* | 998.8697* | 0.9020* |
| Testers | 3 | 0.4415* | 0.5407* | 234.2205* | 41.9781* | 34.1338* | 70.4679* | 33.1344* | 55.7529* | 0.4214* |
| Line × Testers | 27 | 1.2698* | 0.6680* | 141.9736* | 39.0689* | 61.0430* | 49.5083* | 6.9342* | 431.4635* | 0.5608* |
| Errors | 78 | 0.0216 | 0.0424 | 1.8176 | 0.9524 | 2.1378 | 0.8543 | 0.1632 | 4.1422 | 0.0100 |

Significant at 5% level, CL – Calyx length (cm); FPL – Fruit pedicel length (cm); NF/P – Number of fruits per plant; SBI – Shoot borer infestation (%); FBI – Fruit borer infestation (%); LLI – Little leaf incidence (%); ACC – Ascorbic acid content (mg/100g); TPC – Total phenols content (mg/100g); FY/P – Fruit yield per plant (kg).

**Table 4a.** Mean performance and standard heterosis for various quantitative and qualitative characters in eggplant, rainy season 2010-2011.

| Entry | Calyx length (cm) | | Fruit pedicel length (cm) | | Number of fruits per plant | | Shoot borer infestation (%) | |
|---|---|---|---|---|---|---|---|---|
| | Mean value | SH | Mean value | SH | Mean value | SH | Mean value | SH |
| Alavayal Local | 2.45* | | 5.57* | | 25.89 | | 22.86* | |
| Sedapatty Local (Green) | 3.45* | | 6.03* | | 33.06* | | 20.65* | |
| Kariapatty Local | 3.72* | | 4.58 | | 30.85* | | 27.31 | |
| Alagarkovil Local | 3.84* | | 5.42* | | 27.43 | | 25.47 | |
| Palamedu Local | 4.65 | | 5.11 | | 23.42 | | 19.34* | |
| Melur Local | 4.76 | | 4.76 | | 31.49* | | 22.14* | |
| Keerikai Local | 5.54 | | 4.19 | | 27.37 | | 21.40* | |
| Nilakottai Local | 2.42* | | 7.04* | | 25.44 | | 17.89 | |
| Singampunari Local | 3.34* | | 5.37 | | 33.34* | | 22.66* | |
| Sedapatty Local (Blue) | 3.54* | | 4.98 | | 25.17 | | 21.58* | |
| Annamalai | 3.96 | | 4.12 | | 37.92* | | 26.67 | |
| KKM 1 | 3.28* | | 3.44 | | 31.07* | | 27.87 | |
| Punjab Sadabahar | 4.41 | | 4.75 | | 30.94* | | 23.09* | |
| EP 65 | 3.28* | | 5.44* | | 26.99 | | 21.68* | |
| $L_1 \times T_1$ | 1.74* | -56.14** | 4.40 | 6.88* | 35.60* | -6.12 | 22.86* | 6.85* |
| $L_1 \times T_2$ | 1.88* | -52.53** | 4.03 | -2.11 | 24.65 | -35.00** | 20.65* | -19.11** |
| $L_1 \times T_3$ | 2.84 | -28.37** | 5.35* | 30.04** | 20.52 | -45.88** | 27.31 | -30.21** |
| $L_1 \times T_4$ | 2.20* | -44.44** | 4.38 | 6.32* | 28.04 | -26.05** | 25.47 | -11.60** |
| $L_2 \times T_1$ | 2.02* | -48.99** | 3.88 | -5.83 | 43.02* | 13.45** | 19.34* | -0.11 |
| $L_2 \times T_2$ | 2.10* | -46.97** | 3.98 | -3.40 | 26.22 | -30.85** | 22.14* | 26.21** |
| $L_2 \times T_3$ | 2.25* | -43.18** | 4.10 | -0.32 | 36.92* | -2.63 | 21.40* | -9.07** |
| $L_2 \times T_4$ | 1.74* | -56.06** | 4.96* | 20.49** | 29.97 | -20.97** | 17.89 | -15.35** |
| $L_3 \times T_1$ | 2.20* | -44.44** | 4.25 | 3.32 | 33.29 | -12.21** | 22.66* | -15.84** |

**Table 4a.** Contd.

| | | | | | | | | |
|---|---|---|---|---|---|---|---|---|
| $L_3 \times T_2$ | 2.52* | -36.36** | 4.46 | 8.34** | 22.30 | -41.19** | 21.58* | -27.70** |
| $L_3 \times T_3$ | 3.14 | -20.79** | 4.90* | 19.03** | 28.10 | -25.90** | 26.67 | -33.43** |
| $L_3 \times T_4$ | 3.88 | -2.02 | 6.50* | 57.89** | 19.55 | -48.44** | 27.87 | -15.82** |
| $L_4 \times T_1$ | 3.15 | -20.37** | 4.75 | 15.30** | 41.00* | 8.11* | 23.09* | -22.41** |
| $L_4 \times T_2$ | 2.90 | -26.77** | 4.25 | 3.16 | 43.93* | 15.85** | 21.68* | 2.66 |
| $L_4 \times T_3$ | 2.48* | -37.37** | 4.95* | 20.24** | 28.03 | -26.09** | 28.50 | 20.96** |
| $L_4 \times T_4$ | 3.36 | -15.15** | 5.80* | 40.89** | 32.92 | -13.19** | 21.57* | 6.86* |
| $L_5 \times T_1$ | 2.44* | -38.30** | 5.00* | 21.46** | 30.19 | -20.39** | 18.61* | -4.46 |
| $L_5 \times T_2$ | 2.10* | -46.97** | 4.58 | 11.26** | 44.47* | 17.27** | 23.58* | 4.91 |
| $L_5 \times T_3$ | 3.76 | -5.13 | 5.05* | 22.67** | 42.24* | 11.39** | 26.64* | -4.35 |
| $L_5 \times T_4$ | 2.58* | -34.93** | 5.53* | 34.33** | 43.36* | 14.35** | 33.66 | -10.24** |
| $L_6 \times T_1$ | 2.42* | -38.97** | 6.00* | 45.75** | 28.60 | -24.58** | 24.25* | 7.62** |
| $L_6 \times T_2$ | 3.16 | -20.20** | 5.58* | 35.63** | 40.00* | 5.49 | 22.58* | -16.69** |
| $L_6 \times T_3$ | 3.09 | -22.05** | 4.43 | 7.69* | 25.46 | -32.85** | 22.45* | 6.15* |
| $L_6 \times T_4$ | 2.26* | -42.93** | 4.22 | 2.43 | 32.84 | -13.40** | 19.28* | -19.22** |
| $L_7 \times T_1$ | 2.10* | -46.97** | 4.00 | -2.83 | 19.98 | -47.32** | 17.75* | 8.66** |
| $L_7 \times T_2$ | 1.54* | -61.20** | 4.44 | 7.85* | 41.04* | 8.22* | 22.45* | 8.75** |
| $L_7 \times T_3$ | 1.56* | -60.61** | 3.62 | -12.06** | 26.36 | -30.49** | 20.69* | -4.92 |
| $L_7 \times T_4$ | 2.32* | -41.41** | 3.36 | -18.38** | 30.09 | -20.65** | 27.38* | -7.85** |
| $L_8 \times T_1$ | 1.76* | -55.64** | 4.58 | 11.26** | 39.52* | 4.23 | 32.26 | 30.07** |
| $L_8 \times T_2$ | 1.40* | -64.65** | 4.68 | 13.68** | 38.87* | 2.51 | 28.50 | 17.17** |
| $L_8 \times T_3$ | 1.60* | -59.68** | 3.42 | -16.92** | 25.61 | -32.45** | 25.48* | 4.69 |
| $L_8 \times T_4$ | 1.90* | -52.02** | 4.30 | 4.45 | 41.62* | 9.76** | 27.98 | -10.82** |
| $L_9 \times T_1$ | 2.44* | -38.47** | 4.60 | 11.74** | 42.96 * | 13.29** | 25.51* | -2.51 |
| $L_9 \times T_2$ | 1.92* | -51.52** | 4.74 | 15.14** | 36.84* | -2.85 | 23.94* | 29.25** |
| $L_9 \times T_3$ | 1.64* | -58.59** | 5.74* | 39.51** | 21.61 | -43.01** | 28.70 | 18.75** |
| $L_9 \times T_4$ | 2.28* | -42.42** | 4.44 | 7.85* | 27.98 | -26.21** | 22.22* | 4.69 |
| $L_{10} \times T_1$ | 2.46* | -37.96** | 4.86 | 17.98** | 19.63 | -48.24** | 28.31 | 1.55 |
| $L_{10} \times T_2$ | 2.42* | -38.80** | 4.30 | 4.53 | 35.31* | -6.88* | 21.54* | 12.56** |
| $L_{10} \times T_3$ | 2.52* | -36.28** | 5.12* | 24.45** | 23.14 | -38.98** | 28.98 | 8.31** |
| $L_{10} \times T_4$ | 1.70* | -57.07** | 4.50 | 9.39** | 25.56 | -32.60** | 29.00 | 14.16** |
| SEd | 2.34 | - | 0.12 | - | 1.12 | | 0.79 | - |
| CD at 5% | 4.69 | 0.15 | 0.23 | 0.13 | 2.22 | 1.30 | 1.56 | 0.72 |

SH-standard heterosis; *Significance at 5% level; ** Significance at 1% level.

performance is the most important criterion for evaluating hybrids and parents. Selection based on phenotypic expression is easy when the required character is controlled by a few genes and inherited simply. Continuously varying traits of yield and its components, which are under

**Table 4b.** Mean performance and standard heterosis for various quantitative and qualitative characters in eggplant, rainy season 2010-2011.

| Entry | Fruit borer infestation (%) | | Little leaf incidence (%) | | Ascorbic acid (mg/100 g) | | Total phenol content (mg/100 g) | |
|---|---|---|---|---|---|---|---|---|
| | Mean value | SH | Mean value | SH | Mean value | SH | Mean value | SH |
| Alavayal Local | 33.75* | | 18.92 | | 13.37* | | 67.28* | |
| Sedapatty Local (Green) | 37.58* | | 20.74 | | 12.03 | | 51.77 | |
| Kariapatty Local | 38.08* | | 17.83* | | 12.86* | | 50.54 | |
| Alagarkovil Local | 40.41 | | 15.58* | | 13.34* | | 46.04 | |
| Palamedu Local | 35.38* | | 10.42* | | 11.23 | | 58.36 | |
| Melur Local | 39.54 | | 24.75 | | 13.58* | | 82.62* | |
| Keerikai Local | 41.29 | | 25.63 | | 12.13 | | 61.47* | |
| Nilakottai Local | 38.42* | | 16.68* | | 13.55* | | 40.37 | |
| Singampunari Local | 39.58 | | 12.45* | | 10.58 | | 72.23* | |
| Sedapatty Local (Blue) | 40.02 | | 13.86* | | 9.88 | | 48.28 | |
| Annamalai | 38.18* | | 24.08 | | 14.65* | | 77.27* | |
| KKM 1 | 44.57 | | 19.55 | | 12.49 | | 45.33 | |
| Punjab Sadabahar | 42.61 | | 19.76 | | 13.83* | | 29.46 | |
| EP 65 | 28.89* | | 10.05* | | 13.87* | | 56.05* | |
| $L_1 \times T_1$ | 35.17* | -7.88** | 20.74 | 20.16** | 16.47* | 12.42** | 67.28* | -4.68* |
| $L_1 \times T_2$ | 37.96* | -0.58 | 17.83* | -5.97 | 15.42* | 5.28* | 51.77 | -9.08** |
| $L_1 \times T_3$ | 40.28 | 5.50 | 15.58* | -11.78** | 16.74* | 14.29** | 50.54 | -4.09* |
| $L_1 \times T_4$ | 35.36* | -7.39* | 10.42* | 34.65** | 14.15* | -3.39 | 46.04 | -6.51** |
| $L_2 \times T_1$ | 38.48* | 0.79 | 24.75 | 17.28** | 13.61* | -7.10** | 58.36 | -20.53** |
| $L_2 \times T_2$ | 34.29* | -10.20** | 25.63 | 5.69 | 13.95* | -4.78* | 82.62* | 19.16** |
| $L_2 \times T_3$ | 35.74* | -6.38* | 16.68* | -18.05** | 12.48 | -14.81** | 61.47* | 10.05** |
| $L_2 \times T_4$ | 32.05* | -16.06** | 12.45* | 21.54** | 12.53 | -14.47** | 40.37 | -31.50** |
| $L_3 \times T_1$ | 36.98* | -3.14 | 13.86* | -6.80* | 14.47* | -1.23 | 72.23* | 14.29** |
| $L_3 \times T_2$ | 42.38 | 11.00** | 18.92 | -9.25** | 11.22 | -23.41** | 48.28 | -8.89** |
| $L_3 \times T_3$ | 36.47* | -4.48 | 24.08 | -32.56** | 14.33* | -2.16 | 77.27* | 10.88** |
| $L_3 \times T_4$ | 39.56* | 3.62 | 19.55 | 17.33** | 13.18 | -10.01** | 45.33 | -4.97* |
| $L_4 \times T_1$ | 38.82* | 1.67 | 19.76 | -22.23** | 12.74 | -13.06** | 29.46 | 16.58** |
| $L_4 \times T_2$ | 36.76* | -3.71 | 10.05* | -22.58** | 13.95* | -4.78* | 56.05* | 12.25** |
| $L_4 \times T_3$ | 34.40* | -9.91** | 28.93 | -6.45* | 12.22 | -16.59** | 73.65 | 4.51* |
| $L_4 \times T_4$ | 32.96* | -13.67** | 22.64* | -12.82** | 10.62 | -27.51** | 70.25 | 14.33** |
| $L_5 \times T_1$ | 30.33* | -20.56** | 21.24* | 5.65 | 15.57* | 6.28** | 74.11 | -11.42** |
| $L_5 \times T_2$ | 36.65* | -4.01 | 32.42 | 6.52* | 11.56 | -21.07** | 72.24 | 7.76** |
| $L_5 \times T_3$ | 34.78* | -8.91** | 28.24 | -34.61** | 11.60 | -20.80** | 61.40 | -8.20** |
| $L_5 \times T_4$ | 45.57 | 19.36** | 25.45* | -23.33** | 10.61 | -27.58** | 92.07* | 16.48** |
| $L_6 \times T_1$ | 39.98 | 4.71 | 19.73* | -31.80** | 16.46* | 12.33** | 85.03* | -14.15** |

**Table 4b.** Contd.

| | | | | | | | | |
|---|---|---|---|---|---|---|---|---|
| $L_6 \times T_2$ | 31.28* | -18.07** | 29.26 | -7.50* | 11.40 | -22.16** | 52.93 | 9.45** |
| $L_6 \times T_3$ | 45.58 | 19.37** | 22.44* | 3.21 | 11.78 | -19.59** | 88.31* | 10.25** |
| $L_6 \times T_4$ | 34.52* | -9.59** | 21.85* | 4.25 | 10.87 | -25.80** | 70.40 | 26.10** |
| $L_7 \times T_1$ | 38.90* | 1.89 | 16.24* | 18.59** | 14.12* | -3.62 | 85.67* | 28.68** |
| $L_7 \times T_2$ | 34.62* | -9.33** | 28.25 | -7.57* | 11.27 | -23.07** | 73.42 | 30.96** |
| $L_7 \times T_3$ | 36.05* | -5.59 | 18.72* | 29.35** | 15.28* | 4.30* | 90.08* | 8.91** |
| $L_7 \times T_4$ | 35.69* | -6.52* | 18.64* | 1.50 | 11.27 | -23.07** | 86.73* | 27.35** |
| $L_8 \times T_1$ | 29.58* | -22.52** | 22.52* | 10.22** | 10.31 | -29.60** | 80.75* | -10.31** |
| $L_8 \times T_2$ | 45.27 | 18.56** | 20.99* | 18.58** | 14.18* | -3.21 | 88.34* | -24.38** |
| $L_8 \times T_3$ | 48.86 | 27.97** | 25.44* | -7.59* | 10.62 | -27.49** | 68.44 | -8.27** |
| $L_8 \times T_4$ | 40.09 | 5.00 | 25.65* | 33.50** | 9.77 | -33.31** | 83.26* | -13.00** |
| $L_9 \times T_1$ | 44.28 | 15.97** | 15.74* | 26.43** | 16.00* | 9.22** | 70.93 | -28.81** |
| $L_9 \times T_2$ | 40.35 | 5.68 | 18.46* | 10.23** | 14.23* | -2.87 | 90.00* | -19.34** |
| $L_9 \times T_3$ | 45.63 | 19.51* | 16.42* | 18.58** | 14.47* | -1.21 | 66.33 | -16.44** |
| $L_9 \times T_4$ | 40.21 | 5.32 | 22.27* | -7.59* | 10.52 | -28.17** | 84.57* | 12.84** |
| $L_{10} \times T_1$ | 36.58* | -4.19 | 24.85* | 21.03** | 9.63 | -34.24** | 85.19* | 27.35** |
| $L_{10} \times T_2$ | 31.28* | -18.07** | 25.10* | 43.04** | 10.02 | -31.60** | 97.43* | -10.31** |
| $L_{10} \times T_3$ | 34.02* | -10.90** | 28.55 | 10.25** | 11.50 | -21.50** | 99.43* | -24.37** |
| $L_{10} \times T_4$ | 35.82* | -6.17* | 22.25* | 32.45** | 10.61 | -27.55** | 101.19* | -16.08** |
| SEd | 1.19 | - | 0.75 | - | 0.32 | - | 3.29 | - |
| CD at 5% | 2.36 | 1.14 | 1.49 | 0.71 | 0.65 | 0.31 | 7.01 | 1.56 |

SH-standard heterosis; *Significance at 5% level; ** Significance at 1% level.

**Table 4c.** Mean performance and standard heterosis for various quantitative and qualitative characters in eggplant, rainy season 2010-2011.

| Entry | Fruit yield per plant (kg) | |
|---|---|---|
| | Mean value | SH |
| Alavayal Local | 1.72* | |
| Sedapatty Local (Green) | 1.86* | |
| Kariapatty Local | 1.27 | |
| Alagarkovil Local | 1.26 | |
| Palamedu Local | 1.91* | |
| Melur Local | 1.75* | |
| Keerikai Local | 1.79* | |
| Nilakottai Local | 1.27 | |

**Table 4c.** Contd.

| | | |
|---|---|---|
| Singampunari Local | 1.16 | |
| Sedapatty Local (Blue) | 1.37 | |
| Annamalai | 2.12* | |
| KKM 1 | 1.46 | |
| Punjab Sadabahar | 1.56 | |
| EP 65 | 1.36 | |
| $L_1 \times T_1$ | 2.47* | 16.33** |
| $L_1 \times T_2$ | 1.35 | -36.58** |
| $L_1 \times T_3$ | 1.35 | -36.42** |
| $L_1 \times T_4$ | 1.45 | -31.55** |
| $L_2 \times T_1$ | 2.07* | -2.67 |
| $L_2 \times T_2$ | 1.42 | -32.97** |
| $L_2 \times T_3$ | 2.14* | 0.63 |
| $L_2 \times T_4$ | 1.56 | -26.37** |
| $L_3 \times T_1$ | 1.45 | -31.71** |
| $L_3 \times T_2$ | 0.99 | -53.22** |
| $L_3 \times T_3$ | 1.74 | -18.05** |
| $L_3 \times T_4$ | 0.93 | -56.36** |
| $L_4 \times T_1$ | 2.40* | 13.03** |
| $L_4 \times T_2$ | 2.32* | 9.11* |
| $L_4 \times T_3$ | 1.49 | -29.67** |
| $L_4 \times T_4$ | 1.92* | -9.73* |
| $L_5 \times T_1$ | 1.71 | -19.62** |
| $L_5 \times T_2$ | 2.34* | 10.20** |
| $L_5 \times T_3$ | 2.36* | 11.30** |
| $L_5 \times T_4$ | 2.21* | 4.08 |
| $L_6 \times T_1$ | 1.48 | -30.30** |
| $L_6 \times T_2$ | 2.24* | 5.49 |
| $L_6 \times T_3$ | 1.35 | -36.26** |
| $L_6 \times T_4$ | 1.68 | -21.04** |
| $L_7 \times T_1$ | 1.13 | -46.78** |
| $L_7 \times T_2$ | 2.85* | 34.07** |
| $L_7 \times T_3$ | 1.43 | -32.50** |
| $L_7 \times T_4$ | 1.74 | -18.21** |
| $L_8 \times T_1$ | 2.26* | 6.28 |
| $L_8 \times T_2$ | 1.66 | -21.82** |
| $L_8 \times T_3$ | 1.04 | -50.86** |

**Table 4c.** Contd.

| | | |
|---|---|---|
| $L_8 \times T_4$ | 2.06* | -2.83 |
| $L_9 \times T_1$ | 2.29* | 7.69* |
| $L_9 \times T_2$ | 1.73 | -18.68** |
| $L_9 \times T_3$ | 1.22 | -42.54** |
| $L_9 \times T_4$ | 1.42 | -33.28** |
| $L_{10} \times T_1$ | 1.12 | -47.25** |
| $L_{10} \times T_2$ | 1.85 | -13.03** |
| $L_{10} \times T_3$ | 1.43 | -32.81** |
| $L_{10} \times T_4$ | 1.45 | -31.55** |
| SEd | 0.07 | - |
| CD at 5% | 0.15 | 0.08 |

SH-standard heterosis; *Significance at 5% level; ** Significance at 1% level.

influence of polygenes, require selection based on mean performance. It is essential to eliminate undesirable types, which can be achieved by studying mean performance of genotypes as an initial step in the evaluation procedure. Heterosis is a direct property of heterozygosity and due to superior gene content possible in a hybrid contributed by both parents (Mather, 1955). When performance of a hybrid is assessed based on significance of mean performance and heterosis, its superiority over other hybrids could be easily determined.

Evaluation of parents based on mean might result in identification of different sets of promising parents. Chandra et al. (1970) reported that parents with high mean may be able to transmit superior traits to hybrids and the need for combining ability of parents. On the basis of mean values, the line 'Alavayal Local' ($L_1$) had high mean values for fruit pedicel length, calyx length, shoot borer infestation, little leaf incidence, ascorbic acid content, total phenols and fruit yield per plant, and 'Sedapatty local' (Green) ($L_2$) had high mean values for fruit pedicel length, calyx length, number of fruit per plant, shoot borer infestation, fruit borer infestation and fruit yield per plant. Among testers, 'Annamalai' ($T_1$) was the best parent since it had high mean for number of fruit per plant, fruit borer infestation, ascorbic acid content and fruit yield per plant. An analysis of average performance of parents for fruit yield per plant, and other desirable traits, indicated that production of elite hybrids with the correct parents will led to fixes for heterotic effects through isolation of high yielding homozygous lines with better quality and lesser incidence of pest and disease in advance generation.

Chaudhary and Malhotra (2000) reported production of hybrids in crosses involving parents with high average values for yield and its component traits in eggplant. Das and Barua (2001) indicated crosses with two good general combiners to have particular merit in eggplant breeding and suggested bi-parental mating among $F_2$ progenies to evolve better genotypes through combination of desirable attributes. An overview of average performance of parents indicated that lines 'Alavayal local' ($L_1$), 'Sedapatty local' (Green) ($L_2$) and the tester 'Annamalai' ($T_1$) were good parents for further breeding to exploit high yield and low incidence of pest and diseases. The hybrid $L_4xT_1$ had favorable mean values for the characters numbers of fruit per plant, shoot borer infestation, fruit borer infestation, little leaf incidence, total phenol content and fruit yield per plant. This was followed by $L_7 \times T_2$ which had favorable mean performance for yield and quality traits calyx length, number of fruit per plant, fruit borer infestation, total phenols content and fruit yield per plant. The above hybrids could be outstanding for improving growth, yield and quality traits and lower incidence of pest and diseases.

Another important criterion to assess hybrids for heterosis breeding was through standard heterosis. Though, the bases of heterosis are important, Kadambavanasundaram (1980) indicated that heterotic expression over the standard variety should alone be given importance for commercial exploitation of hybrid vigor and the crosses, which showed significantly high value of standard heterosis over 'Annamalai' ($T_1$) for yield and yield components, quality, pest and disease. None of the hybrids expressed favorable standard heterosis for all characters. The hybrids $L_7 \times T_2$ had significant standard heterosis over its parents for calyx length, total phenol content and fruit yield per plant. The hybrid and $L_6 \times T_1$ had better heterosis values for fruit pedicel length, ascorbic acid content and little leaf incidence. For shoot borer infestation and little leaf incidence hybrid $L_3 \times T_3$ can be chosen for reduced pest and disease incidence. The hybrid $L_8 \times T_1$ had superior standard heterosis for fruit borer infestation and these hybrids can be used for heterosis breeding. Hybrids were selected for heterosis breeding based on average performance and standard heterosis for fruit yield per plant. The hybrid $L_4 \times T_1$ was suitable for heterosis breeding since it expressed high values for numbers of fruit per plant, shoot borer infestation, little leaf incidence, total phenols content and fruit yield per plant. Hybrids $L_5 \times T_3$, $L_5 \times T_4$ and $L_7 \times T_2$ could be the next best for high values under those 3 criteria for 4 and 3 traits, respectively. For hybrids based on average performance, SCA effects and standard heterosis, hybrids 'Alagarkovil Local' × 'Annamalai' ($L_4 \times T_1$), 'Palamedu Local' × 'Punjab Sadabahar' ($L_5 \times T_3$), 'Palamedu Local' × 'EP 65' ($L_5 \times T_4$) and 'Keerikai Local' × 'KKM 1' ($L_7 \times T_2$) were suitable for heterosis breeding.

The F1 hybrids 'Keerikai Local' × 'KKM 1' ($L_7 \times T_2$) and 'Alavayal Local' × 'Annamalai' ($L_1 \times T_1$) had the highest yield, respectively; higher than performance of other hybrids, male parent, female parent and commercial varieties. The F1 hybrid 'Keerikai Local' × 'KKM 1' had positive heterosis for calyx length, total phenol content and fruit yield per plant; F1 hybrid 'Alavayal Local' × 'Annamalai' had positive heterosis for fruit pedicel length, ascorbic acid content and fruit yield per plant. Although there were no parental varieties which showed good appearance in all traits, some parents had a high mean value in some characteristics. These are useful in breeding to improve fruit yield and qualities of commercial varieties.

## ACKNOWLEDGEMENTS

The author wishes to thank research institutes viz., Vegetable Research Station, Palur, Agricultural College and Research Institute, Killikulam, Madurai and TNAU, Coimbatore for providing the germplasms to carry out the present investigation. Dr. S. Balakrishnan, Professor (Horticulture), Agricultural College and Research Institute, Madurai is duly acknowledged for giving valuable suggestions for this dissertation research.

## REFERENCES

Abhinav S, Nandan M (2010). Heterosis in relation to combining ability for yield and quality attributes in Brinjal (*Solanum melongena* L.). Elect. J. Plant Breed. 1(4):783-788.

AOAC (2001). Official methods of analysis. Association of Official Analytical Chemists, Washington D.C., U.S.A

Anonymous (2010).Area, production and productivity of brinjal in India during 2009-2010, www.indiastat.com.

Bray HG, Thrope WV (1954). Analysis of phenolic compounds of interest in metabolism. Meth. Biochem. Anal. 1:27-52.

Dhankhar BS, Singh K (1984). Path analysis for fruit yield and its components in brinjal (*Solanum melongena* L.). Haryana J. Hort. Sci., 12: 38-41.

Chandra V, Singh B, Singh A, Kapoor LD (1970). Variation in the solasodine content of fruits of *Solanum khasianum* at different stages of development in Lucknow. Indian For. 96:352-360.

Chaudhary B (1976).Vegetables (4th Edn.), National Book Trust, New Delhi, pp. 50-58.

Chaudhary DR, Malhotra SK (2000). Combining ability of physiological and growth parameters in eggplant (*Solanum melongena* L.). Indian J. Agric. Res. 34(1):55-58.

Chowdhury MJ, Ahmad S, Nazim UM, Quamruzzaman AKM, Patwary MMA (2010). Expression of heterosis for productive traits in F1 brinjal (*Solanum melongena* L.) hybrids. Agriculturists 8(2):8-13.

Das G, Barua NS (2001). Heterosis and combining ability for yield and its components in eggplant. Ann. Agric. New Series. 22(3):399-403.

Dhillion BS (1975). The application of partial diallel crosses in plant breeding – A review. Crop Improv. 2:17.

Fonseca S, Patterson FL (1968). Hybrid vigour in a seven parent diallel crosses in common winter wheat (*Triticum aestivum* L.). Crop Sci. 8:85-88.

Kadambavanasundaram M (1980). Heterotic system in cultivated species of *Gossypium*. An appraisal (Abst). Genetic and crop improvement of heterotic systems. In: Pre-congress scientific meeting of XV international congress of genetics, TNAU, Coimbatore. P. 20.

Pachiyappan R, Saravanan K, Kumar R (2012). Heterosis is yield and yield components in eggplant (*Solanum melongena* L.). Int. J. Curr. Agric. Sci. 2(6):17-19.

Panse and Sukhatme (1967).Statistical methods for Agricultural workers.

Panse VG (1942).The inheritance of quatitative characters and plant breeding:http://www.ias.ac.in/jarch/jgenet/40/283.pdf pp. 284-302.

Prabakaran S (2010). Evaluation of local types of brinjal (*Solanum melongena* L.). M.Sc., (Hort.) Thesis, Agricultural College and Research Institute, TNAU, Madurai.

Turner JR (1953). A study on heterosis in upland cotton II. Combining ability and inbreeding effects. Agron. J. 45:487-490.

# Effect of kinetin (Kn) and naphthalene acetic acid (NAA) on the micropropagation of *Matthiola incana* using shoot tips, and callus induction and root formation on the leaf explants

**Behzad Kaviani[1], Afshin Ahmadi Hesar[1], Alireza Tarang[2], Sahar Bohlooli Zanjani[2], Davood Hashemabadi[1] and Mohammad Hossein Ansari[3]**

[1]Department of Horticultural Science, Rasht Branch, Islamic Azad University, Rasht, Iran.
[2]North Biotechnology Institute, Rasht, Guilan, Iran.
[3]Department of Agronomy, Rasht Branch, Islamic Azad University, Rasht, Iran

**Shoot proliferation, callus induction and root formation on callus are possible when kinetin (Kn) and naphthalene acetic acid (NAA) are added to Murashige and Skoog's (MS) medium. Seeds of *Matthiola incana* (an ornamental plant) were germinated on solid MS medium without plant growth regulators. Shoot tips and leaf micro-cuttings from four-week-old *in vitro* germinated seedlings were subcultured on solid MS medium containing Kn (0, 0.5, 1 and 2 mg/L) and NAA (0, 0.5, 1 and 2 mg/L) for shoot tips explants and Kn (0, 0.5 and 1 mg/L) and NAA (0, 0.5 and 1 mg/L) for leaf explants. Shoot tips media supplemented with 2 mg/L Kn without NAA and 2 mg/L NAA without Kn resulted in the best shoot length (1.20 cm) and root number (1.90), respectively. The callus was induced from leaf media after four weeks of culture, except for medium containing 0.5 and 1 mg/L NAA. The development of roots was observed from callus in MS medium containing suitable concentration of Kn and NAA. MS mediums containing 0.5 mg/L Kn (100%) and 0.5 mg/L Kn + 0.5 mg/L NAA (100%) were most effective in induction of callus on leaf micro-cuttings. The largest number (1.83) and the highest length (15.7 mm) of roots were obtained in MS medium supplemented with 1 mg/L Kn + 0.5 mg/L NAA. NAA did not stimulate callus induction and root formation when it was applied alone. Also, this hormone prevented root formation originated from callus with concentration of 1 mg/L along with 0.5 and 1 mg/L Kn in medium.**

**Key words:** Brassicaceae, micropropagation, organogenesis, ornamental plants, plant growth regulators.

## INTRODUCTION

The ornamental species *Matthiola incana*, belonging to Brassicaceae, is a pot plant. The Brassicaceae is a fairly large family with many economically important taxa, but from viewpoint of tissue culture, it has been little studied. Natural propagation of *M. incana* takes place by seed. The economic value of ornamental plants has increased significantly worldwide and is increasing annually by 8 to 10% (Jain and Ochatt, 2010). The techniques for *in vitro* propagation of ornamental plants and tissue culture laboratory equipment are being continuously improved to meet the demand of the floriculture breeding and industry (Rout et al., 2006). Tissue culture has become a routine technique in agricultural and horticultural development which has revolutionized the ornamental industry and most popular application of this technique is micropropagation (Maira et al., 2010; Bhattacharya and Bhattacharyya, 2010). Micropropagation through tissue culture permits the regeneration of large numbers of disease free plants from small pieces (explants) of stock plants in a relatively short period and, crucially, without seasonal restrictions (Preil et al., 1988). In general, the number of publications on different aspects of the culture of *M. incana* is limited, with emphasis on micropropagation through somatic explants (Gautam et al., 1983). In the field of ornamental plants, tissue culture has allowed mass propagation of superior genotypes and plant improvement, thus enabling the commercialization of healthy and uniform planting material (Winkelmann et al., 2006; Nhut et al., 2006). The success of the micropropagation method depends on several factors like genotype, media, plant growth regulators and type of explants, which should be observed during the process (Pati et al., 2005; Nhut et al., 2010). In general, three modes of *in vitro* plant regeneration have been in practice: Organogenesis, embryogenesis and axillary proliferation. In tissue culture, cytokinins and auxins play a crucial role as promoters of cell division and act in the induction and development of meristematic centers leading to the formation of organs (Peeters et al., 1991). The most frequently used growth regulators for micropropagation of ornamental plants by organogenesis, embryogenesis and axillary proliferation are naphthalenacetic acid (NAA), and benzyl adenine (BA) (Jain and Ochatt, 2010). Kn has been applied for micropropagation of many plants (Jain and Ochatt, 2010). In this paper, potential of shoot tips and leaf explants of *in vitro* grown *M. incana* seedling to proliferation, and induction of callus and root by Kn and NAA has been discussed.

## MATERIALS AND METHODS

Seeds of *M. incana* were prepared from Mohaghegh-e-Ardabili University, Iran. The seeds were washed thoroughly under running tap water for 20 min and disinfected with a 20% NaOCl aqueous solution and Tween-20 for 10 min then rinsed three times in sterile distilled water (10 min each). At the end, seeds were sterilized for 2 min in 70% ethanol followed by three times rinses with sterile distilled water (15 min each). Five seeds were cultivated in culture flasks on MS (Murashige and Skoog, 1962) basal medium without growth regulators. Micro-cuttings (shoot tips and leaves) were isolated from 4-week-old plants and cultivated on MS media supplemented with 0, 0.5, 1 and 2 mg/L Kn, and 0, 0.5, 1 and 2 mg/L NAA for shoot tips, also, 0, 0.5 and 1 mg/L Kn, and 0, 0.5 and 1 mg/L NAA for leaves. The media were adjusted to pH 5.7 to 5.8 and solidified with 7 g/L agar-agar. The media were pH adjusted before autoclaving at 121°C, 1 atmosphere for 20 min. The cultures were incubated in growth chamber whose environmental conditions were adjusted to 25±2°C and 75 to 80% relative humidity, under a photosynthetic photon density flux 50 μmol/m$^2$/s with a photoperiod of 14 h per day. Some characters such as callus, fresh weight, number of root, and root length were calculated after 30 days. The experimental design was R.C.B.D. Each experiment was carried out in three replicates and each replicate includes five specimens. Data were subjected to ANOVA (analysis of variance) and significant differences between treatments means were determined by LSD test.

## RESULTS AND DISCUSSION

The plant growth regulators are widely used for callus, rooting and shoot induction in tissue culture studies. Therefore, we studied the effect of Kn and NAA on shoot proliferation, callus production and rooting of *M. incana*, an ornamental plant. The medium supplemented with 2 mg/L Kn without NAA resulted in the best shoot length (1.20 cm) (Table 1). Data analysis showed that the effect of Kn, NAA and Kn × NAA were significant on the length of shoot and ($p \leq 0.01$) (Table 2). When the shoot tips were inoculated in the medium containing 2 mg/L NAA without Kn, the best result was observed for root number (1.90) (Table 1). Analysis of variance showed that the effect of Kn was no significant on the root number, while the effect of NAA and Kn × NAA on the root number was significant ($p \leq 0.05$) (Table 2). Similar to our findings, many researchers showed that Kn induced multiple shoot formation (Sajina et al., 1997b; Mathai et al., 1997; Luo et al., 2009). Some studies showed the positive effect of NAA on rooting (Gautam et al., 1983; Hammaudeh et al., 1998; Lee-Epinosa et al., 2008). The results on leaf explants revealed that the largest number and highest length of root were obtained in MS basal medium containing 0.5 mg/L Kn + 1 mg/L NAA. Our data revealed that there are differences in the effect of the different concentrations of Kn and NAA on the root number and length. The most roots length (15.67 mm) and the most number of roots (1.833) were found when we used 0.5 mg/L Kn + 1 mg/L NAA (Table 3). This result was comparatively better than the growth of control. Data analysis showed that the effect of Kn and NAA was significant on the length and number of root ($p \leq 0.01$) (Table 4). Interaction effect of Kn and NAA was significant on the length and number of root ($p \leq 0.01$ and $p \leq 0.05$, respectively) (Table 4). The highest percent of callus induction (100%) was seen in explants grown in MS medium containing 0.5 mg/L NAA and 0.5 mg/L Kn + 0.5 mg/L NAA (Table 3). Data analysis showed that the effect of Kn and NAA were significant on the callus formation ($p \leq 0.01$) (Table 2). The effect of Kn + NAA was no significant on the callus formation (Table 4). The most fresh weight between explants was obtained in explants grown in MS medium supplemented with 0.5 mg/L NAA (0.833 g) and 0.5 mg/L Kn + 1 NAA (0.817 g) (Table 3). Data analysis showed that the effect of Kn was significant on the fresh weight ($p \leq 0.01$) (Table 4). No the effect of NAA and Kn + NAA were significant on the fresh weight (Table 4).

All living parts of plants can be used as explants, but in case of ornamental plants, leaf especially obtained from *in vitro* grown plantlets has more extensively been applied.

**Table 1.** Effect of different concentrations of Kn and NAA on the shoot length and root number of *Matthiola incana*.

| Plant growth regulators (mg/L) | Traits | |
|---|---|---|
| | Shoot length (mm) | Root No. |
| 0 Kn | $8.46^{a}$ | $0.85^{a}$ |
| 0.5 Kn | $6.58^{b}$ | $0.42^{a}$ |
| 1 Kn | $7.37^{ab}$ | $0.75^{a}$ |
| 2 Kn | $8.58^{a}$ | $0.81^{a}$ |
| 0 NAA | $9.26^{a}$ | $0.50^{b}$ |
| 0.5 NAA | $7.25^{b}$ | $0.50^{b}$ |
| 1 NAA | $5.76^{c}$ | $0.76^{ab}$ |
| 2 NAA | $8.72^{a}$ | $1.05^{a}$ |
| 0 Kn + 0 NAA | $6.95^{c}$ | $0.36^{cd}$ |
| 0 Kn + 0.5 NAA | $7.65^{b}$ | $0.25^{d}$ |
| 0 Kn + 1 NAA | $9.45^{a}$ | $1.25^{ab}$ |
| 0 Kn + 2 NAA | $9.48^{a}$ | $1.90^{a}$ |
| 0.5 Kn + 0 NAA | $9.20^{a}$ | $0.56^{cd}$ |
| 0.5 Kn + 0.5 NAA | $7.50^{c}$ | $0.38^{cd}$ |
| 0.5 Kn + 1 NAA | $3.65^{h}$ | $0.45^{d}$ |
| 0.5 Kn + 2 NAA | $6.50^{d}$ | $0.75^{bc}$ |
| 1 Kn + 0 NAA | $8.92^{a}$ | $0.70^{cd}$ |
| 1 Kn + 0.5 NAA | $5.85^{e}$ | $1.60^{ab}$ |
| 1 Kn + 1 NAA | $4.75^{g}$ | $0.40^{d}$ |
| 1 Kn + 2 NAA | $10.00^{a}$ | $0.75^{bc}$ |
| 2 Kn + 0 NAA | $12.00^{a}$ | $0.45^{cd}$ |
| 2 Kn + 0.5 NAA | $8.55^{b}$ | $0.20^{d}$ |
| 2 Kn + 1 NAA | $5.28^{f}$ | $1.80^{a}$ |
| 2 Kn + 2 NAA | $8.92^{a}$ | $0.85^{bc}$ |

In each column, means with the similar letters are not significantly different at 5% level of probability using LSD test.

**Table 2.** Analysis of variance (ANOVA) for the effect of different concentrations of Kn and NAA on the shoot length and root number of M. incana.

| Source of variations | df | M.S. | |
|---|---|---|---|
| | | Shoot length | Root No. |
| Kn | 3 | 0.174** | $0.770^{ns}$ |
| NAA | 3 | 0.477** | 1.120* |
| Kn × NAA | 9 | 0.175** | 2.470** |
| Error | 64 | 0.03782 | 0.402 |
| c.v. | | 25.18 | 9.8 |

**Significant at $\alpha = 1\%$, *significant at $\alpha = 5\%$, $^{ns}$not significant.

We used from leaf explants taken from *in vitro* germinated seeds of *M. incana*. Many researchers applied leaves of ornamental plants as explants (Ibrahim and Debergh, 2000; Pati et al., 2004; Tyagi et al., 2010; Godo et al., 2010; Eeckaut et al., 2010; Radice, 2010).

Organogenesis in explants during micropropagation takes place either directly or after callus formation. Studies on many ornamental plants showed both kinds of organogenesis (Jain and Ochatt, 2010). There are many reports on organogenesis via callus formation (Pati et al., 2010; Jain and Ochatt, 2010). Studies of Maira et al. (2010) on *Anthurium andreanum* Lind cv Rubrun revealed that the four-week-old in plants obtained from micro-cuttings, showed callus proliferation at the stem base. The development of plantlets was observed from callus tissue. *In vitro* leaf explants in *Rosa damascena* and some other ornamental plants were used for direct organogenesis (Leffering and Kok, 1990; Ibrahim and Debergh, 2001; Dubois and de Vries, 1995). Nencheva (2010) showed direct organogenesis from pedicel explants of Chrysanthemum.

Cytokinins and auxins are usually known to promote the formation of callus and root in many excited and *in vitro* cultured organs (Jain and Ochatt, 2010). Proper type and concentration of these hormones are different for each species. We observed that callus was formed on the explants in many treatments. NAA did not stimulate much callus induction and root formation when it was applied alone (Table 3). Similar to our findings, many researchers showed that cytokinins and auxins induced callus induction and root formation in ornamental plants (Fuller and Fuller, 1995; Sangavai and Chellapandi, 2008; Hashemabadi and Kaviani, 2010; Dorion et al., 2010; Pati et al., 2010; Ochatt et al., 2010; Jain and Ochatt, 2010). Callus induction and root formation was performed for most Rhododendron genotypes by indole-3-acetic acid (IAA), NAA, indole-3-butyric acid (IBA) and 2,4-Dichlorophenoxy acetic acid (2,4-D) (Eeckaut et al., 2010). Rout et al. (1990) observed that the addition of benzylaminopurine (BAP) (2.0-3.0 mg/L) as the only growth regulator in the culture medium resulted in feeble callusing at the cut ends of the explants and the shoot elongation was considerably slow.

Rooting is a crucial step to the success of micropropagation. Without an effective root system plant acclimatization will be difficult and the rate of plant propagation may be severely affected (Gomes et al., 2010). The ideal concentrations of cytokinins and auxins differ from species to species and need to be established accurately to achieve the effective rates of multiplication (Gomes et al., 2010). The most types of cytokinins and auxins applied for root formation on callus or organs are BA, Kn and IAA, and NAA, IBA and 2,4-D, respectively. Some studies showed the positive effect of cytokinins on rooting (Gomes et al., 2010). A review of the literature clearly points out to a negative effect of cytokinins on shoot rooting (Van Staden, 2008), although a positive role has been occasionally referred (Nemeth, 1979; Bennett et al., 1994). Studies of Godo et al. (2010) and Wong and Bhalla (2010) on *Lysionotus pauciflorus* Maxim. and *Scaevola*, respectively, showed that the regenerated shoots rooted easily on medium without any plant growth regulators. Current study showed the positive effect of Kn and NAA on root formation. Contrary to our findings, root formation was inhibited in the medium culture of *Lilium longiflorum* Georgia containing BA (Han et al., 2004). Nayak et al. (2010) showed that the

**Table 3.** Effect of different concentrations of Kn and NAA on the root length and number, callugenesis percent and fresh weight of *M. incana*.

| Plant growth regulators (mg/L) | Traits | | | |
|---|---|---|---|---|
| | Root length (mm) | Root No. | Callugenesis (%) | Fresh weight (g) |
| 0 Kn | 7.44[a] | 1.08[a] | 75[a] | 0.757[a] |
| 0.5 Kn | 8.22[a] | 0.96[a] | 52.1[b] | 0.663[a] |
| 1 Kn | 1.30[b] | 0.12[b] | 29.7[c] | 0.49[b] |
| 0 NAA | 2.44[b] | 0.277[b] | 15[c] | 0.68[a] |
| 0.5 NAA | 5.77[a] | 0.722[a] | 84.8[a] | 0.65[a] |
| 1 NAA | 7.44[a] | 1.055[a] | 57[b] | 0.581[a] |
| 0 Kn + 0 NAA | 7.33[d] | 0.833[e] | 45[e] | 0.757[b] |
| 0.5 Kn + 0 NAA | 1[g] | 0.21[h] | 7[g] | 0.517[e] |
| 1 Kn + 0 NAA | 1[g] | 0.3[f] | 8[g] | 0.47[f] |
| 0 Kn + 0.5 NAA | 2[f] | 1.1[c] | 100[a] | 0.833[a] |
| 0.5 Kn + 0.5 NAA | 8.33[c] | 1.067[d] | 100[a] | 0.657[c] |
| 1 Kn + 0.5 NAA | 9[b] | 0.2[i] | 54.7[d] | 0.550[d] |
| 0 Kn + 1 NAA | 6.67[e] | 1.333[b] | 80[b] | 0.683[c] |
| 0.5 Kn + 1 NAA | 15.67[a] | 1.833[a] | 56.3[c] | 0.817[a] |
| 1 Kn + 1 NAA | 1[g] | 0.22[g] | 34.7[f] | 0.450[f] |

In each column, means with the similar letters are not significantly different at 5% level of probability using LSD test.

**Table 4.** Analysis of variance (ANOVA) for the effect of different concentrations of Kn and NAA on the root length and number, callugenesis percent and fresh weight of *M. incana*.

| Source of variations | df | M.S. | | | |
|---|---|---|---|---|---|
| | | Root length | Root No. | Callugenesis | Fresh weight |
| Kn | 2 | 185.44** | 3.202** | 4601.59** | 0.166** |
| NAA | 2 | 58.33** | 1.370** | 11139.37** | 0.0231[ns] |
| Kn × NAA | 4 | 64.61** | 0.680* | 510.44[ns] | 0.0348[ns] |
| Error | 15 | 7.48 | 0.192 | 258.44 | 0.0123 |
| Total | 23 | | | | |
| C.V. | | 52.37 | 64.04 | 30.74 | 17.42 |

**Significant at α = 1%, *Significant at α = 5%, [ns]Non sense.

lowest rooting of *Bambusa arundinacea* was observed in medium without Kn. Fuller and Fuller (1995) demonstrated that the least and most percentage of explants regeneration with root percent (5.0 and 65.0%) in *Brassica* spp. obtained in culture medium without IBA and Kn, and 2 mg/L IBA without Kn, respectively. The studies of Gautam et al. (1983) on *in vitro* regeneration of plantlets from somatic explants of *M. incana* showed only a few shoots developed on explants reared on MS medium supplemented with 0.1 mg/L Kn. Also, NAA (1 and 4 mg/L) induced profuse rooting in explants. Nhut et al. (2010) demonstrated adventitious shoots of *Begonia tuberous* can be rooted on MS medium supplemented with 0.5 mg/L BA + 0.1 mg/L NAA. Root was induced on nodal segments of *Vanda teres* on medium containing 2 mg/L Kn + 0.5 mg/L NAA (Alam et al., 2010). Tyagi et al. (2010) showed root induction at the cut ends of shoots obtained from leaf explants of *Crataeva adansonii* on MS basal medium devoid of growth regulators. Shoot cuttings induce roots on MS medium with 1 mg/L NAA in 4 to 5 weeks, and in *Dianthus caryophyllus* L. with NAA and IBA (Casas et al., 2010). IAA (0.5 to 1 mg/L) helped rooting in *Pelargonium × hortorum* (Dorion et al., 2010). Studies of Ruffoni et al. (2010) on *Myrtus communis* showed that rooting was better in medium containing IAA than control, BA and BA + IAA. Ochatt et al. (2010) demonstrated that for rooting of *Lathyrus odoratus* L. micro-shoots, they are explanted onto medium with 0.5 to 1 mg/L NAA for 3 weeks. In conclusion, kind and concentration of growth regulators and kind of species are the most important factors in production of callus and root. Current study showed positive effect of Kn and NAA on callus induction and root formation on *in vitro* grown leave explants of *M. incana*, if we use suitable concentrations of them, alone or in combination.

## REFERENCES


Alam MF, Sinha P, Hakim ML (2010). Micropropagation of *Vanda teres* (Roxb.) Lindle. In: Jain SM, Ochatt SJ (eds) Protocols for *In Vitro* Propagation of Ornamental Plants. Springer protocols. Humana Press. pp. 21-28.

Bhattacharya S, Bhattacharyya S (2010). In vitro propagation of Jasminum officinale L.: a woody ornamental vine yielding aromatic oil from flowers. In: Jain SM, Ochatt SJ (eds) Protocols for In Vitro Propagation of Ornamental Plants. Springer protocols. Humana press. pp. 117-126.

Casas JL, Olmos E, Piqueras A (2010). In vitro propagation of carnation (Dianthus caryophyllus L.). In: Jain SM, Ochatt SJ (eds) Protocols for In *Vitro* Propagation of Ornamental Plants. Springer protocols. Humana Press. pp. 109-116.

Dorion N, Jouira HB, Gallard A, Hassanein A, Nassour M, Grapin A (2010). Methods for *in vitro* propagation of *Pelargonium × hortorum* and others: from meristems to protoplasts. In: Jain SM, Ochatt SJ (eds) Protocols for *In Vitro* Propagation of Ornamental Plants. Springer protocols. Humana Press. pp. 197-212.

Dubois LAM, de Vries DP (1995). Preliminary report on direct regeneration of adventitious buds on leaf explants of in vitro grown glass house rose cultivars. Gartenbauwissenschaft 60:249-253.

Eeckaut T, Janssens K, Keyser ED, Riek JD (2010). Micropropagation of Rhododendron. In: Jain SM, Ochatt SJ (eds) Protocols for *In Vitro* Propagation of Ornamental Plants. Springer Protocols. Humana Press., pp. 141-152.

Fuller MP, Fuller FM (1995). Plant tissue culture using Brassica seedlings. J. Biol. Edu. 20(1):53-59.

Gautam VK, Mittal A, Nanda K, Gupta SC (1983). *In vitro* regeneration of plantlets from somatic explants of *Matthiola incana*. Plant. Sci. Lett. 29:25-32.

Godo T, Lu Y, Mii M (2010). Micropropagation of *Lysionotus pauciflorus* Maxim. (Gesneriaceae). In: Jain SM, Ochatt SJ (eds) Protocols for *In Vitro* Propagation of Ornamental Plants. Springer Protocols. Humana Press. pp. 127-140.

Gomes F, Simões M, Lopes ML, Canhoto M (2010). Effect of plant growth regulators and genotype on the micropropagation of adult trees of *Arbutus unedo* L. (strawberry tree). New Biotech.

Hammaudeh HY, Suwwan MA, Abu Quoud HA, Shibli RA (1998). Micropropagation and regeneration of Honeoye strawberry. Dirasat Agric. Sci. 25:170-178.

Han BH, Yu HJ, Yae BW, Peak KY (2004). In vitro micropropagation of *Lilium longiflorum* 'Georgia' by shoot formation as influenced by addition of liquid medium. Sci Hortic 103:39-49.

Hashemabadi D, Kaviani B (2010). *In vitro* proliferation of an important medicinal plant Aloe-A method for rapid production. Austr. J .Crop Sci. 4(4):216-222.

Ibrahim R, Debergh PC (2001). Factors controlling high efficiency adventitious bud formation and plant regeneration from *in vitro* leaf explants of roses (Rosa hybrida L.). Sci. Hortic. 88:41-57.

Ibrahim R, Debergh PC (2000). Improvement of adventitious bud formation and plantlet regeneration from *in vitro* leaflet explants of roses (*Rosa hybrida* L.). Acta Hortic. 520:271-280.

Jain SM, Ochatt SJ (2010). Protocols for *in vitro* propagation of ornamental plants. Springer Protocols. Humana Press.

Lee-Epinosa HE, Murguia-Gonzalez J, Garcia-Rosas B, Cordova-Contreras AL, Laguna C (2008). In vitro clonal propagation of vanilla (Vanilla planifolia Andrews). Hortic. Sci. 43:454-458.

Leffering R, Kok E (1990). Regeneration of rose via leaf segments. Prophyta 8:244.

Luo JP, Wawrosch C, Kopp B (2009). Enhanced micropropagation of *Dendrobium huoshanense* C.Z.Tang et S.J.Cheng through protocorm-like bodies: the effects of cytokinins, carbohydrate sources and cold pretreatment. Sci. Hortic. 123:258-262.

Maira O, Alexander M, Vargas TE (2010). Micropropagation and organogenesis of *Anthurium andreanum* Lind cv Rubrun. In: Jain SM, Ochatt SJ (eds) Protocols for *In Vitro* Propagation of Ornamental Plants. Springer Protocols. Humana Press. pp. 3-14.

Mathai MP, Zacharia JC, Samsudeen K, Rema J, Nirmal Babu K, Ravindran PN (1997). Micropropagation of *Cinnamomum verum* (Bercht and Presl.). Proceedings of the National Seminar on Biotechnology of Spices and Aromatic Plants, April 24-25, Calicut, India, pp. 35-38.

Murashige T, Skoog F (1962). A revised medium for rapid growth and bioassays with tobacco tissue culture. Physiol. Plant 15:473-497.

Nayak S, Hatwar B, Jain A (2010). Effect of cytokinin and auxins on meristem culture of *Bambusa arundinacea*. Der Pharmacia Lett. 2(1):408-414.

Nencheva D (2010). *In vitro* propagation of Chrysanthemum. In: Jain SM, Ochatt SJ (eds) Protocols for *In Vitro* Propagation of Ornamental Plants. Springer protocols. Humana Press. pp. 177-186.

Nhut DT, Don NT, Vu NH, Thien NQ, Thuy DTT, Duy N, Teixeira da Silva JA (2006). Advanced technology in micropropagation of some important plants. In: Teixeira da Silva JA (ed) Floriculture Ornamental and Plant Biotechnology, vol. II. Global Science Books, UK, pp. 325-335.

Nhut DT, Hai NT, Phan MX (2010). A highly efficient protocol for micropropagation of *Begonia tuberous*. In: Jain SM, Ochatt SJ (eds) Protocols for *In Vitro* Propagation of Ornamental Plants. Springer protocols. Humana Press., pp 15-20.

Ochatt SJ, Conreux C, Jacas L (2010). *In vitro* production of sweet peas (*Lathyrus odoratus* L.) via axillary shoots. In: Jain SM, Ochatt SJ (eds) Protocols for *In Vitro* Propagation of Ornamental Plants. Springer protocols. Humana Press. pp. 293-302.

Pati PK, Kaur N, Sharma M, Ahuja PS (2010). *In vitro* propagation of rose. In: Jain SM, Ochatt SJ (eds) Protocols for *In Vitro* Propagation of Ornamental Plants. Springer protocols. Humana Press. pp. 163-176.

Pati PK, Rath SP, Sharma M, Sood A, Ahuja PS (2005). *In vitro* propagation of rose-a review. Biotech. Adv. 94-114.

Pati PK, Sharma M, Sood A, Ahuja PS (2004). Direct shoot regeneration from leaf explants of *Rosa damascena* Mill. In Vitro Cell Dev. Biol. Plant. 40:192-195.

Peeters AJM, Gerards W, Barendse GWM, Wullems (1991) *In vitro* flower bud formation in tobacco: interaction of hormones. Plant Physiol. 97:402-408.

Preil W, Florak P, Wix U, Back A (1988). Towards mass propagation by use of bioreactors. Acta Hortic. 226:99-107.

Radice S (2010). Micropropagation of *Codiaeum variegatum* (L.) Blume and regeneration induction via adventitious buds and somatic embryogenesis. In: Jain SM, Ochatt SJ (eds) Protocols for *In Vitro* Propagation of Ornamental Plants. Springer protocols. Humana Press. pp 187-186.

Rout GR, Debata BK, Das P (1990). *In vitro* clonal multiplication of roses. Proc. Natl. Acad. Sci. India 60:311-318.

Rout GR, Mohapatra A, Mohan Jain S (2006). Tissue culture of ornamental pot plant: a critical review on present scenario and future prospects. Biotechnol Adv. 24(6):531-560.

Ruffoni B, Mascarello C, Savona M (2010). *In vitro* propagation of ornamental Myrtus (*Myrtus communis*). In: Jain SM, Ochatt SJ (eds) Protocols for *In Vitro* Propagation of Ornamental Plants. Springer protocols. Humana Press. pp. 257-270.

Sajina A, Geetha SP, Minoo D, Rema J, Nirmal Babu K, Sadanandan AK, Ravindran PN (1997b). Micropropagation of some important herbal species. In: Biotechnology of Spices, Medicinal and Aromatic Plants, Edison S, Ramana AV, Sasikumar B, Nirmal Babu K, Santhosh JE (eds.). Indian Society for Spices, Calicut, India. pp. 79-86.

Sangavai C, Chellapandi P (2008). *In vitro* propagation of a tuberose plant (*Polianthes tuberosa* L.). Electronic J. Biol. 4(3):98-101.

Tyagi P, Sharma PK, Kothari SL (2010). Micropropagation of *Crataeva adansonii* D.C. Prodr: an ornamental avenue tree. In: Jain SM, Ochatt SJ (eds) Protocols for *In Vitro* Propagation of Ornamental Plants. Springer protocols. Humana Press. pp. 39-46.

Van Staden D (2008). Plant growth regulators, II: cytokinins, their analogues and inhibitors. In: Plant Propagation by Tissue Culture (edn 3) (George EF, et al eds), pp. 205-226, Springer.

Winkelmann T, Geier T, Preil W (2006). Commercial *in vitro* plant production in Germany in 1985-2004. Plant Cell Tiss. Org. Cult.

86:319-327.
Wong CE, Bhalla PL (2010). *In vitro* propagation of Australian native ornamental plant, *Scaevola*. In: Jain SM, Ochatt SJ (eds) Protocols for *In Vitro* Propagation of Ornamental Plants. Springer protocols. Humana Press. pp. 235-242.

16

# Effects of container volume and pruning on morpho-functional characters of *Salix elaeagnos* Scop. under water stress for Mediterranean riparian ecosystems restoration

**B. Sagrera, C. Biel and R. Savé**

IRTA, Environmental Torre Marimon E-08140 Caldes de Montbui, Spain.

**Restoration success using nursery grown plants in riparian environments of Mediterranean areas is limited by summer high mortality rate due to severe drought stress. The aim of this work is to develop a morphological / physiological modified plant which could establish itself in low water availability areas. Four morphologic formats of *Salix elaeagnos* Scop. were tested depending on two factors: Container volume (3 and 6 L) and pruning (pruned and not pruned). Twenty four individuals of four different formats were grown in a nursery for at least three months and outplanted in a per liter filled assay pool simulating a riverbed restoration for 12 weeks. Morpho-physiological destructive measurements of 32 plants (8 per format) were done in the beginning (T0), half, 6 weeks, (T1) and ending (T2) of the assay. It was found that in T0, there were important biomass differences, being the 6 L volume and not pruned format (6NP) the highest. During T1, water availability was high and it was advantageous for 6NP which showed absolute and relative highest growth rates. However at T2, water availability declined and relative rates of growth and new roots production were higher for the 6 L volume and pruned format (6P). Three liters formats suffered harder water stress than 6 L ones. 6 L containers better performance is explained for its high volume and depth, which leads to reaching underground water level easily. Results shown as pruning, high root /canopy ratio can be beneficial when water conditions are or will be unfavorable in river restorations now and under potential new climate change conditions.**

**Key words:** Agronomical practices, climate change, drought stress, leaf water content, relative growth rate, relative soil.

## INTRODUCTION

Under Mediterranean weather and climatic conditions (Mitrakos, 1980) and due to socioeconomical development, geomorphological characteristics of water streams, from middle to last years of XX century rivers and creeks and consequently their associated vegetation where destroyed or degraded (http://www.restorerivers.eu/; http://www.ecrr.org/). River restoration refers to a large variety of ecological, physical, spatial and management measures and practices. These are aimed at restoring the natural state and functioning of the river system in support of biodiversity, control of flows, recreation, flood safety and landscape development. By restoring natural conditions, river restoration improves the resilience of streams systems and provides the framework for the sustainable multifunctional use of estuaries, rivers and streams. River restoration is an integral part of sustainable water management and is in direct support of the aims of the EU Water Framework Directive (2000), and national and

*Corresponding author. E-mail: robert.save@irta.cat

regional water management policies.

Planting riparian trees and shrubs is one of the most effective methods to recover river's ecological quality. It provides important benefits: Soil fixation, erosion reduction, high regeneration power after floods, slope stabilization, sediment retention and habitat formation enhancing biodiversity. One of the point that must be taken in consideration in river restoration is plant material, which must offer some important characteristics in order to promote a good ecological restoration, as to be autochthonous, available at nurseries and resistant to transplant (Cortina et al., 2006). *Salix* L. species are among the most common plant groups living in areas directly disturbed by floods. They are exceptionally adapted to this environment, since their mechanical properties make it possible to bear moderate freshets, and in case to be uprooted, swept or fragmented by more heavy freshets, they have full guarantee of sprouting new shoots (Karrenberg et al., 2002). The economic importance of Salix is currently increasing and emerging in a wide array of practical applications to restore damaged ecosystems (Kuzovkina and Quigley, 2005). Rosemary willow (*Salix elaeagnos* Scop.) is a common shrub, which is adaptable to perturbations and it shows high survivorship and growth rates. Hence, it is good generalist for riparian restoration specially indicated for disturbed areas along the river banks (Francis et al., 2005). However, restoration success is not always optimal, due to low survivorship rates (Cortina et al., 2006). Plant establishment to field conditions is the most critic period, were higher mortality appears. In Mediterranean environments, the water stress associated to the first summer season is the main reason for plant dieback (Cortina et al., 2006; Chirino et al., 2008). Rivers and especially little Mediterranean streams can undergo high flow oscillations, becoming completely dry during long periods of time and hard droughts (Medici et al., 2008), which promotes the fact that water source for plants is the underground water table (Horton and Clark, 2001). Moreover, soil characteristics in medium and low stretches of certain rivers and streams show high gravelliness resulting in low water retention. Hence, phreatic level is a limiting factor in water resource availability (ACA, 2009). The potential climate change attributable to global change can increase local and general temperature (IPCC, 2007). These yearly small changes in temperature may have great influence in the atmospheric carbon balance (Valentini et al., 2000). This increase will not be the same around the world (IPCC 2007). Global change is the combination of many stresses in the same space and at the same time, which can cause synergic effects on vegetation (Llebot, 2010). Some models are generating scenarios of climate change that are showing us that some regions in the world will be affected by average duration of dry periods (4 to 6 months) and the length between periods (more 12 months), being these episodes from 3 to 8 times more frequent than at present (Sheffield and Wood, 2008). According to these models applied to aquifer recharge, basic river flow will suffer a hard reduction, and therefore also groundwater level (Medici et al., 2008; IPCC, 2007; ACA, 2009; ACCUA, 2011). With increasing aridity, the riparian vegetation of Mediterranean-type streams becomes shorter, more scattered, more restricted to the side of the active channel, and markedly different from the upland regions (Gasith and Resh, 1999). There is a restoration need in drier conditions of rivers and streams. Therefore, it is very necessary to develop new strategies to produce plants which can establish themselves in arid conditions.

In general, classical plant production in nurseries has focused in aerial development as the main commercial attribute (Cortina et al., 2006). The commonly used container volume in Mediterranean ecosystem restoration has been relatively small and shallow (max 3 L – 18 cm) (Pastor et al., 1999). Container characteristics are important determining outplanting performance (Tsakaldimi et al., 2005). It has been proved that larger and deeper container sizes increase root mass in certain riparian species including *Salix* spp. (Houle and Babeux, 1998; Dreesen et al., 2002) and other typical Mediteranean species of *Pinus* and *Quercus* genus (Chirino et al., 2008; Dominguez-Lerena et al., 2006; Lamhamedi et al., 1998; Permán et al.,, 2006). However, there are not similar studies for riparian species under Mediterranean climate. Since it is considered that water availability is a limiting factor and water absorption potential is determined by root system size and depth (Carlson, 1986; Lloret et al., 1999), an optimum container is needed to develop such a root system.

Some agronomical treatments at nursery and / or field conditions as pruning plant aerial parts may stimulate photosynthesis because it helps keeping water balance in some species, reducing leaf total area (Elfadl and Luukkanen, 2003). In *Salix* species, individuals are perfectly able to compensate aerial biomass loss and survivorship is not affected (Hjaelten, 1999; Guillet and Bergström, 2006). High resprouting capacity of Salicaeace potentially points to pruning as a good practice to reduce transpiration (Hjaelten and Price, 1996; Karrenberg et al., 2002; Sennerby-Forsse and Christersson, 1994) and it might be an advantage under water stress conditions.

Quality description of forestry plant is based in two attribute types: Material attributes, which are directly measurable; and behavior attributes, which reflect plant response under certain environment conditions (Oliet et al., 2003; Ritchie and Landis, 2010). Behavior attributes can be considered synthetically because they can summarize in one or few parameters several morpho-physiological quality characters (Burdett, 1990). For example: Root growth potential, relative growth rate, photosynthesis, water use efficiency etc. (Folk and Grossnickle, 1997).

The hypothesis held in this paper is that a bigger and deeper container will improve outplanted individuals capacity to reach underground water, and aerial part reduction will diminish transpiration and thus will improve drought resistance. To prove these hypotheses a restoration will be simulated where different treated Rosemary willow (*S. elaeagnos*) will be outplanted. Morphological characters which are good predictors for survivorship and growth, and are important for nursery production will be controlled, especially those related to water balance. The aims of the work were:

1. To develop a morphologic *S. elaeagnos* format with high water stress resistance and improved performance in restoration projects than standard formats.
2. To determinate the effect of container size and pruning on survivorship and growth under water stress conditions.

## MATERIAL AND METHODS

### Experimental design

#### *Nursery phase*

Cuttings 10 cm long of *S. elaeagnos* Scop. (common name: rosemary willow) were collected from several adult individuals found in Ripoll river in a St. Llorenç Savall, Catalunya (41°N, 2°E, 466 masl). They were planted in 300 $cm^3$ plastic container or pots in February 2009. They were grown outdoor in the nursery Tres Turons SCP, in Castellar del Vallès, Catalunya, Spain (41°N, 2°E, 350 msnm). Growth substrate was standard peat moss (Vermhurg CC20, Gramoflor, Germany) and slow-release fertilizer of 3 to 4 month was added (1.5 $g/dm^3$, Osmocote Plus© 15-8-11 N:P:K). Irrigation by sprinkler using Ripoll river water (electrical conductivity: 0.67 dS/m) was about 90 $L/m^2h$ for one and a half hour, every other day. From rooted cuttings four treatments were prepared in summer 2010, counting 24 individuals each treatment with a density of 25 containers per $m^2$. The treatments were:

3NP: Container SMH 3 (Soparco-Odena, France; volume 3 L, diameter 18 cm, depth 15 cm). No pruning (standard commercial format). , 3P: Container SMH 3 and pruning.
6NP: Container C1600 (IEM Plastics, USA; volume 6.5 L, square cross-section 18 cm, depth 34 cm) and no pruning.
6P: Container C1600 and pruning.

In January 2011, plants of 3P and 6P treatments were pruned to 20 cm height. Slow-release fertilizer of 8 to 9 month was added when outplanting (1.5 $g/dm^3$, Osmocote Plus© 15-8-11 N:P:K).

#### *Outplanting phase*

In January 2011, 96 plants were placed in a round plastic pool(diameter 3.5 m, depth 0.7 m). The growth medium was perlite (Europerl© A13, World Minerals Europe), to achieve maximum homogeneity and to facilitate root processing. Distance among plants was approximately 0.5 m giving a density of 10.39 plants per $m^2$. Treatments were mixed, simulating a real riparian restoration. However all non-pruned plants (NP) were placed in the north face to avoid shading to the pruned ones (P). Plants grew during a 12 weeks period. In the beginning of the field assay, pool was filled up to 40 cm with river water (conductivity: 0.67 dS/m), no fertilizer was added. Rainfall was registered through the essay.

### Non-destructive measurements

#### *Morphologic development*

Height was measured as the vertical distance between the substrate and the highest shoot. Diameter at root collar was measured. When plants had more than one stem, equivalent diameter was calculated. Relative growth rate (RGR) of height and diameter was calculated as:

$$RGR = [(Ln(H_{tn}) - Ln(H_{tn-1})] / (t_n - t_{n-1})$$

Where H was the height or diameter and t is the day of measure.

#### *Chlorophyll content*

Relative chlorophyll content was measured with a SPAD-502, (Konica Minolta© , Japan) in 5 leaves per plant (only for the third plant block).

#### *Soil water content*

Two capacitive probes EC-5 (Decagon Devices©, USA) were permanently installed at different depths: 30 and 0 cm (bottom). Water saturation values were weekly registered with a Procheck data logger (Decagon Devices©, USA). Water level from the bottom was also registered.

### Destructive measurements

The 96 initial individuals were divided in three blocks of 32, each one formed by 8 individuals of each format. Block 0 was processed in the laboratory in the 0 week, so it was not planted in the assay pool and was considered as the control block at nursery outlet. Blocs 1 and 2 were planted in the pool were they grow for 6 and 12 weeks, respectively. Therefore the parameters measured in the laboratory were performed for each block in the beginning (T0), middle (T1) and end (T2) of the assay. All the dry weight measures were taken to drying the vegetal material in a air forced oven for 48 h at 65°C and using an analytic scale MS204S (Mettler Toledo©, Spain).

#### *Biomass*

Stems, leaves and roots were dried and weighted separately. Relation between underground and aboveground biomass was calculated (Root: Shoot ratio). Biomass growth was analyzed in terms of relative growth (biomass at T1 or T2 per biomass unit at T0). Biomass enhancement rate (BER) was calculated as the increase of the new formats respect the control format 3NP (Poorter and Naves, 2003).

#### *Leaf morphology*

Five leaves per plant were individually scanned and weighted, and were processed with WinFolia© software (Regents Instruments, Canada) software to obtain leaf area. Specific weight ($g/m^2$), specific leaf area ($m^2/g$) and total leaf area (TLA=specific leaf area × total leaves dry weight) were calculated.

#### *Root growth potential (RGP)*

This parameter was only measured in blocks 1 and 2, which had been planted in the assay pool. All the roots exploring new medium (perlite) were considered new roots. These were cut and weighted. A subsample was taken from new roots, which was scanned and weighted, to be processed with WinRhizo© software (Regent Instruments, Canada). From root length values of the subsample, specific length (m/g) and RGP was calculated, considering t as the total length of new roots (RGP = specific length × new roots dry weight) (Folk and Grossnickle, 1997).

#### *Relative water content (RWC)*

Five leaves randomly collected from each plant were weighted before and after placing them in a water solution for one day at dark and 5°C (saturation weight or maximum swelling). Finally, dry weight was obtained and relative water content was calculated as:

RWC = [(fresh weight – dry weight) / (saturation weight – dry weight)] × 100

#### Data analysis

Statistical analyses to find differences in data between time and plant and container format, were carried out using R free software (version 2.11.1, 2010). A type II factorial ANOVA test followed of a multiple means Tukey comparison test was performed considering three factors: container size (3 and 6 L), pruning (NP and P) and time (T0, T1 and T2).

## RESULTS AND DISCUSSION

### Water evolution during the pool assay

On T0 pool was water filled up to 40 cm from the bottom. Thus, both sensors bottom (0 cm) and sensor 30 cm from bottom were under water, displaying water saturation values of 59.9%. During the first 6 weeks, water level went down steadily, rising 16 cm on week 6 (T1). At this time, point sensor Bottom displayed 58.2% while sensor 30 cm did 1.1%. Since week 2, when water level dropped to 23 cm, sensor 30 cm displayed 2%. However, accumulated rainfall during first 6 weeks was considerable: 117 mm, but not sufficient to compensate evapotranspiration and consequently increase water storage. Between week 7 and 12, water level keep on dropping regularly. From week 10 onwards water level was under 1 cm and sensors detected no humidity (Figure 1a). Rainfall was 34 mm between T1 and T2 period.

Water conditions have been unfavorable even though accumulated rainfall (151 mm in 12 weeks). Perlite showed low capillarity because when water level dropped under the sensors, medium displayed water saturation values near to zero (Figure 1b). Therefore, fast elongation of new roots to reach phreatic permanent flood level has been decisive on water availability for plants (Francis et al., 2005). It might be considered that original substrate (peat coming from nursery), has retained rainfall water, mitigating the effects of water stress. However, calculations considering effective rainfall collection surface, water retention capacity of peat, and transpiration measurements in *Salix* species on field conditions, that this effect would be negligible. Thus results obtained in this work will be analyzed considering only pool water level, which has been much higher during T1 period than T2.

### Plant material at T0

The plant formats showed big differences when the assay began (T0). Average height of NP willows was 131.6 cm while P was 46.5 cm. It means that NP plants were 193% higher than P ($p<0.001$), but no differences in diameter where found. About total dry weight, NP plants showed 124% more biomass than P ($p=0.004$), while 6 L volume plants where 50% heavier than 3 L ones ($p=0.002$). Stem dry weight is the big part of the total biomass. Plants NP have 220% more stems ($p<0.001$) than P ones. About root dry weight, formats NP were 66% heavier than P. Format 6NP was the heaviest (27.1 g) significantly higher than 3P, the lowest one (10.2 g, Figure 2). Therefore it was not container volume that determined root weight at T0, but pruning ($p=0.003$). About Root:Shoot ratio, P plants showed higher values than NP ones and significant interaction ($p=0.006$, results not shown) was found between volume and pruning factors.

About leaf biomass, 6 L treatments had 133% more leaves than 3 L ($p<0.001$) and NP plants were 81% higher than P plants ($p=0.004$). Formats P presented lower leaf specific weight to NP ($p=0.02$). Total leaf area was 75% higher in 6 L plants ($p<0.001$). Finally, leaf relative water content (RWC) was very similar between treatments. The only significant difference was between 3P format, the highest (87.35%) and 3NP format, the lowest (78.19%). As had been shown previously, the four treatments have shaped different formats during nursery phase. Therefore, evolution of these formats during the assay will be analyzed considering its nature at planting time.

### Height and diameter growth

All plants survived during the assay. However, to analyze growth beyond survivorship, information is given about plant quality. It is related with the new environment, genetic potential and morphological and physiological status of plants at transplant time (Mexal and Landis, 1990).

In general, plants had small weekly growth in height and diameter. About height relative growth rate (RGRh), P willows showed a significant higher RGRh than NP, especially during the first 6 weeks (n=68). In this period, highest RGRh recorded was on 3P, followed by 6P. Treatments NP showed significant lower RGRh (Table 1). The need to recover aerial biomass and to raise photosynthetic structure could explain the higher growth

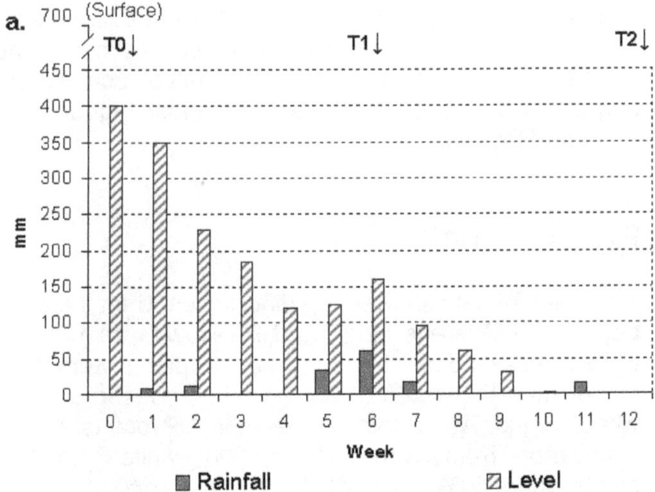


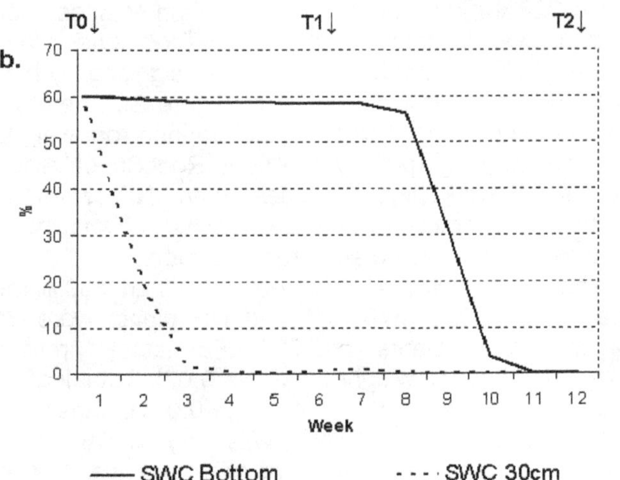


**Figure 1.** Water evolution of the assay pool and rainfall (a). Soil water content of bottom and 30 cm from bottom (b). Plant extracting time of each block is indicated with an arrow above corresponding to weeks.

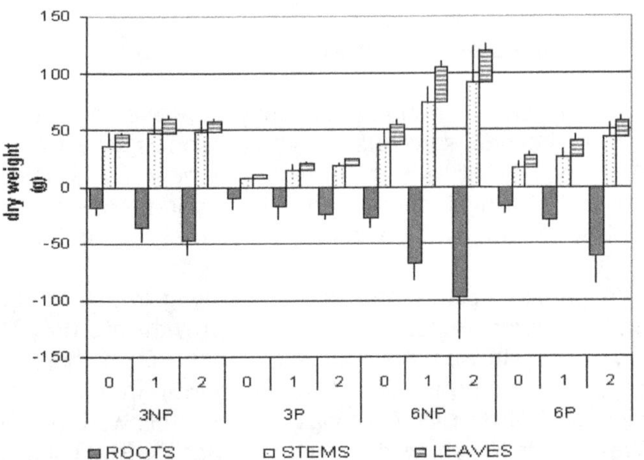


**Figure 2.** Dry weight of different fractions in each sampling time and format (mean ± standard deviation, n=8).

rate on pruned plants together with the high water availability at substrate level respect canopy surface. There are morphologic differences between 6P and 3P which have been observed in the field that should be highlighted, although no measures have been taken. Format 3P had few shoots which tend to elongate. This observation agrees with the work by Hjaelten and Price (1996) on *S. lasiolepis*. However, format 6P had many shoots which tend to multiply. Therefore there is an intrinsic effect in the interaction pruning-volume that might determine resulting aerial plant morphology.

About diameter, although 6 L plants showed higher RGRd in the whole period, it was not statistically significant due to high variability. It might be explained by two factors. First, the shrub nature of rosemary willow and its high sprouting capacity causes it to produce between one and seven stems per individual. Thus, equivalent diameter measures are subjected to high dispersion rates and increase error. On the other hand, stems are not uniform and it makes hard to take accurate measures every week. Possibly bigger container determines a root morphology which allows higher diameter growth (Dominguez-Lerena et al., 2006), even though it is not statistically sustained.

Height and diameter are broadly used attributes in nurseries for quality control. However these attributes do not always have a predictive capacity on survivorship and growth in reforestation (Cortina et al., 2006). Study species characteristics, a multistem shrub, advise against using these parameters to define plant quality.

## Biomass growth

NP plants showed a significant increase on every parameter in T1, but not in T2 (Figure 2, Table 4). Format 6NP presented the highest relative and absolute increases in T1. Format 3NP did not have any significant increase and it even lose 36% of leaves at T2 (Table 2). Format 3P grew significantly in weight in every parameter at T1 and it showed some important relative increases in T2 (aerial parts 123%, $p<0.001$). However, it was the format presenting less biomass in all sampling days and parameters (Figure 2). On the other hand format 6P showed low relative increases in T1 while in T2 increases significantly in every parameter, being the format which obtained the highest relative growth rates in total (176%, $p<0.001$) roots (269%, $p<0.001$) and stems (171%, $p<0.001$), even though leaves dry weight did not increase through the time (Table 2). All the formats tended to allocate biomass to underground fraction, increasing R:S ratio through the time.

Overall, in T2 willows 6L presented 69% more total biomass than 3L ones, being 6NP the highest, 3NP and 6P similar in the middle and 3P the lowest. Willows NP showed more biomass than P ones, but this difference was bigger in T1 (124%) than in T2 (62%). This tendency was very similar in every biomass parameter (Figure 2).

**Table 1.** Relative growth rate in height and diameter for the four treatments between T0 and T1, and total (between T0 and T2).

| Treatment | RGR height 0-6 weeks | RGR height total | RGR diameter 0-6 weeks | RGR diameter total |
|---|---|---|---|---|
| 3NP | 0.008±0.0042[c] | 0.008±0.0035[c] | 0.023±0.0179 | 0.024±0.0173 |
| 3P | 0.063±0.0077[a] | 0.05±0.0079[a] | 0.029±0.0166 | 0.028±0.0171 |
| 6NP | 0.013±0.0036[c] | 0.017±0.0052[bc] | 0.057±0.0158 | 0.049±0.0137 |
| 6P | 0.034±0.0063[b] | 0.03±0.0064[b] | 0.059±0.0176 | 0.047±0.0159 |

Values followed by different letters are significantly different in each column. (Mean ± standard deviation, 0-6 weeks n=102; 0-12 weeks n=108).

**Table 2.** Biomass relative increase in T1 and T2 related to T0.

| Parameter | 3NP | | 3P | | 6NP | | 6P | |
|---|---|---|---|---|---|---|---|---|
| Time | T1 | T2 | T1 | T2 | T1 | T2 | T1 | T2 |
| Aerial parts (%) | 31 | 27 | 88 | **123*** ** | **89** | 115*** | 45 | 112** |
| Roots (%) | 104 | 165*** | 74 | 137*** | **152** | 262*** | 70 | **269*** ** |
| Leaves (%) | 43 | 7* | 64 | **69*** | **70** | 51* | 17 | 30 |
| Stems (%) | 28 | 32 | 99 | 150** | **98** | 147*** | 66 | **171*** ** |
| Total | 51 | 66* | 81 | 130* | **110** | 163* | 55 | **176*** ** |

Remarkable values of each time is in bold (n=8; p<0.001 '***' p<0.01 '**' p<0.05 '*').

It means that P plants had a total relative growth rate 23% higher than NP, and 6 L plants 51% higher than 3 L ones. Thus, format 6NP was the one accumulating more biomass, but the highest relative biomass increase was in 6P.

Dry weight of different plant fractions displays information about response of every format to transplant, just like its capacity to assimilate carbon and allocate it to functional parts (Chapin, 1991). During first 6 weeks every format gains biomass in all fractions, but specially 6NP. In this period water soil availability was relatively high and therefore having higher biomass at point zero may be advantageous when initial conditions are favorable (Cortina et al., 2006). However, when checking relative growth rates through 12 weeks, tendency changes.

Format 6P reaches the most important biomass increases in most of the parameters. Thus, pruning may reduce water stress effect decreasing transpiring surface (Savé et al., 1993).

## Biomass enhancement rate (BER)

Biomass enhancement rate was used to compare the weight increase of the three format tested against standard, 3NP format.

A substantial increase in biomass was observed in format 6NP respect the control one, 3NP. Already at T0 it had produced significantly more leaves, and it showed highly positive BER values in all parameters at T1 and T2, reaching a 106% higher increase in total dry weight at T2 (Figure 4).

Format 6P showed negative stem BER values in T0. However it tended increase in BER, reaching values around zero on roots and stems, or even significantly positive leaf BER values (53%, p=0.02). Therefore, format 6P does not overcome 3NP in biomass, but it gains more biomass during three months. If tendency would carry on, 6P but have positive BER values in few weeks after the end of the assay. Therefore 6NP clearly improves biomass growth from the standard format, but 6P shows the best tendency and it must be considered too.

## Leaf morphology and physiology

Chlorophyll leaf content (SPAD) weekly measured did not change in any treatment through time. It stood around 32 in all formats except 3P, which was significantly lower to the rest (p<0.001) with a value of 27.

About total leaf area, it has been observed that volume determines a higher leaf area for 6 L plants (p<0.001). Tendency through the time is to keep similar values in 6 L but 3 L plants even decreased significantly in T2 (Table 4). Leaf specific weight of all treatments tends to rise through the time, but in different ways. In T1 all specific weight values rise significantly except for 3NP format. Volume factor is significant (p<0.001) being 6 L treatments the highest leaf specific weights (141 $g/m^2$). On the other hand at T2 a significant interaction among factors appears (p=0.041). Formats 3L rise, and finally become the ones with the highest specific weight (185 $g/m^2$, p<0.001, Table 4). Moreover, RWC decreases in every format reaching values around 66%, without group

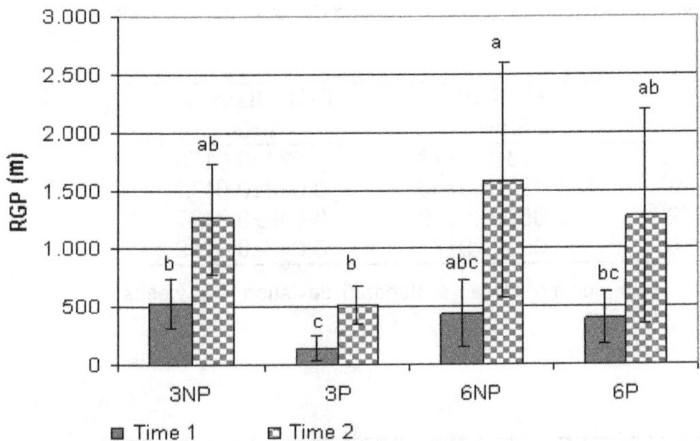


**Figure 3.** Root growth potential (RGP) in root total length (m) for each format in Times 1 and 2. (mean ± standard deviation, n=8). Different letters show significant differences between times and formats.

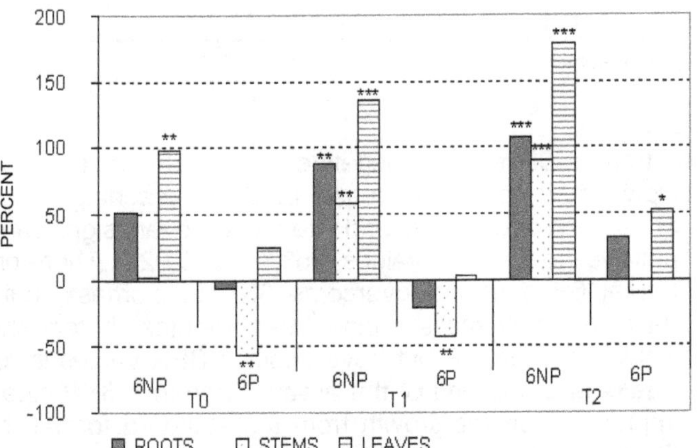


**Figure 4.** Biomass enhancement rate (BER) from format 6NP and 6P respect 3NP. (n=8; p<0.001 '***' p<0.01 '**' p<0.05 '*'). Format 3P has been omitted because all BER values were negative.

**Table 3.** Dry weight of new roots related to root initial dry weight.

| Treatment | T1 | T2 |
|---|---|---|
| 3NP | **0.08** | 0.24 |
| 3P | 0.04 | 0.21 |
| 6NP | 0.05 | 0.2 |
| 6P | 0.06 | **0.32** |

Remarkable values for each time in bold.

differences, which indicates an important stress level (Bradford and Hsiao, 1982). These data point out that all formats are under water stress. There is a double response to this stress: In one hand reduction of leaf area due to leaf loss, and on the other hand hardening of remaining ones. Therefore it is possible that transplanting shock and water shortage conditions going worse force plants to restrict transpiration and finally stop growth. However response is different depending on the format, and the intrinsic properties that arise from the combination volume-pruning, as it is shown on the significant interactions detected in T2 in each water status parameter (leaves dry weight, p=0.009; RWC, p=0.022; specific leaf weight, p=0.041). Hence, plants which have more substrate (6 L) were less affected by drought stress as suggested by lower specific weights and higher relative growth rates. Conversely, in 3 L formats, shock is longer and no recuperation signs appear.

## Root growth potential (RGP)

Root growth potential gives information of the capacity to outcome transplant impact and to explore the new medium (Oliet et al., 2003). In the current study few differences are found among formats except in 3P. Format 3NP showed the highest total length of new roots in T1 (527 m) but only significantly different than 3P. In T2, the highest RGP is for 6NP format (1583 m), but again only significantly different than 3P, but not from 3NP and 6P (Figure 3). Since root specific length values were not significantly different between times and formats, new roots dry weight show a very similar tendency than RGP. However specific length decreases significantly between times for P formats (p=0.001 for 3P and p= 0.015 for 6P, Table 4).

The intrinsic variability of the attribute should be considered (Sutton, 1990). RGP depends largely of plant reserves at transplant time, and most of these are found in the roots (Dickmann and Pregitzer, 1992; Kozlowski, 1992). Hence, normalizing RGP to root dry weight at T0 might be indicative of the capacity to produce new roots from the existing (Noland et al., 1996). Considering this index, in T2 format 6P is higher than all the others T2 which stay with similar values (Table 3). However these values only point a tendency and they are not sustained by any statistic test.

There is another consideration about container differences, which has been observed but not measured. Format 3NP at T0 shows very compressed roots in the container, so new roots growth will be almost entirely in the new medium. Conversely, 6 L formats (and especially 6P) can still produce new roots in the original medium after transplant, so that counted roots to calculate RGP are only those standing out of the original substrate. This phenomenon may have produced an underestimation.

Exploring new medium guarantees to reach the phreatic level quickly, and it provides sufficient water availability (Francis et al., 2005) at least during the first weeks. It was observed that at T2 all treatments have produced long roots to the pool bottom, so even though it was no directly measured, most likely all plants reached the bottom with at least one root. This observation agrees with the work by Francis et al. (2005), where high root

**Table 4.** Evolution of each parameter related to previous time.

| Parameter | 3NP | | | 3P | | | 6NP | | | 6P | | |
|---|---|---|---|---|---|---|---|---|---|---|---|---|
| | T1 | T2 | Pr(>F) | T1 | T2 | Pr(>F) | T1 | T2 | Pr(>F) | T1 | T2 | Pr(>F) |
| Aerial part dw(g) | = | = | 0.156 | ↑ | = | <0.001*** | ↑ | = | <0.001*** | = | ↑ | 0.001** |
| Roots dw(g) | ↑ | = | <0.001*** | ↑= | ↑= | 0.018* | ↑ | = | 0.029* | = | ↑ | 0.028* |
| Leaves dw(g) | ↑ | ↓= | 0.03* | ↑ | = | 0.015* | ↑ | = | 0.001 | = | = | 0.388 |
| Stems dw(g) | = | = | 0.165 | ↑ | = | <0.001*** | ↑ | = | <0.001*** | = | ↑ | 0.072*** |
| Total dw(g) | ↑ | = | 0.01* | ↑= | ↑= | 0.002** | ↑ | = | 0.074. | = | ↑ | <0.001*** |
| R:S | ↑ | ↑ | <0.001*** | = | = | 0.419 | ↑ | ↑ | <0.001*** | = | ↑ | <0.001*** |
| Total leaf area ($m^2$) | = | ↓ | 0.085. | ↑= | ↓ | 0.032* | = | = | 0.869 | = | = | 0.071 |
| LSW ($g/m^2$) | = | ↑ | <0.001*** | ↑ | ↑ | <0.001*** | ↑ | = | <0.001*** | ↑ | = | <0.001*** |
| RWC (%) | ↓ | ↑= | 0.006** | ↓ | = | 0.048* | ↓= | ↓= | <0.001*** | ↓ | = | 0.003** |
| New roots dw(g) | | ↑ | <0.001*** | | ↑ | <0.001*** | | ↑ | 0.002** | | ↑ | <0.001*** |
| RGP (m) | | ↑ | 0.002** | | ↑ | 0.078. | | ↑ | 0.008** | | ↑ | 0.02* |
| NRSL (m/g) | | = | 0.155 | | ↓ | 0.002** | | = | 0.547 | | ↓ | 0.015* |

Values are p-value from ANOVA test within treatment. Arrows show significant increase (↑) or dicrease (↓) of the parameter from preciding time. =, Points no significant variation. Post-hoc Tukey contrast used (n=8; p<0.001 '***' p<0.01 '**' p<0.05 '*'). R:S, root:shoot, LSW, leaf specific weight; RWC, relative water content; RGP, root growth potential; NRSL, new roots specific length.

elongation rates are described in *S. elaeagnos* in response to a decreasing water level in similar experimental condition to the current study. Parameter RGP correlates with survivorship and growth (McKay, 1999). However, sufficient water availability could have mitigated this possible relation (Simpson and Ritchie, 1997) during first weeks. Overall, from RGP data it is shown that 3P has a poor response and 6P is capable to produce more roots in relation to the initial root volume than any other treatment. Plants tend to allocate biomassto roots, or at least root growth is higher than aerial in all formats but 3P. This tendency is more pronounced during last 6 weeks, when more water deficit is given. This observation agrees with other studies which demonstrate that R:S ratio increases in drought situations (Timmer and Miller, 1991; Villar-Salvador et al., 2004; Ovaska et al., 1993). However this response can be explained by carbohydrate accumulation to roots as a reservoir tissue (Von Fircks and Sennerby-Forsse, 1998) as a water stress consequence. Hence, water availability is probably neither dependant of root volume or R:S ratio, but deep rooting capacity (Padilla and Pugnaire, 2007).

Results demonstrate that a bigger container is beneficial, because it allows higher growth and better drought resistance at nursery and in the first steps after transplant. The main attribute to explain better performance of 6 L containers is probably depth. Several studies in forestry confirm this hypothesis (Padilla and Pugnaire, 2007; Permán et al., 2006; Chirino et al., 2008). However, other factors may be also important, such as a less dense and deformed root system in 6 L, which could help overcoming transplant shock (Dominguez-Lerena et al., 2006). Pruning is a practice that forces to mobilize root reservoirs to sprout (Von Fircks and Sennerby-Forsse 1998; Carpenter et al., 2008). Therefore it is very important to have enough root volume as it is demonstrated in the current study by 6P better performance in front of 3P. Pruning advantage appears when water availability decreases. It seems that plants with reduced aerial biomass have better resistance and is capable to keep on growing (South and Blake, 1994; Carpenter et al., 2008; Savé et al., 1993). Therefore, it is possible that in more extreme conditions, pruned plants would show better survivorship rates.

Formats 6 L perform better in the experiment. Much more considerations about these formats suitability should be done, such as weed competency, herbivory, flood resistance or economic feasibility. Results do not demonstrate physiological advantage of 6P on 6NP, but, pruning can permit high plant density at nursery, which improves sources use efficiency (Cortina et al., 2006). Different pruning levels could be tested in a further study. Moreover, it would be need to run a longer assay completing a summer period, testing plant material in drier conditions. Observed tendencies could not be indicative of the long term restoration success, but they are objective indications for the early stages of landscape restoration (Cortina et al., 2006). Larger samples would be needed to offset certain attributes variability, such as RGP. These morphologic treatments should be tested in other species to contrast it as a general hypothesis for nursery plants for restoration. However, results point 6P format as the optimum one for restoration in potentially dry riparian environments, with a 3 month guarantee from nursery exit and with good survivorship expectations.

## Conclusion

According to climatic change models applied to Mediterranean region, rivers will suffer a reduction on water input and therefore groundwater level will be harder to reach by plants. Riparian restoration of degraded river systems usually involves the planting of native pioneer species such as *S. elaeagnos* in order to restore modified floodplain communities. Plant capacity to reach underground water is determinant on survivorship and growth. Nursery practices are determinant on restoration exit, in this way, big container volume and specially depth, promote positive attributes for plants at root level used in plant restoration in wide environmental conditions. The size of aerial parts of plants can provide avoidance mechanism against environmental stresses, in spite of growth reduction.

## ACKNOWLEDGEMENT

This research was partially economically supported by Project CONSOLIDER MONTES (CSC2008-00040).

### REFERENCES

ACA (Agència Catalana de l'Aigua) (2009). Aigua i canvi climàtic. Diagnosi dels impactes previstos a Catalunya. (http://cercador.gencat.cat/cercador/AppJava/index.jsp?q=Aigua+i+canvi+climatic)

ACCUA (Adaptacions al Canvi Climàtic en l'Ús de l'Aigua) Informe final. (2011). http://www.creaf.uab.cat/accua/ACCUA_tecnica_internet.pdf)

Bradford KJ, Hsiao TC (1982). Physiological responses to moderate water stress. In: Encyclopedia of Plant Physiology, Vol. 12. (Lange, O.D., Nobel, P.S., Osmond, C.B., Ziegler, H., Eds.). Springer Verlag. Berlin. Heidelberg. New York. pp. 263-324.

Burdett AN (1990). Physiological processes in plantation establishment and the development of specifications for forest planning stock. Can. J. For. Res. 20:415-427.

Carlson WC (1986). Root System Considerations in the Quality of Loblolly Pine Seedlings. Southern. J. Appl..For.10(2):87-92.

Carpenter LT, Pezeshki SR, Shields FD (2008). Responses of nonstructural carbohydrates to shoot removal and soil moisture treatments in Salix nigra. Trees 22(5):737-748.

Chapin FS (1991). Integrated responses of plants to stress. Bioscience 41:29–36.

Chirino E, Vilagrosa A, Hernandez EI, Matos A, Vallejo VR (2008). Effects of a deep container on morpho-functional characteristics and root colonization in Quercus suber L. seedlings for reforestation in Mediterranean climate. For. Ecol. Manage. 256:779–785.

Cortina J, Peñuelas JL, Puértolas J, Savé R, Vilagrosa A (2006). Actual knowhow about seedlings quality for landscape restoration under arid and semiarid Mediterranean conditions. Organismo Autónomo Parques Nacionales, Ministerio de Medio Ambiente de España. p.191.

Dickmann D, Pregitzer KS (1992). The structure and dynamics ofwoody plant root systems. In Ecophysiology of Short Forest Crops. Eds. Mitchell CP, Ford-Robertson JB, Hinckley T, Sennerby-Forsse H. Elsevier Applied Science, London. pp. 95-123.

Directive 2000/60/EC of the European Parliament and of the Council of 23 October (2000). establishing a framework for Community action in the field of water policy. Official J. L 327 , 22/12/2000 P. 0001 – 0073

Dominguez-Lerena S, Herrero-Sierra N, Carrasco-Manzano I, Ocaña-Bueno L, Peñuelas-Rubira JL, Mexal JG (2006). Container characteristics influence Pinus pinea seedling development in the nursey and field. For. Ecol. Manage. 221:63-71.

Dreesen D, Harrington J, Subirge T, Stewart P, Fenchel G (2002). Riparian Restoration in the Southwest – Species Selection, Propagation, Planting Methods, and Case Studies. Forest Service, Rocky Mountain Research Station, Proceedings RMRS-P-24. p.370 .

Elfadl M, Luukkanen O (2003). Effect of pruning on *Prosopis juliflora:* considerations for tropical dryland agroforestry. J. Arid Environ. 53:441–455.

Folk RS, Grossnickle SC. (1997). Determining field performance potential with the use of limiting enviromental conditions. New For. 13:121-138.

Francis RA, Gurnell AM, Petts GE, Edwards PJ (2005). Survival and growth responses of Populus nigra, Salix elaeagnos and Alnus incana cuttings to varying levels of hydric stress. For. Ecol. Manag. 210:291–301.

Gasith A, Resh V (1999). Streams in Mediterranean climate regions: abiotic influences and biotic responses to predictable seasonal events. Ann. Rev. Ecol. Syst. 30:51-81.

Guillet C, Bergström R (2006). Compensatory growth of fast-growing willow (Salix) coppice in response to simulated large herbivore browsing. Oikos 113:33–42.

Hjaelten J (1999). Willow response to pruning: The effect on plant growth, survival and susceptibility to leaf gallers. Ecoscience 6(1):62-67.

Hjaelten J, Price PW (1996). The effect of pruning on willow growth and sawfly population densities. Oikos 77:549-555.

Houle G, Babeux P (1998). The effects of collection date, IBA, plant gender, nutrient availability, and rooting volume on adventitious root and lateral shoot formation by *Salix planifolia* stem cuttings from the Ungava Bay area (Quebec, Canada). Can. J. Bot. 76:1687–1692.

Horton JL, Clark JL (2001). Water table decline alters growth and survival of Salix gooddingii and Tamarix chinensis seedlings. For. Ecol. Manag. 140:239-247.

IPCC (2007). Climate Change 2007: Synthesis Report. Contribution of Working Groups I, II and III to the Fourth Assessment Report of the Intergovernmental Panel on Climate Change [Core Writing Team, Pachauri, R.K and Reisinger, A. (eds.)]. IPCC, Geneva, Switzerland, p.104.

Karrenberg S, Edwards PJ, Kollmann J (2002). The life history of Salicaceae living in the active zone of floodplains. Freshwater Biol. 47:733–748.

Kozlowski TT (1992). Carbohydrate sources and sinks in woody plants. Bot. Rev. 58:108-222.

Kuzovkina YA, Quigley MF (2005). Willows beyond wetlands: uses of *Salix* L. species for environmental projects. Water, Air Soil Pollution 162:183–204.

Lamhamedi M, Bernier P, Hébert C, Jobidon C (1998). Physiological and growth responses of three sizes of containerized Picea mariana seedlings outplanted with and without vegetation control. For. Ecol. Manag. 110:13-23.

Llebot JE (Ed.) (2010). Segon informe sobre el canvi climàtic a Catalunya. Generalitat de Catalunya i Institut d'Estudis Catalans, Barcelona (Spain).

Lloret F, Casanovas C, Peñuelas J (1999). Seedling survival of Mediterranean shrubland species relation to root:shoot ratio, seed size and water and nitrogen use. Funct. Ecol. 13:210-216.

McKay HM (1999). Root electrolyte leakage and root growth potential as indicators of spruce and larch establishment. Silva Fennica 32(3):241-252.

Medici C, Butturini A, Bernal S, Vázquez E, Sabater F, Vélez JI,Francés F (2008). Modelling the non-linear hydrological behaviour of a small Mediterranean forested catchment. Hydrol. Proces. 22(18):3814-3828.

Mexal JG, Landis TD (1990). Target seedling concepts: Heigth and diameter. In: Proceedings of combined Meeting of the Western Forest Nursery Associations. pp17-35. Roseburg, Oregon, August 13-17.

Mitrakos KA (1980). A theory for Mediterranean plant life. Acta Oecol. 1:245–252.

Noland TL, Mohammed GH, Scott M (1996). The dependance of root growth potential on light level, photosynthetic rate, and root starch content in jack pine seedlings. New For. 13:105-119.

Oliet J, Planelles R, Artero F, Martínez Montes E, Álvarez Linarejos L, Alejano R, López Arias M (2003). El potencial de crecimiento radical en planta de vivero de Pinus halepensis Mill. Influencia de la

fertilización. Invest. Agrar.: Sist. Recu. For. 12(1):51-61.
Ovaska J, Ruuska S, Rintamaki E. (1993). Combined effects of partial defoliation and nutrient availability on cloned Betula pendula saplings. II. Changes in net photosynthesis and related biochemical properties. J. Exp. Bot. 44:1395–1402.
Padilla F, Pugnaire F (2007). Rooting depth and soil moisture control Mediterranean woody seedling survival during drought. Funct. Ecol. 21:489-495.
Pastor JN, Burés S, Savé R, Marfà O, Pages JM (1999). Transplant adaptation in landscape ornamental shrubs in relation with substrate physical properties and container sizes. Acta. Hortic. 481:137-144.
Permán J, Voltas J, Gil-Pelegrin E (2006). Morphological and functional variability in the root systems of Quercus ilex L. subject to confinement: consequences for afforestation. Ann. For. Sci. 63:425-430.
Poorter H, Naves ML (2003). Plant growth and competition at elevated $CO_2$: on winners, losers and functional groups. New Phytol. 157:175–198.
Ritchie GA, Landis T (2010). Assessing Plant Quality, In: The Container Tree Nursery Manual, Chapter. 2:17-80. USDA Agric. Handbook p. 674.
Savé R, Alegre L, Pery M, Terradas J (1993). Ecophysiology of after-fire resprouts of Arbutus unedo L. Orsis 8:107-119.
Sennerby-Forsse L, Christersson L (1994). The role of energy forestry in alternative energy planning, waste recycling and agriculture in Sweden. World Resour. Rev. 6:395--405.
Simpson DG, Ritchie GA (1997). Does RGP predict field performance. A debate. New For. 13(1-3):253-277.
Sheffield J, Wood EF (2008). Projected changes in drought occurrence under future global warming from multi – model, multiscenario, IPCC AR4 simulations. Clim. Dyn. 31:79-105.
Sutton R (1990). Root growth capacity in coniferous forest trees. Hotscience 25:259-266.
Timmer V, Miller B (1991). Effects of contrasting fertilization and moisture regimes on biomass, nutrients, and water relations of container grown red pine seedlings. New For. 5:335-348.
Tsakaldimi M, Zagas T, Tsitsoni T, Ganatsas P (2005). Root morphology, stem growth and field performance of seedlings of two Mediterranean evergreen oak species raised in different container types. Plant Soil 278:85–93.
South DB, Blake JI (1994). Top-pruning increases survival of pine seedlings. Alabama Agricultural Experiment Station. Highlight. Agric. Res. 41(2):9.
Valentini R, Matteucci G, Dolman AJ (2000). Respiration as the main determinant of carbon balance in European forest. Nature 404: 861-865.
Villar-Salvador P, Planelles R, Enríquez E, Peñuelas J (2004) Nursery cultivation regimes, plant functional attributes, and field performance relationships in the Mediterranean oak Quercus ilex L. For. Ecol. Manage. 196:257-266.
Von Fircks Y, Sennerby-Forsse L (1998). Seasonal fluctuations of starch in root and stem tissues of coppiced Salix viminalis plants grown under two nitrogen regimes. Tree Physiol. 18:243-249.

# Effect of bulb density, nitrogen application time and deheading on growth, yield and relative economics of daffodil cv. Tunis (*Narcissus* sp.)

**I. M. Khan, F. U. Khan, M. Salmani, M. H. Khan, M. A. Mir and Amir Hassan**

Division of Floriculture, Medicinal and Aromatic Plants, Sher-e- Kashmir University of Agricultural Science and Technology of Kashmir, Shalimar, Srinagar, India.

**A two years field experiment was carried out to study the effect of bulb density, nitrogen application time and deheading on the growth and bulb yield of daffodil cv. Tunis was conducted during winter seasons of 2009 to 2010 and 2010 to 2011 on silty clay loam soil, low in available Nitrogen, medium in available phosphorus and potassium with neutral pH. The study revealed that the bulb density at 15.0 t ha$^{-1}$ proved significantly superior recording higher values of growth characters and resulted in 30.30 and 57.47% higher bulb yield than 12.50 and 10.00 t ha$^{-1}$, respectively. Application of nitrogen in two splits at 1st week of March and April significantly improved the growth characters and registered 9.88 and 21.56% higher bulb yield than single split of nitrogen at 1st week of March, respectively. Deheading at tight bud stage did not affect the top growth characters *viz.*, number and dry weight of leaves and leaf area however, significantly improved bulb dry weight and increased the bulb yield by 5 to 15 % over no deheading. Bulbs planted at 15.0 t ha$^{-1}$ and supplied with Nitrogen in 2 splits at 1st week of March and April followed by deheading at tight bud stage recorded the highest benefit cost of Rs 0.51(per rupee) and net returns of Rs 6, 30,508.0 ha$^{-1}$.**

**Key words:** B.C ratio, bulb weight, nitrogen splits, deheading, tight bud stage.

## INTRODUCTION

Daffodil (*Narcissus* sp.), the most important early season on blooming flower, is known for its unmatched beauty and fragrance. Like other bulbous crops, the flower and bulb production in daffodil is influenced greatly by agronomic and environmental factors. Unfortunately the work done on these aspects is lacking in the country. As such, it is essential to standardize the cultivation of daffodils so that they can be grown in the farmers field by adopting proper management practices because cultural and management practices play a vital role in the qualitative traits of flowering plants. One of the important management practices is the maintenance of proper bulb density for obtaining maximum quality bulbs per unit area. Planting of quality bulbs would improve the quality of flowers. No doubt, nitrogen is the limiting nutrient element for growth and development of bulbs but there is need to identity the time for its application so that it is need utilized by the crop to maximum level. Besides, it is a known fact that the flowers and bulbs are two major sinks in bulb crops and the removal of one sink may influence the other sink (John and Khan, 2003). Deheading at tight bud stage would lead to maximum accumulation of photosynthates in bulbs thereby would improve the bulb yield (Xia et al., 2005).

In view of this, the present study was initiated during winter seasons of 2009 to 2010 and 2010 to 2011.

## MATERIALS AND METHODS

A field experiment to study the effect of bulb density, nitrogen application time and deheading on growth, bulb production in daffodil cv. Tunis was conducted during winter seasons of 2009 to 2010 and 2010 to 2011 at the Research Farm, Division of Floriculture, Medicinal and Aromatic plants, Sher-e- Kashmir University of Agriculture Sciences and Technology of Kashmir, Shalimar on a silty clay loam soil low in available nitrogen (269.0 kg $ha^{-1}$), medium in available phosphorus (14.6 kg $ha^{1}$) and potassium (160.0 kg $ha^{-1}$) with neutral pH (7.5). The treatments consisting of 3 factors *viz.*, 3 bulb densities (10.00, 12.50 and 15.00 kg $ha^{-1}$), 3 nitrogen application time (2 splits in 2nd week of Nov and 1st week of March, 2 splits in 1st week of March and April and single split in 1st week of March), 2 deheadings (no deheading and deheading at tight bud stage) were laid out in a randomized block design replicated thrice. The fertilizers *viz.*, phosphorus at the rate of 150 kg $ha^{-1}$ and potassium at the rate of 75 kg $ha^{-1}$ through single super phosphate and murate of potash, respectively were applied as basal before planting of bulbs. Nitrogen at the rate of 150 kg $ha^{-1}$ through urea was applied as per treatments. Other cultural and management operations were carried as per recommended package of practices. Uniform sized bulbs at the rate as per treatment were planted in 2nd week of November, 2009. Deheading operation as per treatment was carried at tight bud stage as per treatment. Observations on various parameters *viz.*, number and dry weight of leaves $palnt^{-1}$, leaf area per plant at 30, 60 and 90 DAM during 2009 to 2010 and 2010 to 2011, bulb dry weight at 30, 60, and 90 DAM during 2010 to 2011 and bulb yield were recorded from 5 randomly selected plants from each plot. Bulbs were excavated on 25.07.2011. Data was analysed by the method given by Panse and Sukhatme (1985). Benefit cost ratio and net returns were determined on the basis of cost of cultivation.

## RESULTS AND DISCUSSION

The growth characters viz., number and dry weight of leaves and leaf area recorded significant improvement with bulb density of 15.00t $ha^{-1}$ over 10.00 t $ha^{-1}$ during 2010 to 2011 at 30, 60 and 90 days after March 1st (DAM), however, all these parameters remained non significant during 2009 to 10 (Table 1). Number of bulbs and bulb yield was significantly highest at bulb density of 15.0 t $ha^{-1}$ (Table 2).Whereas bulb dry weight at 30, 60 and 90 DAM was significantly highest bulb density of 10.0 t $ha^{-1}$. Bulb yield at 15.0 t $ha^{-1}$ 30.30 and 57.47% more than 12.5 and 10.0 t $ha^{-1}$, respectively, Leopold and Kriedmann (1980) indicated that within limits, plants inherent capacity for growth is direct consequence of how successfully it exploits the local environment. Such factors as light, temperature, moisture and mineral nutrition assumes major importance. The higher density affords more leaves, greater leaf area and more bulb dry weight per unit area and therefore greater potential for overall yield. When higher density does not produce competition for above factors, the photosynthetic rate of individual leaves would tend to increase Kumar et al. (2003) and Nazki et al. (2005) also support the results. Singh and Singh (2005) found that medium bulb spacing of 20 × 25 cm produced maximum growth and number of bulbs in tuberose as compared to bulb spacings of 20 × 20 cm and 30 × 30 cm. Similarly, Patel et al. (2006) reported that for higher yield of bulbs, tuberose could be planted at closer spacing instead of wider spacing.

Application of nitrogen in splits at 1st week of March and April resulted in significant improvement in growth characters that is, number and dry weight at 30, 60 and 90 DAM and number of bulbs $ha^{-1}$ during 2010 to 2011 in comparison to single split of nitrogen at 1st week of March and 2 splits of nitrogen at 2nd week of November and 1st week of March and the treatments recorded 9.88 and 21.56% higher bulb yield than single nitrogen split in 1st week of March and 2 splits of nitrogen in 2nd week of November and 1st week of March, respectively. The treatment ensured regular supply of nitrogen is small quantities at critical growth stages. Thus, proper utilization of applied nitrogen might be responsible for better growth. Goss (1973) also reported similar findings. Besides, Nitrogen is an essential constituent of proteins, chlorophyll and amino acids. Regular and timely supply of nitrogen to bulbs might have increased the weight of bulbs and hence increase in total bulb yield. Singh et al. (2005) and Hermandez et al. (2008) found that application of nitrogen in 2 splits *viz.,* half at planting and half at spike initiation increased the number of leaves $plant^{-1}$, leaf length, plant height, reduced the days to sprouting and lesser number of stems of inferior quality of tuberose bulbs. These results also support the present findings.

Deheading did not affect the growth characters *viz.*, number of leaves, leaf dry weight and leaf area. However, deheading at tight bud stage significantly improved bulb dry weight at 30, 60 and 90 DAM and number of bulbs than no deheading. The treatment recorded 5.15% increase in total bulb yield over no deheading. Increase in bulb dry with and total bulb yield under deheading at tight bud stage is obviously a result of more resource allocation to the under ground sinks which could other wise have been used by the developing flowers (Wang and Breen, 1984). Any relocation of photosynthates as a consequence of elimination of floral sink mostly takes place to the main bulb. The results agree with those of John and Khan (2003).

The study revealed that the bulbs planted at the rate of 15.0 t $ha^{-1}$ and supplied with 150 kg $ha^{-1}$ nitrogen in 2 splits at 1st week of March and April followed by deheading at tight bud stage recorded the highest benefit cost ratio of Rs 0.51 and net returns of 6,30,508.16 $ha^{-1}$ (Table 3). Thus the study leads to the conclusion that for producing economic highest bulb yield in daffodil cv. Tunis bulb at the rate of 15.00 t $ha^{-1}$ be planted and supplied with 150 kg N $ha^{-1}$ in 2 splits at 1st week of March and April followed by deheading at tight bud stage.

**Table 1.** Growth characters in daffodil cv. Tunis as affected by bulb density, nitrogen application time and deheading.

| Treatment | | Days after March 1st (DAM) | | | | | | | | | | | |
|---|---|---|---|---|---|---|---|---|---|---|---|---|---|
| | | Number of leaves plant$^{-1}$ (g) | | | | | | Leaf area palnt$^{-1}$ (m$^2$) | | | | | |
| | | 30 | | 60 | | 90 | | 30 | | 60 | | 90 | |
| | | 2009-2010 | 2010-2011 | 2009-2010 | 2010-2011 | 2009-2010 | 2010-2011 | 2009-2010 | 2010-2011 | 2009-2010 | 2010-2011 | 2009-2010 | 2010-2011 |
| **Planted bulb weight (t ha$^{-1}$)** | | | | | | | | | | | | | |
| $W_1$ | 10.00 | 3.20 | 3.77 | 4.18 | 4.47 | 3.88 | 4.17 | 24.08 | 28.30 | 34.90 | 38.00 | 28.36 | 31.26 |
| $W_2$ | 12.50 | 3.28 | 3.92 | 4.26 | 4.64 | 3.96 | 4.34 | 26.78 | 31.60 | 35.05 | 34.26 | 28.55 | 33.05 |
| $W_3$ | 15.00 | 3.34 | 4.13 | 4.31 | 4.87 | 4.01 | 4.59 | 27.98 | 32.54 | 27.98 | 32.54 | 29.16 | 33.66 |
| CD (P= 0.05) | | NS | 0.25 | NS | 0.39 | NS | 0.37 | NS | 0.85 | NS | 1.58 | NS | 1.74 |
| **Nitrogen application time (150 kg ha$^{-1}$)** | | | | | | | | | | | | | |
| $N_1$ | Two splits (2nd week of Nov. and 1st week of March) | 3.24 | 3.88 | 4.24 | 4.52 | 3.98 | 4.22 | 25.81 | 30.89 | 35.00 | 39.78 | 28.50 | 31.80 |
| $N_2$ | Two splits (1st week of March and 1st week of April) | 3.30 | 4.10 | 4.31 | 4.98 | 4.02 | 4.68 | 26.98 | 31.70 | 35.40 | 42.74 | 29.21 | 34.61 |
| $N_3$ | Single split (1st week of March) | 3.29 | 3.84 | 4.20 | 4.50 | 3.85 | 4.20 | 35.85 | 29.85 | 35.20 | 37.17 | 28.36 | 31.66 |
| CD (P= 0.05) | | NS | 0.25 | NS | 0.39 | NS | 0.37 | NS | 1.85 | NS | 1.58 | NS | 1.74 |
| $D_0$ | No deheading | 3.26 | 3.91 | 4.24 | 4.71 | 3.98 | 4.39 | 26.36 | 31.22 | 25.30 | 40.12 | 29.05 | 32.94 |
| $D_1$ | Deheading at tight bud stage | 3.28 | 3.97 | 4.26 | 4.62 | 3.92 | 4.34 | 26.20 | 80.40 | 35.10 | 39.74 | 28.33 | 32.44 |
| CD (P= 0.05) | | NS | NS | NS | NS | NS | NS | NS | NS | NS | NS | NS | NS |

| Treatment | | Leaf dry weight plant$^{-1}$ (g) | | | | | |
|---|---|---|---|---|---|---|---|
| | | 30 | | 60 | | 90 | |
| | | 2009-2010 | 2010-2011 | 2009-2010 | 2010-2011 | 2009-2010 | 2010-2011 |
| **Planted bulb weight (t ha$^{-1}$)** | | | | | | | |
| $W_1$ | 10.00 | 0.79 | 0.93 | 0.90 | 1.04 | 0.99 | 1.14 |
| $W_2$ | 12.50 | 0.82 | 1.00 | 0.92 | 1.11 | 1.00 | 1.18 |
| $W_3$ | 15.00 | 0.82 | 1.04 | 0.94 | 1.15 | 1.04 | 1.25 |
| **CD (P= 0.05)** | | **NS** | **0.05** | **NS** | **0.06** | **NS** | **0.08** |
| **Nitrogen application time (150 kg ha$^{-1}$)** | | | | | | | |
| $N_1$ | Two splits (2nd week of Nov. and 1st week of March) | 0.81 | 0.97 | 0.91 | 1.08 | 1.00 | 1.16 |
| $N_2$ | Two splits (1st week of March and 1st week of April) | 0.82 | 1.02 | 0.95 | 1.17 | 1.05 | 0.28 |
| $N_3$ | Single split (1st week of March) | 0.80 | 0.98 | 0.90 | 1.05 | 0.98 | 1.30 |
| CD (P= 0.05) | | NS | 0.05 | NS | 0.06 | NS | 0.08 |
| **Deheading** | | | | | | | |
| $D_0$ | No deheading | 0.81 | 1.00 | 0.93 | 1.11 | 1.03 | 1.21 |

**Table 1.** Contd.

| | | | | | | |
|---|---|---|---|---|---|---|
| $D_1$ Deheading at tight bud stage | 0.81 | 0.98 | 0.91 | 1.09 | 0.99 | 1.17 |
| CD (P= 0.05) | NS | NS | NS | NS | NS | NS |

**Table 2.** Bulb dry weight (after March 1st) and total bulb yield daffodil cv. Tunis as affected by bulb density, nitrogen application time and deheading.

| Treatment | Bulb dry weight plant$^{-1}$ (g) | | | Bulb weight plant$^{-1}$ (g) | Number of bulbs ha$^{-1}$ (lakhs) | Total bulb yield (q/ha$^{-1}$) |
|---|---|---|---|---|---|---|
| | 30 | 60 | 90 | | | |
| **Planted bulb weight (t ha$^{-1}$)** | | | | | | |
| $W_1$ 10.00 | 10.85 | 13.261 | 15.94 | 39.02 | 6.101 | 14.86 |
| $W_2$ 12.50 | 10.03 | 12.74 | 15.06 | 38.85 | 7.141 | 17.16 |
| $W_3$ 15.00 | 9.70 | 11.50 | 13.37 | 37.29 | 9.864 | 22.36 |
| CD (P= 0.05) | 0.29 | 0.38 | 0.36 | 0.97 | 0.72 | 0.39 |
| **Nitrogen application time (150 kg ha$^{-1}$)** | | | | | | |
| $N_1$ Two splits (2nd week of Nov. and 1st week of March) | 10.19 | 12.07 | 14.15 | 37.02 | 7.262 | 16.37 |
| $N_2$ Two splits (1st week of March and 1st week of April) | 10.65 | 13.06 | 15.70 | 39.77 | 8.130 | 19.90 |
| $N_3$ Single split (1st week of March) | 10.34 | 12.33 | 14.52 | 38.37 | 7.711 | 18.11 |
| **CD (P= 0.05)** | **0.29** | **0.38** | **0.36** | **0.97** | **0.72** | **0.39** |
| **Deheading** | | | | | | |
| $D_0$ No deheading | 10.18 | 12.20 | 14.20 | 37.74 | 7.655 | 17.67 |
| $D_1$ Deheading at tight bud stage | 10.60 | 12.78 | 15.38 | 39.02 | 7.747 | 18.58 |
| CD (P= 0.05) | 0.23 | 8.31 | 0.29 | 0.79 | NS | 0.32 |

**Table 3.** Relative economics (hectare Basis) of daffodil cv. Tunis as affected by bulb density, nitrogen application time and deheading.

| Treatment combination | Gross returns (ha$^{-1}$) | Cost of cultivation (ha$^{-1}$) | Net returns (ha$^{-1}$) | Benefit cost ratio (returns per rupee invested) |
|---|---|---|---|---|
| $W_1 N_1 D_0$ | 10,02,240.00 | 8,29,061.84 | 1,73,178.16 | 0.21 |
| $W_1 N_1 D_1$ | 10,27,800.00 | 8,29,061.84 | 1,98,738.16 | 0.24 |
| $W_1 N_2 D_0$ | 11,22,660.00 | 8,29,061.84 | 2,93,598.16 | 0.35 |
| $W_1 N_2 D_1$ | 11,39,220.00 | 8,29,061.84 | 3,10,158.16 | 0.37 |
| $W_1 N_3 D_0$ | 10,92,060.00 | 8,29,061.84 | 2,62,998.16 | 0.32 |
| $W_1 N_3 D_1$ | 11,08,620.00 | 8,29,061.84 | 2,79,558.16 | 0.34 |
| $W_2 N_1 D_0$ | 11,96,460.00 | 10,29,611 .84 | 1,66,848.16 | 0.16 |
| $W_2 N_1 D_1$ | 12,13,020.00 | 10,29, 611.84 | 1,83,40.08 | 0.18 |

**Table 3.** Contd.

| | | | | |
|---|---|---|---|---|
| $W_2 N_2 D_0$ | 13,52,700.00 | 10,29, 611.84 | 3,23,088.16 | 0.31 |
| $W_2 N_2 D_1$ | 13,69,280.00 | 10,29, 611.84 | 3,39,668.16 | 0.33 |
| $W_2 N_3 D_0$ | 12,77,280.00 | 10,29, 611.84 | 2,47,668.16 | 0.24 |
| $W_2 N_3 D_1$ | 12,93,840.00 | 10,29, 611.84 | 2,64,228.16 | 0.25 |
| $W_3 N_1 D_0$ | 16,88,220.00 | 12,30,151.84 | 4,58,068.16 | 0.37 |
| $W_3 N_1 D_1$ | 17,04,780.00 | 12,30 ,151.84 | 4,74,628.16 | 0.38 |
| $W_3 N_2 D_0$ | 18,44,460.00 | 12,30 ,15 1.84 | 6,14,308.16 | 0.50 |
| $W_3 N_2 D_1$ | 18,60,660.00 | 12,30 ,151.84 | 6,30,508.16 | 0.51 |
| $W_3 N_3 D_0$ | 17,69,040.00 | 12,30 ,15 1.84 | 5,38,888.16 | 0.44 |
| $W_3 N_3 D_1$ | 17,85,600.00 | 12,30 ,15 1.84 | 5,55,448.16 | 0.45 |

Where: $W_{1\,=\,10.00}$ t $ha^{-1}$, $W_{2\,=12.50}$ t $ha^{-1}$, $W_{3\,=15.00}$ t $ha^{-1}$, $N_1$ =2 splits in $2^{nd}$ week of November and 1st week of March, $N_2$ = 2 splits in 1st week of March and April, $N_3$ = single split in 1st week of March, $D_0$ = No deheading, $D_1$ = Deheading at tight bud stage.

## REFERENCES


Goss GA (1973). Physiology of plants and their cells Pergamon Press, INC, New York.

Hermandez DMI, Marrero GV, Gonzalez HM, Salgado PJM, Geda VA (2008). Nitrogen levels and their fractioning in gladiolus cultivation for Fessalitic red soils. Pesquisa Agropecuaria Erasileira. 43(1):21-27.

John AQ, Khan FU (2003). Effect of flower and leaf removal on bulb production in tulip cv. Cassini. SKUAST J. Res. 5:190-193.

Kumar R, Chaturvedi OP, Misra RL (2003). Effect of N and P on growth and flowering of gladiolus. J. Ornamental Hortic. 6(2):100-103.

Leopold AC, Kriedemann PE (1980). Plant growth and development. 2nd Ed. Tata Mc Graw-Hill Publications, New Delhi.

Nazki IT, Khan FU, Qadri ZA, Paul TM, Sheikh MQ (2005). Effect of crop duration, planting density and bulb grade on foliar, floral and bulb growth in *Naciassus tazetta* linn. cv. Paper White Grandiflorus. J. Ornamental Hortic. 8(3):222-224.

Panse VG, Sukhatme PV (1985). Statistical Method for Agricultural Workers. (2nd Eds.). Indian council of Agricultural Research, New Delhi. P. 381.

Patel MM, Parmar PB, Parmar BR (2006). Effect of nitrogen, phosphorus and spacing on growth and flowering in tuberose (*Polyanthus tuberosa* Linn.) cultivar Single. J. Ornamental Hortic. 9(4):286-289.

Singh SK, Singh RK (2005). Response of different levels of nitrogen and plant density on the performance of tuberose (*Polyanthus tuberosa* L.) cv. Double Plant Arch. 4(2):515-517.

Singh SRP, Dhiraj K, Singh VK, Dwivedi R (2005). Effect of NPK fertilizers on growth and flowering of tubrose cv. Single. Haryana J. Hortic. Sci. 34(1/2):84.

Wang YT, Breen PJ (1984). respiration and weight changes of Easter lily during development. Hortic. Sci. 19:702-703.

Xia YP, Zheng HJ, Huang CH (2005). Studies on the bulb development and its physiological mechanism in lilium oriental hybrids. Acta Hortic. 29(3):273-281.

# Studies on interspecific hybridization in *Cyamopsis* species

**Anju Ahlawat**[1], **S. K. Pahuja**[2] **and H. R. Dhingra**[1]

[1]Department of Botany and Plant Physiology, CCS Haryana Agricultural University, Hisar-125004, India.
[2]Forage Section, Department of Genetics and Plant Breeding, CCS Haryana Agricultural University, Hisar-125004, India.

**Reproductive characters of three species of *Cyamopsis* were studied to find out barriers to interspecific crosses between *Cyamopsis tetragonoloba* × *Cyamopsis serrata* and *C. tetragonoloba* × *C. senegalensis* which may serve as a stepping stone for development of extra early varieties of guar. Pollen grains of *C. tetragonoloba* and *C. senegalensis* showed more than 95% of viability while those of *C. serrata* had 87% viability. Nutritive requirement for *in vitro* germination of pollen revealed that pollen of *C. tetragonoloba* required 25% sucrose + 100 ppm boric acid + 300 ppm calcium nitrate while *C. senegalensis* pollen needed 35% sucrose with same basal medium. On the other hand, *C. serrata* pollen required 35% maltose + 6% PEG 6000 along with above dose of boric acid and calcium nitrate. Moreover, pollen germination in *C. serrata* was initiated after 30 h of incubation and its pollen tubes were slow growing attaining 174.7 µm length in 48 h. The length of style of *C. tetragonoloba* and *C. serrata* was nearly identical (2.6 mm) while *C. senegalensis* possessed longest style (3.8 mm). Interspecific hybridization between *C. tetragonoloba* x *C. serrata* was successful through the use of stub smeared with pollen germination medium (PGM) and as a consequence 10.43% of pod setting was observed. Colour and shape of hybrid seeds was similar to the female parent (*C. tetragonoloba*), hybrid plants showed early flowering just like male parent (*C. serrata*) whereas the plant height was intermediate between the two parents.**

**Key words:** Interspecific hybrids, pollen, *in vitro* germination, stub pollination, *in vivo* tube growth.

## INTRODUCTION

India and Pakistan are the main producers of cluster bean, accounting for 80% production of the world's total, while Thar, Punjab Dry Areas in Pakistan and Rajasthan occupies the largest area (82.1%) under guar cultivation in India. In addition to its cultivation in India and Pakistan, the crop is also grown as a cash crop in other parts of the world (Pathak et al., 2010). In India, 3.34 million hectares of the farmable land was under guar cultivation during the year 2006/07 (Ministry of Agri. and Co-op GOI, 2010). It is cultivated in arid zones of Rajasthan, some parts of Gujarat,Haryana and Madhya Pradesh. The productivity of guar ranges from 474 kg/ha in Rajasthan to 1200 kg/ha in Haryana. The most important growing area centres on Jodhpur in Rajasthan, India where demand for guar for fracking produced an agricultural boom as of 2012 (Gardiner 2012).

Guar grows well under a wide range of soil conditions and is tolerant of low fertility, soil salinity and alkalinity. It performs best on fertile, medium-textured and sandy loam alluvial soils but does not tolerate heavy black soils

(Wong and Parmar 1997).

In the recent past, guar cultivation has become an attractive option with the farmers due to availability of high yielding varieties with high gum (30 to 35% of whole seed) content (galactomannans) in its endosperm which has great value as an enhancer of viscosity in food industry, like stiffner in soft ice-cream, a stabilizer for cheese, instant pudding and whipped cream substitutes and as a metal binder. It is widely used from paper and cosmetic to mining and explosive industry (Whistler and Hymowitz, 1979). Al-Hafedh and Siddiqui (1998) reported that the C. *tetragonoloba* L. beans had 32.81% crude proteins, 3.18% crude fats, 4.19% ash and 10.87% crude fibers.

Guar meal contains about 12% gum residue (7% in the germ fraction and 13% in the hulls) (Lee et al., 2005), which increases viscosity in the intestine, resulting in lower digestibilities and growth performance (Lee et al., 2009). Its uses in tissue culture media as a gelling agent has also been reported (Jain et al., 2005). *Cyamopsis tetragonoloba* L. is a well-known traditional plant used in folklore medicine.

A critical requirement for crop improvement in general, is the introduction of new genetic material in the cultivated lines of interest, whether through conventional or non-conventional breeding or *in-vitro* techniques. *C. tetragonoloba,* an erect herb with indeterminate growth and broad trifoliate leaves, matures in 80 to 120 days. However, one of its wild relatives, that is, *C. serrata* is an extra early maturing (40-50 days),slow growing with narrow trifoliate leaves while the other wild species, that is, *C. senegalensis* is also slow growing with narrow pentafoliate leaves and matures in 120 to 130 days (Menon, 1973). Both these wild relatives possess some desirable attributes like drought resistance, photo-and thermo-insensitivity and disease resistance. Interspecific hybridization among the *C. tetragonoloba* and its wild relatives is anticipated to produce hybrid with trait of early maturity, disease resistance and photo-and thermo-insensitivity.

According to Harlan and De Wet (1971) gene pool concept, *C. tetragonoloba* is included in primary gene pool (GP-1) while *C. serrata* and *C. senegalensis* are included in secondary gene pool (GP-2). According to them species included in GP-2 can be crossed with GP-1 with some fertility in F1's and thus gene transfer is feasible.

Unfortunately, conventional plant breeding technique has so far failed to yield desired results (Mathiyazhagan, 2009). Such a failure may be due to presence of pre- and/or post-fertilization barriers. To combat such barriers, it is essential to have detailed knowledge of reproductive biology of all the species of *Cyamopsis* in question. It was therefore, contemplated to follow a systematic approach to identify pre- as well as post- fertilization barriers in interspecific crossing of *Cyamopsis*. Supplementing conventional plant breeding with unconventional less popular methods along with plant biotechnological techniques is anticipated to go headway in resolving the issue. Present investigation was thus undertaken to study some relevant reproductive characters in three different species of *Cyamopsis* and work out cross-ability among these by conventional and unconventional less popular methods.

## MATERIALS AND METHODS

Plants of three species of *Cyamopsis* viz. *C. tetragonoloba* cv. HG563, *C. serrata* and *C. senegalensis* were raised in cemented pots (from authenticated seeds collected from forage department of CCS HAU, Hisar) in the screen house of the Department of Botany and Plant Physiology. Pots were filled with mixture of soil and farm yard manure. Before sowing, the seeds of uniform size were soaked in liquid broth of Rhizobium strain 1305 for 10 min. Five seeds were sown in each pot at uniform depth and distance. After 25 days of sowing thinning was done to leave three plants of uniform size in each pot. Irrigation with canal water was given as and when required. At flowering, the following reproductive characters of each species were recorded.

### Number of pollen/flower

Flower buds from each species were collected a day before anther dehiscence and employed for quantification of pollen grains produced per flower. For this, twenty anthers were suspended in 2 ml of 50% diluted glycerine (50% V/v glycerine with water) containing a few drops of safranine. Anthers were crushed with the help of glass rod and the suspension was passed through a brass sieve with a mesh of 48 sq/cm$^2$ (Kapoor and Nair, 1974). Number of pollen grains per flower drop was counted by haemocytometer.

### Pollen viability and *In vitro* pollen germination

Viability of pollen grains was assessed by 2, 3, 5 triphenyl tetrazolium chloride (TTC) test (Hauser and Morrison, 1964). Flower buds were collected from 3 randomly selected plants in the early morning (6.30 a.m.) on the day of anthesis and pollen of these floral buds was mixed thoroughly on glazed paper and used immediately for viability test and *in vitro* germination on the semi solid medium contained in petri dishes. Preliminary studies revealed that sugar type and its concentration and other adjuvants required for pollen germination varied with species. After preliminary trials, germination medium consisting of 25% sucrose (for *C. tetragonoloba*), 35% sucrose (for *C.senegalensis*), 35% maltose +6% PEG 6000 (for *C. serrata*) along with 100 ppm boric acid, 300 ppm calcium nitrate and 0.8% agar were used for *in vitro* germination and tube growth. After pollen inoculation, Petri plates were incubated at 30±2ºC for 4 h in dark in a BOD incubator with three replicates per treatment. However, inoculated Petri plates of *C. senegalensis* and *C.serrata* were incubated for 30 and 48 h respectively. After pollen germination, the pollen activity was terminated by flooding the surfaces of the media with killing and fixing solution of following composition (Sass, 1951): Formaldehyde = 5 ml; Glacial acetic acid = 3 ml; Water = 72 ml; Glycerine = 20 ml.

Pollen producing a tube length of a size greater than double of its diameter was designated as germinated. Twenty readings for pollen germination and thirty for tube length from different microscopic fields of each petri plate were made from area with uniform distribution of pollen and fairly good population.

**Pistil and yield related characters**

The above collected flower buds were used to record shape of the stigma and length of the ovary and style by micrometry. Pistils were cut open under a stereoscopic microscope and number of ovules per pistil from at least twenty pistils was recorded. At maturity, thirty pods from each species were collected randomly and used to measure length and breadth of pods, number of seeds per pod and test weight of 100 healthy; uniform sized seeds from each species was recorded in three replicates per species were used.

***In vivo* pollen tube growth**

Self pollinated pistils from flowers of three species of *Cyamopsis* were collected at 24, 48 and 72 h of anther dehiscence and fixed in acetic alcohol (Acetic acid : Ethanol; 1 : 3) for 4 h and processed for aniline blue test (Dumas and Knox, 1983). The observations for germination of pollen grains on stigmatic surface and extent of tube growth in stylar tissue and penetration of the ovule by tube were made under florescent microscope. Fifteen random pistils for each species were used for these studies.

**Biochemical composition of stigma + style**

***Total soluble carbohydrates***

The total soluble carbohydrate content (mg $g^{-1}$FW) was estimated by the method of Yemm and Willis (1954).

***Extraction***

Extraction of soluble carbohydrates was done according to Barnett and Naylers (1966) procedure. 50 mg of fresh material of stigma + style was finely ground in 80% alcohol by using pestle and mortar. Total soluble carbohydrates were extracted in 2 ml of 80% ethanol (v/v) on a water bath at 50±1°C for 15 min. It was then cooled and centrifuged at 5000 x g for 5 min. The supernatent (extract) was kept aside and the pellet re-extracted twice with 80% ethanol. Total volume of extract was made to 5 ml with 80% ethanol. This extract was used for the analysis of total soluble carbohydrates while the pellet was used for extraction of the total soluble proteins.

***Reagents***

Anthrone reagent: 0.4% anthrone in concentrated sulphuric acid.

**Procedure**

Aliquot (0.1 ml) of ethanol extract was evaporated to dryness in a test tube. After cooling, the residue was dissolved in 1 ml of distilled water and to it 4 ml of anthrone reagent was added. The mixture was then boiled in a water bath for 10 min. After cooling, absorbance was recorded at the wavelength of 620 nm against a reagent blank with the help of UV-Vis spectrophotometer. Standard curve was prepared using graded concentration of D-glucose (20-100 μg/ml).

**Estimation of soluble proteins**

***Sample preparation***

Pellet left after soluble carbohydrate extraction was extracted in 1.25 ml chilled Tris buffer (0.1 M, pH 8.0) containing 0.1% polyvinyl pyrrolidone (PVP). It was centrifuged at 10000 rpm for 15 min. The supernatant containing the proteins was taken in a test tube and pellet was discarded and processed for the quantification of proteins by the method of Bradford (1976).

***Reagents***

Commassie Brilliant Blue G-250 reagent was used. 100 mg of CBBG-250 reagent was dissolved in 50 ml of 95% ethanol. To this solution 100 ml 85% (w/v) phosphoric acid was added and final volume was made to 200 ml with double distilled water. The solution was filtered through Whatman No. 1 filter paper, and final volume was made to 1 L and stored at 4°C in amber colour bottle.

***Procedure***

To 100 μl of the aliquot taken in test tube, 5 ml of the CBBG-250 reagent was added and mixed thoroughly either by inversion or vortexing. The optimal density (O.D.) was measured at 595 nm after 15 min and before 1 h against reagent blank. Standard curve was prepared using graded concentration of bovine serum albumin (20-100 μg/ml).

**Interspecific hybridization**

Since the three species of *Cyamopsis* employed in the present study differed in their flowering schedule; these were grown in a staggered manner to synchronize their flowering. The flowers of the female designate parent were emasculated in the evening (between 04:00 p.m-06:00 p.m.) prior to their anther dehiscence. Generally only two floral buds were emasculated on a raceme to permit their proper development. The crossing of *C. tetragonoloba* was carried out on the following morning between 07:00a.m. and 08:30 a.m. with pollen grains of *C. serrata* or *C. senegalensis.* Pollinated flowers were harvested after 24, 48 and 72 h fixed in acetic alcohol and processed for *in vivo* germination and pollen tube growth by aniline blue method as described earlier. The allogamous pistils left *in situ* were allowed to set pods. Percent pod set and number of seeds/pod were recorded. In addition to the conventional breeding method, non-conventional methods like stub pollination with or without smearing with pollen germination medium (PGM), *in vivo* placental pollination and placental pollination followed by *in vitro* pistil culture on MS medium supplemented with naphthalene acetic acid (NAA), indole acetic acid (IAA) and benzylaminopurine (BAP) also were attempted.

**Stub pollination**

The stigma of emasculated pistils of *C. tetragonoloba* was excised and was smeared with cool and molten pollen germination medium (PGM) with the help of camel hair brush and then pollen of *C. serrata* and *C. senegalensis* were applied separately on the stigma of emasculated flowers. Pollinated pistils were collected after 24, 48 and 72 h and processed in the same way as explained above for study *in vivo* pollen germination and tube growth. All self pollinated flowers below the selected buds were removed thereby ensuring that all the lowest buds on the raceme are always emasculated ones. To avoid damage to the raceme, upper buds were not removed until 3 day after emasculation; however, upper buds blooming during this period were removed immediately. The whole inflorescence was bagged to check any undesired pollination.

**Pollination through perforation in the basal part of style**

With the help of sterilized syringe needle, a hole was made in the

**Table 1.** Comparison of floral and male reproductive characters in three different species of *Cyamopsis*.

| S/N | Parameter | *C. tetragonoloba* | *C. serrata* | *C. senegalensis* |
|---|---|---|---|---|
| 1 | No. of pollen/flower | 4765.94 ± 241.23 | 4684.26 ± 506.55 | 4582.66 ± 453.49 |
| 2 | Pollen size (μm ) | 34.90 ± 0.23 | 38.47 ± 0.30 | 54.46 ± 0.52 |
| 3 | Pollen viability (%) | 95.77 | 87.01 | 98.12 |
| 4 | *In vitro* pollen germination (%) | 94.42 | 25.59 | 70.01 |
| 5 | Pollen tube length (μm) | 1204.9 | 174.7 | 949.2 |
| 6 | Pistil length (cm) | 0.67 ±0.02 | 0.53 ±0.01 | 0.77 ±0.02 |
| 7 | Shape of stigma | Capitate | Subapical crescent | Capitate |
| 8 | Number of ovules/pistil | 7-8 | 8-9 | 7-8 |
| 9 | Number of pods/ cluster | 6.75±0.23 | 12.00±0.10 | 9.50±0.52 |
| 10 | Length x breadth of pod (cm) | 5.95 x 0.54 | 4.51 x 0.36 | 5.1 0x 0.40 |
| 11 | Number of seeds/pod | 7.58±0.19 | 8.75±0.12 | 7.45±0.35 |
| 12 | Test weight of 100 seeds (g) | 2.78±0.21 | 1.81±0.19 | 1.84±0.30 |

upper part of ovary. Another hole opposite to this was also made to allow release of air. Pollen suspension of *C. serrata* and *C. senegalensis* in the liquid PGM was injected separately into the ovary through a hole. The hole was plugged with petroleum jelly and pistils were collected after 24 and 48 h and processed as explained earlier.

**Placental pollination**

A cut with the help of sharp needle was made along the ventral suture of the pistil to open up the pistil and expose the placenta. Pollen grains were dusted on the placenta and pistil was rolled back into their normal configuration under *in situ* conditions. Pistils were collected after 24 and 48 h of placental pollination and processed as explained earlier. In other set of experiment, pistils were excised from the plant, sterilized with 95% ethanol on the hood of laminar flow, cut open to expose placenta, pollinated with the desired pollen *viz C. serrata* or *C. senegalensis* and inoculated on MS basal medium supplemented with BAP, IAA, kinetin, NAA, adenine sulphate and casein hydrolysate (CH).

## RESULTS AND DISCUSSION

### Reproductive characters and *in vivo* pollen germination

Table 1 clearly evinces that flowers of all species of *Cyamopsis* studied produced nearly identical number (nearly 4500) of pollen grains per flower. Auto fertile species, like legumes in general, are known to produce few small pollen grains in small flowers (Diaz and Macnair, 1999) and such a reduced male allocation in the species studies is reflective of cost effective allocation to male function (Berlin, 1988) Leguminous pollen are know to contain carbohydrates as well as lipids (Baker and Baker, 1979). These storage compounds not only are source of energy to growing pollen tubes but may play a role in protecting pollen from desiccation e.g. sugars (Pacini 1996) or in synthesis of membranes eg. lipids. Among different species studied, pollen diameter of *C. senegalensis* was maximum (54.46 ±1.10 μm) whereas the value was least in *C. tetragonoloba* cv. HG 563 (34.90 ±0.63 μm).The diameter of *C. serrata* pollen grains was 38.47 ±0.80 μm. Pollen viability as assessed by 2,3,5-triphenyl tetrazolium chloride (TTC) test was more than 95% in *C. senegalensis* and *C. tetraganoloba* whereas the value was comparatively lower (87.01%) in *C. serrata*. Pollen grains of *C. tetragonoloba* showed maximum germination (94.42%) whereas pollen grains of *C. senegalensis* yielded 70.01% germination. On the other hand, 25.59% pollen grains of *C. serrata* germinated after 48 h of incubation. Among the tested species, pollen grains of C. tetragonoloba cv. HG 563 produced the longest tube (1204.9 μm) after 4 h of incubation while those of *C. serrata* produced the smallest pollen tube (174.7μm) after 48 h of incubation. Tube length of *C. senegalensis* pollen was 949.2 μm after 30 h of incubation (Figure 3). It seems that at anthesis, the embryosac probably is not fully developed and following pollination, it attains maturity in 1 to 2 day as has been reported for Pistacia vera (Shuraki and Sedgley, 1997). Standardization of *in vitro* germination would be helpful in interspecific hybridization in a number of ways. It is pre-requisite for stub pollination smeared with PGM, *in vitro* pollination and fertilization etc. Interestingly, nutritive requirement and lag period for *in vitro* germination varied significantly in the three species. Pollen grains of *C. tetragonoloba* required 25% sucrose +100 ppm boric acid + 300 ppm calcium nitrate + 0.8% agar.

The pistil plays a crucial role in the reproductive biology of flowering plants. Studies on the pollen-pistil interaction in leguminous taxa are limited in spite importance of legumes in agricultural production. Stigma of guar is most receptive during 7:30 to 9:00 am while pollen grains are reported to remain viable throughout the day (Anonymous, 1984). Among the three species of *Cyamopsis* studied, the pistil and style length was maximum (77, 38 mm) in *C. senegalensis* and minimum in (53, 25 mm) *C. serrata. C. tetragonoloba* and *C.*

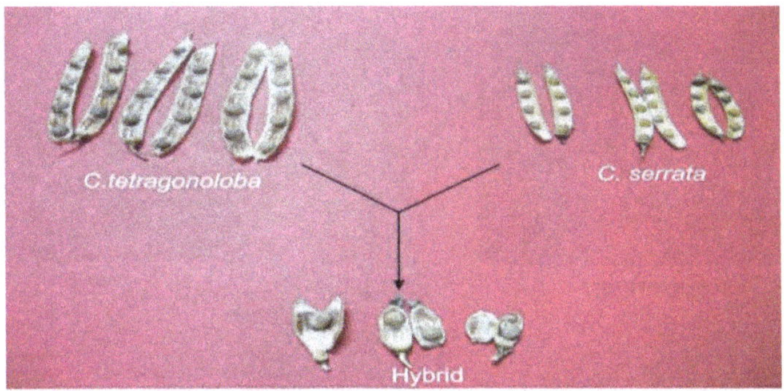


**Figure 1.** Comparison of morphological features of pods of *C. tetragonoloba, C. Serrata* and their $F_1$ hybrid.

*senegalensis* possessed capitate type of stigma whereas *C. serrata* is characterized by subapical crescent shaped stigma (Figure 4). Fabaceae, in general, is characterized by wet stigma. The stigma is covered with surface cells that often lyse to release viscuous secretion containing proteins, amino acids, lipids, polysaccharides and pigments. These secretions not only support retention and germination of pollen grains but protect stigma against desiccation. The role of proteins in "Wet" stigmas is not clear (Esau, 1977). Mattson et al., (1974) emphasized on the role of proteins in hydration of pollen grains. Proteins, which constitute one of the main constituent of stigma + styles of angiosperms was nearly identical quantitatively in all the three species of *Cyamopsis* studied (8 mg/100 mg FW). Total soluble carbohydrate content of Stigma + styles of *C. tetragonoloba* and *C. serrata* was nearly identical (5-6 mg/100 mg FW), whereas those of *C. senegalensis* consisted of minimum quantity (2.4 mg/100 mg FW) of soluble carbohydrates. Similar to the observations of Cruden (2009) in Fabaceae, pollen size was not correlated with style length in different species of *Cyamopsis* studied. The pistil of *C. tetragonoloba* and *C. senegalensis* possessed nearly identical number of ovules (7-8) while *C. serrata* is characterized by 8 to 9 ovules per pistil. Number of seeds per pod ranged from 7 to 9 and did not reveal any significant difference in the wild and cultivated species of *Cyamopsis (*Table 1*)*. Among the three species, 100 seed weight of *C. tetragonoloba* was maximum (2.78 g) whereas the value was nearly identical in *C. serrata* and *C. senegalenis* (1.80 g).

Selfing in *C. tetraganoloba* resulted in good percentage of *in vivo* pollen germination and pollen tubes could be traced up to the base of the ovary. Micropylar entry of pollen tube in the ovule was evident. Interestingly, in *C. senegalensis* and *C. serrata* no pollen germination was evident after 1 day of anthesis. Pollen germination and tubes became evident on/after two days of anthesis and grew until the 3 day after the anthesis.

## Interspecific hybridization

Interspecific hybridization holds great promise in broadening the genetic base of domesticated plant species and success in interspecific hybridization depends upon the extent cross compatibility between the cultigen and its wild relative. Interspecific hybridization between *C. tetragonoloba* x *C. serrata* and *C. tetragonoloba* x *C. senegalensis* was attempted using conventional and non-conventional breeding methods. Studies deploying conventional method of plant breeding revealed no pod setting in the above said crosses. Among an array of non conventional plant breeding methods tried smearing of stub of *C. tetragonoloba* with agarified pollen germination medium (PGM) prior to pollination was successful (Figure 1). Among 792 crosses attempted, 83 pods were recovered for the cross *C. tetragonoloba* x *C. serrata* which amounted to 10.47% pod set. The hybrid pods were nearly 1.70 cm long which contained 2.28 seeds per pod. Seed colour and shape of $F_1$ hybrid was similar to the female parent, that is, *C. tetragonoloba (*Figure *2)*. However, no success was achieved in 429 crosses attempted between *C. tetragonoloba* x *C. senegalensis* even with stub pollination combined with PGM application, although number of pollen grains sticking on the stub increased.

Among the other methods tried *viz.* pollination through perforation in the basal part of style, *in vivo* placental pollination and placental pollination followed by *in vitro* pistil culture did not reveal any pod set; the pistils turned brown and abscised after about 2 to 3 days of pollination from the plant. Failure of placental pollination may be ascribed to withering of ovules after pollen application as has been described for Fabaceae genera (Zenkteler, 1980). In case of placental pollination followed by *in vitro* pistil culture, callusing was observed in all growth combinations tried.

The $F_1$ seeds obtained from crossing (2008-09) *C. tetragonoloba* × *C. serrata* were sown during the year 2009-10. Morphological and phenological features

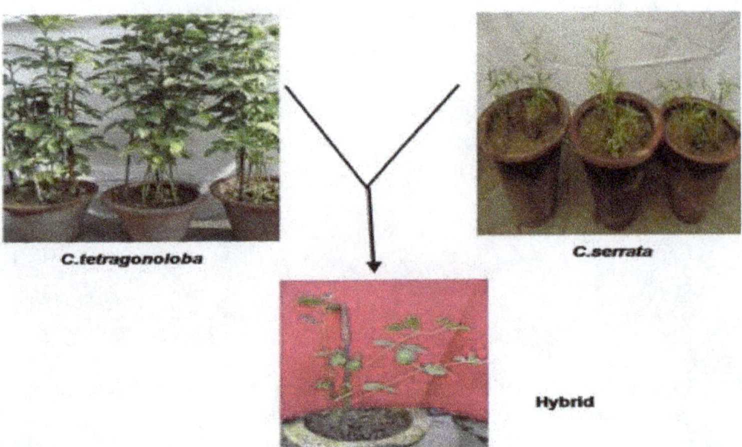


**Figure 2.** Comparison of plant morphological features of *C. tetragonoloba*, *C. Serrata* and their $F_1$ hybrid.

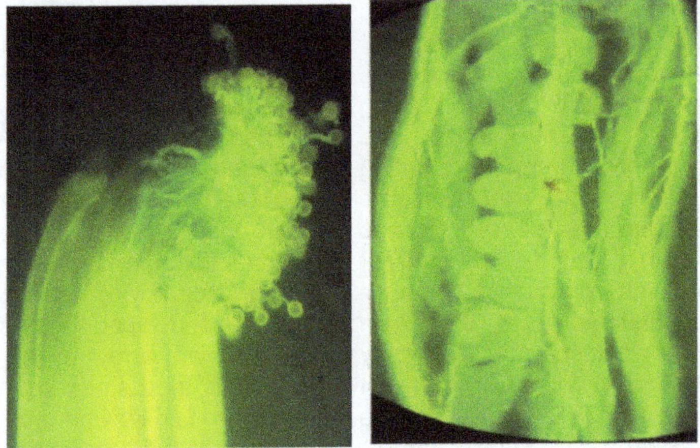

**Figure 3.** L-R *In vivo* germination of *C. serrata* pollen on the pistil stub of *C. tetragonoloba* smeared with pollen germination medium. Pollen tube entering the ovules is evident.

revealed that the plants showed flowering after 21 days of sowing as that of *C. Serrata*, the wild parent. The flowers were pinkish in colour like the female parent (*C. tetragonoloba)* but the shape of pods was akin to the male parent (*C. serrata*). The height of the plant and pod size was 45.5 cm and 4.32 × 0.50 cm, respectively which is intermediate between the two parents. The hybrid plants produced 14.8 pods per plant, 3 to 4 pods per cluster, 6.47 seeds per pod and $3^{rd}$ or $4^{th}$ leaf turned out to be the first trifoliate leaf in contrast to $5^{th}$ to $7^{th}$ leaf in the parent plants. All these morphological and phenological characters are suggestive of the hybrid nature of the plants (Figure 3). Inheritance of seed size in this cross revealed its association with the female parent due to its large size over *C. serrata* however, yield potential of $F_1$ hybrid was low.

It is thus evident that the differences in the nutritive requirements and wide variations in lag period during pollen germination of three species of *Cyamopsis* are the potent pre-fertilization barriers in rearing interspecific hybrids by conventional breeding methods. Smearing of pistil stub of *C. tetragonoloba* with molten and cool pollen germination medium followed by manual pollination with *C. serrata* pollen induced germination and subsequent tube growth culminating in seed set. Since hybrid plant showed early flowering over *C. tetragonoloba*, the transfer of earliness trait from *C. serrata* to the cultivated background is possible by the above method. Further, the attempts can be made to test the fidelity of the interspecific hybrid using molecular markers which may further be tested in the field and transgressive segregants can be selected which may help in identifying extra early varieties of guar. This may prove to be a stepping stone for raising two crops of guar in one year under north Indian conditions which ultimately help in increasing production and productivity of guar.

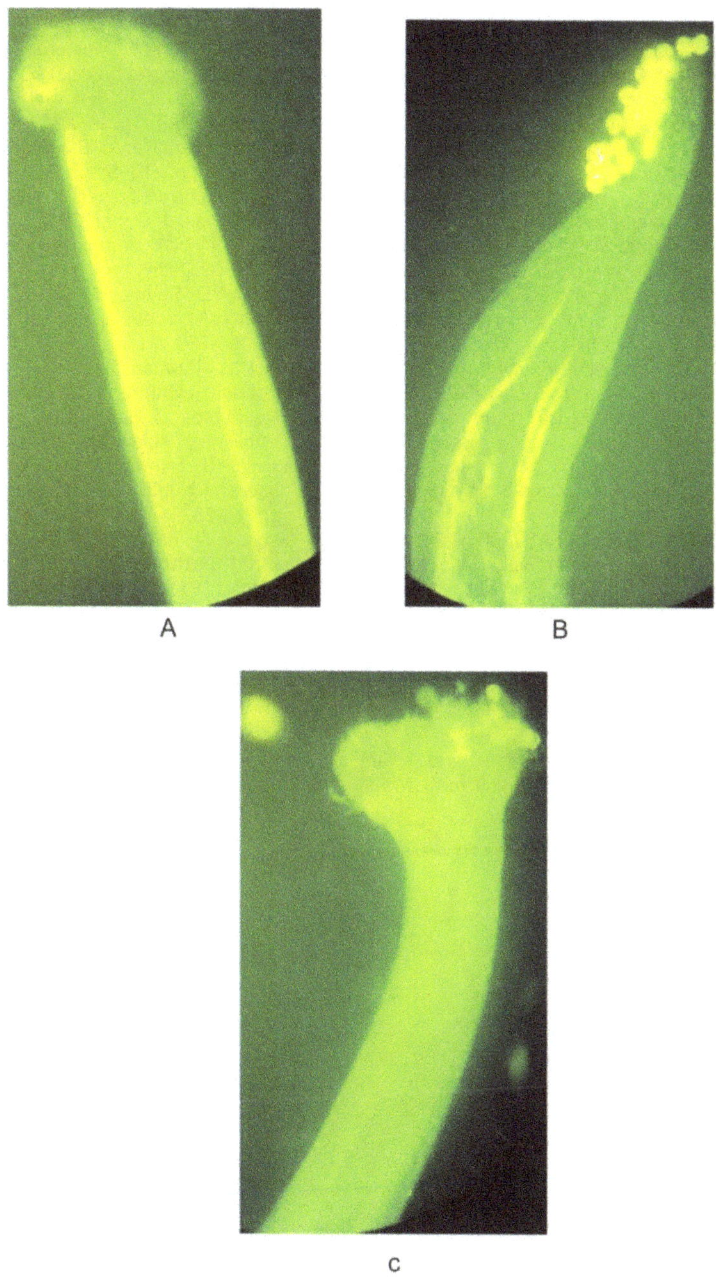


**Figure 4.** Shaps of stigma in different species of *Cyamopsis- Capitate C. tetragonoloba* (A), sub-apical crescent *C. cerrate* (B) and Capitate- *C. senegalensis* (C).

## ACKNOWLEDGEMENT

The author wishes to thank Head Department of Botany and Plant Physiology CCS HAU, Hisar, India. The doctoral fellowship provided by CCS HAU, Hisar is also acknowledged.

### REFERENCES

Anonymous (1984). Final Report (1978-79 to 1983-84). ICAR cess funds scheme on "Genetic improvement in guar for seed yield gum, and resistance to disease". Department of Plant Breeding, CCS HAU, Hisar.

Al-Hafedh YS, Siddiqui AQ (1998). Evaluation of *C. tetragonoloba* seed as a protein source in Nile yilapia, *Oreochromia niloticue* (L.) practical diets. Aquacult. Res. 29:701-708.

Baker HG , Baker I (1979). Starch in angiosperms pollen and its evolutionary significance. Am.J.Bot. 66:591-600.

Berlin RI (1988). Paternity in plant.In: Plant Reproductive Ecology: Patterns and Strategies. Lovett-Doust L (ed): Oxford University Press, Oxford, UK. pp. 30-59.

Bradford M (1976). A rapid and sensitive method for the quantification of microgram quantities of protein utilizing the principle of protein dye binding. Anal. Biochem. 72:248.

Cruden RW (2009). Pollen grain size, stigma depth and style length: the relationship revisited. Plant Systemat. Evol. 278:223-238.

Diaz A, Macnair MR (1999). Pollen tube competition as a mechanism of prezygotic reproductive isolation between *Mimulus nasutus* and its presumed progenitor *M.guttatus*. New Phytologist 144:471-478

Dumas C, Knox RB (1983). Callose and determination of pistil viability and incompatibility. Theorat. Appl. Genet. 67:1-10http://ecocrop.fao.org/ecocrop/srv/en/home

Esau K (1977). Anatomy of seed plants. Wiley, New York.Gardiner Harris July 16, (2012). In Tiny Bean, India's Dirt-Poor Farmers Strike Gas-Drilling Gold. *The New York Times*.

Harlan JR, de Wet JMJ (1971). Towards a rational classification of cultivated plants. *Taxon* 20:509-517.

Hauser EJP, Morrison JH (1964). Cytochemical reduction of nitroblue tetrazolium as an index of pollen viability. Am. J. Bot. 51:748-753

Jain R, Anjaiah V, Babbar SB (2005). Guar gum, a cheap substitute for agar in microbial culture media. Lett. Appl. Microbiol. 41:345-349.

Kapoor S, Nair PKK (1974). Pollen production in some Indian vegetable crops. Geobios. 1:71-73.

Lee JT, Connor-Appleton S, Bailey CA, Cartwright AL (2005). Effects of guar meal by-product with and without beta-mannanase hemicell on broiler performance. Poult. Sci. 84(8):1261-1267.

Lee JT, Bailey CA, Cartwright AL (2009). In vitro viscosity as a function of guar meal and beta-mannanase content of feeds. Int. J. Poult. Sci, 8(8):715-719.

Mattson O, Knox RB, Heslop- Harrison J, Heslop- Harrison Y (1974). Protein pellicle of stigmatic papillae as a probable recognition site in incompatibility reactions. Nature 247:298-300.

Mathiyazhagan S (2009). Interspecific hybridization and plant regeneration in guar [Cyamopsis tetragonoloba (L.) Taub]. M.Sc. Thesis, CCS HAU, Hisar.

Menon U (1973). A comprehensive review of crop improvement and utilization of clusterbean [Cyamopsis tetragonoloba (L) Taub.].Monograph Series-2, Department of Agriculture Rajasthan. P. 51.

Pacini E (1996). Types and meaning of pollen carbohydrate reserves. Sexual Plant Rerod. 9:362-366.

Pathak Singh SK, Singh M, Henry A (2010). Molecular assessment ofgeneticdiversity in clusterbean (*Cyamopsis tetragonoloba*) genotypes. J. Genet. 89:243-246.

Sass JE (1951). Botanical microtechnique. Oxford and IBH Publishing Co. New Delhi.

Shuraki YD, Sedgley M (1997).Pollen tube pathway and stimulation of embryo sac development in *Pistacia vera* (Anacardiaceae). Ann. Bot. 79:361-369.

Wong LJ, Parmar C (1997). Cyamopsis tetragonoloba (L.) TaubertIn: Faridah Hanum, I and van der Maesen LJG (Editors). Plant Resources of South-East Asia No. 11: Auxiliary plants. Backhuys Publisher, Leiden, The Netherlands, pp. 109-113.

Whistler RL, Hymowitz T (1979). Guar. Agronomy, production, Industrial use and Nutrition. Purdue University press wheat Lafayette, Indiana pp. 16-28,

Yemm EW, Wills AJ (1954). The estimation of carbohydrates in plant extract by anthrone. Biochem. J. 57:508-514.

Zenkteler M (1990). *In vitro* fertilization and wide hybridization in higher plants. Plant Sci. 9:267-279.

# 19

# Growth and yield of sunnhemp (*Crotalaria juncea* L.) as influenced by spacing and topping practices

**M. K. Tripathi[1], Babita Chaudhary[1], S. R. Singh[2] and H. R. Bhandari[1]**

[1]Sunnhemp Research Station (CRIJAF, ICAR), Pratapgarh, U.P., India-230001, India.
[2]Central Research Institute for Jute and Allied Fibres (ICAR), Barrackpore, W. B., India.

**Field experiment were conducted during rainy (kharif) season of 2009 and 2010 to study the effects of various spacing and topping practices on growth and seed yield of sunnhemp. Spacing (30 × 10 cm, 30 × 20 cm, 45 × 10 cm and 45 × 20 cm) in main plots and topping (no topping, topping at 30 days after sowing and topping at 45 days after sowing) in sub plots were studied in split plot design with three replications. The individual plant performance with respect to plant height, basal diameter, dry matter accumulation, number of branches and yield attributes was found maximum under the spacing of 45 × 20 cm whereas total biomass and seed yield per unit area was obtained highest with the spacing of 30 × 10 cm. Topping at 30 days after sowing gave higher yield attributes and seed yield being at par with topping at 45 days after sowing. Thus sunnhemp grown at the spacing of 30 × 10 cm coupled with topping at 30 days after sowing realized higher yield.**

**Key words:** Sunnhemp, spacing, topping, seed yield.

## INTRODUCTION

Sunnhemp (*Crotalaria juncea* L., *Fabaceae*), a native of India, is a fast growing annual crop. It is an important source of natural fibre. Traditionally its fibre is used in preparation of ropes, twines, fishing nets, tat-patties, handmade paper etc. (Tripathi et al., 2012). It has been identified as the most promising indigenous raw material for manufacturing of high quality tissue paper, cigarette paper and paper for currency. It is one of the most outstanding green manure crops suited to almost all parts of the India (Ram and Singh, 2011). In Hawaii, sunnhemp (Tropic Sun) has added 150 to165 kg of nitrogen per hectare to the soil when grown for 60 days and then incorporated in test plots (Rotar and Joy, 1983). Sunnhemp possesses medicinal properties and is also used as forage to a limited extent. In present day context of climate change, it holds high promise on account of its ability to fix atmospheric nitrogen, adding organic matter into the soil, suppressing weeds, reducing soil erosion and controlling root knot nematode populations thereby reducing dependence of agriculture on chemical inputs.

Despite having many valuable uses, acreage under this crop has drastically reduced in past decades. The unavailability of good quality seeds is one of the important reasons for reduced popularity of sunnhemp (Chittapur and Kulkarni, 2003). Seed is the critical input in any agricultural system and high yield of quality seed can be obtained only with improved agro-techniques. Abundant research has been done aiming at standardization of different factors for seed production in many crops but most of the agronomic practices have still not been standardized for seed crop of sunnhemp.

Spacing is one of the factors affecting seed yield of different crops. It influences growth rate and crop yield as a result of inter plant competition for different inputs needed for growth and development. Thus investigation on spacing arrangements becomes mandatory for understanding the mechanism of yield enhancement. Apical topping breaks the apical dominance and induces development of lateral branches thereby increase the site for pod development. The practice of topping has proved to be effective in increasing the yield levels of different crops (Bhattacharjee and Mitra, 1999; Sajjan et al., 2002; Jagannatham et al., 2008; Singh et al., 2011). Very meager information on spacing along with topping is available in seed crop of sunnhemp. Keeping these points in view the present investigation was carried out to study the effect of spacing and topping management practices on growth and seed yield of sunnhemp.

## MATERIALS AND METHODS

The field experiment was conducted at Sunnhemp Research Station (CRIJAF-ICAR), Pratapgarh, Uttar Pradesh (25° 34' N latitude and 81° 59' E longitude) during the rainy (Kharif) seasons of 2009 to 2010 and 2010 to 2011 on a sandy loam soil with pH 7.6, organic carbon 2.7 g/kg, bulk density 1.42 g/cc and available nitrogen, phosphorus and potassium 225, 12.2 and 189 kg/ha, respectively. The research station is located at a height of 137 m above sea level. The experiment was laid out in split plot design keeping spacing treatments in main plots and topping in sub plots with three replications. The experiment comprised of four levels of spacing (30 × 10 cm, 30 × 20 cm, 45 × 10 cm and 45 × 20 cm) and three levels of topping (no topping, topping at 30 days after sowing and topping at 45 days after sowing) constituting 12 treatment combinations. The crop was sown on 05.08.2009 and 06.08.2010 during 1$^{st}$ and 2$^{nd}$ years of study, respectively. The sunnhemp variety Shailesh (SH-4) was sown manually in plot size of 4.5 × 5.0 m during both the years. The station received total annual rainfall of 634 mm and 932 mm during the year 2009 to 2010 and 2010 to 2011, respectively. The nutrients NPK were applied at the rate of 20, 40 and 20 kg per ha. The full dose of nitrogen, phosphorus and potassium were applied through urea, single super phosphate and muriate of potash at the time of sowing as basal dose. The crop was thinned 10 days after sowing to maintain plant to plant spacing according to treatments. Remaining package of practices was adopted as per recommendation. The crop was harvested at maturity at the age of 140 days. Observations were recorded both on plant as well as on plot basis at the time of harvest. The plant based data comprised of plant height, basal diameter, number of primary and secondary branches per plant, dry matter accumulation per plant, number of pods per plant, pod weight per plant and seed yield per plant. Plot based data included biomass, seed and stalk yield. Individual plant data were recorded for five random plants and averaged. The statistical analysis of the experimental data was carried out using Indostat.

## RESULTS AND DISCUSSION

### Effect of spacing on growth characteristics

Spacing had significant impact on growth characteristics of seed crop of sunnhemp during both the years of experimentation (Table 1). Basal diameter, number of primary and secondary branches and dry matter accumulation per plant were observed highest under the influence of wider spacing (40 × 20 cm) whereas opposite trend was observed in case of plant height. The highest plant height was observed under the effect of closer spacing (30 × 10 cm) and vice versa although the treatment effect was non significant. The plant height decreased gradually with the increase in spacing. The closer spacing between plants caused comparatively lesser availability of space around the plants for lateral development therefore, forced them to grow vertically. Similar results on plant height were reported by Ram and Singh (2011). The basal diameter (9.12 mm) recorded under the spacing of 45 × 20 cm was superior over rest of the spacing. The spacing of 30 × 20 cm and 45 × 10 cm were at par with respect to their effect on basal diameter. The minimum basal diameter (7.66 mm) was recorded under the spacing of 30 × 10 cm. Rajendra et al. (2008) also reported the positive effect of wider spacing on basal diameter in mesta.

The highest number of primary branches per plant (7.14) was observed in spacing of 45 × 20 cm followed by spacing of 30 × 20 cm (6.45) whereas minimum was recorded in 30 × 10 cm spacing. Similarly in case of secondary branches also the maximum value (17.01) was obtained in 45 × 20 cm of spacing followed by 30 × 20 cm of spacing and being least under 30 × 10 cm spacing (12.69). The variation in dry matter accumulation per plant was also found significant due to different spacing treatments. The highest dry matter accumulation per plant (73.11 g) was noticed under the spacing of 45 × 20 cm which was significantly superior to other spacing treatments whereas lowest was found under the effect of 30 × 10 cm (45.94 g). The better growth characteristics of plant like basal diameter and branches under wider spacing may be attributed to presence of sufficient space that reduced inter plant competition, availability of sufficient moisture, nutrient and light for their development under reduced inter plant competition and high dry matter accumulation was probably due to profuse branching and better growth and development of the plants. The similar findings in sunnhemp crop were reported by Ulemale et al. (2003).

### Effect of topping on growth characteristics

The data presented in Table 1 indicated that basal diameter, number of primary and secondary branches per plant and dry matter accumulation per plant increased significantly with the practice of topping compared to no topping whereas opposite trend was noted in plant height during both the years. The results revealed that plant height decreased significantly due to topping in comparison to no topping. The lowest plant height (196.14 cm) was noticed when the topping was done at 30 days after sowing followed by topping at 45 days after sowing. Although un-topped plants achieved maximum plant height (219.94 cm) maximum dry weight per plant was recorded when plants were topped at 30 days after sowing perhaps due to more number of branches associated with this treatment. Decrease in plant height

**Table 1.** Effect of spacing and topping on growth parameters of seed crop of sunnhemp.

| Treatments | Plant height(cm) | | Basal diameter(mm) | | Primary branch/plant | | Secondary branch/plant | | Dry matter/plant (g) | |
|---|---|---|---|---|---|---|---|---|---|---|
| | 2009-2010 | 2010-2011 | 2009-2010 | 2010-2011 | 2009-2010 | 2010-2011 | 2009-2010 | 2010-2011 | 2009-2010 | 2010-2011 |
| **Spacing** | | | | | | | | | | |
| 30 × 10 cm | 206.31 | 218.92 | 7.60 | 7.72 | 5.40 | 5.51 | 12.67 | 12.72 | 45.22 | 46.67 |
| 30 × 20 cm | 201.07 | 212.13 | 8.42 | 8.81 | 6.40 | 6.50 | 15.19 | 15.46 | 58.56 | 60.44 |
| 45 × 10cm | 203.21 | 215.87 | 8.34 | 8.45 | 6.19 | 6.12 | 14.46 | 14.28 | 52.44 | 53.78 |
| 45 × 20 cm | 200.04 | 208.04 | 8.97 | 9.28 | 7.16 | 7.12 | 16.97 | 17.06 | 72.00 | 74.22 |
| S.Em. (±) | 5.52 | 5.19 | 0.21 | 0.15 | 0.24 | 0.17 | 0.56 | 0.69 | 3.39 | 2.63 |
| CD at 5% | NS | NS | 0.73 | 0.51 | 0.83 | 0.59 | 1.94 | 2.39 | 11.73 | 9.11 |
| **Topping** | | | | | | | | | | |
| No topping | 215.48 | 224.40 | 8.12 | 8.22 | 5.92 | 5.97 | 13.95 | 14.35 | 53.83 | 54.08 |
| Topping at 30 DAS | 191.35 | 200.93 | 8.66 | 8.95 | 6.62 | 6.62 | 15.73 | 15.46 | 60.50 | 63.33 |
| Topping at 45 DAS | 201.15 | 215.90 | 8.23 | 8.52 | 6.33 | 6.34 | 14.78 | 14.82 | 56.83 | 58.92 |
| S.Em. (±) | 5.12 | 2.84 | 0.14 | 0.15 | 0.18 | 0.14 | 0.41 | 0.27 | 1.48 | 2.29 |
| CD at 5% | 15.34 | 8.50 | 0.42 | 0.46 | 0.54 | 0.42 | 1.24 | 0.82 | 4.43 | 6.87 |

as a result of topping was reported by Obasi and Msaakpa (2005) in cotton. This shows that topping at earlier stages of crop growth inhibits vertical growth. This is in accordance to the report of Bhattacharjee and Mitra (1999) where they found maximum plant height in un-topped plants. They reported topping at 45 days after sowing in jute to be optimum for seed yield and component traits. Significantly maximum basal diameter (8.66 mm) was recorded by topping at 30 days after sowing followed by topping at 45 days after sowing (8.23 mm) compared to no topping. Lowest basal diameter (8.12 mm) was recorded in un-topped plants. Similarly topping at 30 days after sowing recorded maximum number of primary (6.62) and secondary branches (15.59) per plant being at par with topping at 45 days after sowing but significantly superior to no topping. This is in agreement to the reports of Marie et al. (2007) in okra. The topping at 45 days after sowing was at par with that of no topping. Toping practices exerted significant variations in dry matter accumulation by the plants. The highest dry matter accumulation per plant (61.91g) was observed by the practice of topping at 30 days after sowing followed by topping at 45 days after sowing. The lowest dry matter accumulation per plant (53.95 g) was noted in un-topped plants. Similar findings on growth characteristics was reported by Kathiresan and Duraisamy (2001) in sesbania, Lakshmi et al. (1995) in mesta and Marie et al. (2007) in okra.

## Effect of spacing on yield attributes

Most of the yield attributing characters of seed crop of sunnhemp was significantly influenced due to different spacing treatments except test weight (1000 seed weight) during both the years (Table 2). The number of pods per plant, pod weight per plant, seed yield per plant and number of seeds per pod were noted maximum under wider spacing of 45 × 20 cm followed by 30 × 20 cm of spacing and being lowest in the spacing of 30 × 10 cm. Number of pods per plant is one of the most effective elements in producing seed yield. Spacing of 45 × 20 cm recorded significantly maximum number of pods per plant (74.04) being statistically on par with that at 30 × 20 cm of spacing during second year whereas lowest values (47.09) were noted in the spacing of 30 × 10 cm. The enhancement in number of pods per plant under the influence of wider spacing might be on account of maximum number of primary and secondary branches per plant at the same spacing which resulted into increase in pod weight

**Table 2.** Effect of spacing and topping on yield attributing characters of seed crop of sunnhemp.

| Treatments | Pods/plant | | Pod weight/plant (g) | | Seed yield/plant (g) | | Seeds/pod | | Test weight (g) | |
|---|---|---|---|---|---|---|---|---|---|---|
| | 2009-2010 | 2010-2011 | 2009-2010 | 2010-2011 | 2009-2010 | 2010-2011 | 2009-2010 | 2010-2011 | 2009-2010 | 2010-2011 |
| **Spacing** | | | | | | | | | | |
| 30 × 10 cm | 44.33 | 49.86 | 14.89 | 16.56 | 9.42 | 10.36 | 7.34 | 7.29 | 37.58 | 37.78 |
| 30 × 20 cm | 57.84 | 67.48 | 20.67 | 22.42 | 12.67 | 13.02 | 8.20 | 8.08 | 38.34 | 38.67 |
| 45 × 10cm | 52.00 | 60.27 | 18.89 | 20.57 | 10.87 | 11.34 | 7.79 | 7.47 | 38.18 | 38.28 |
| 45 × 20 cm | 69.64 | 78.44 | 24.22 | 26.02 | 14.82 | 15.18 | 8.68 | 8.36 | 38.79 | 39.08 |
| S.Em. (±) | 2.52 | 3.50 | 1.27 | 1.18 | 0.71 | 0.45 | 0.25 | 0.20 | 0.36 | 0.32 |
| CD at 5% | 8.73 | 12.13 | 4.4 | 4.07 | 2.45 | 1.56 | 0.86 | 0.70 | NS | NS |
| **Topping** | | | | | | | | | | |
| No topping | 51.93 | 58.00 | 18.00 | 18.75 | 10.93 | 11.55 | 7.84 | 7.47 | 37.80 | 38.22 |
| Topping at 30 DAS | 59.70 | 70.80 | 21.67 | 23.63 | 12.92 | 13.50 | 8.11 | 8.12 | 38.59 | 38.71 |
| Topping at 45DAS | 56.23 | 63.23 | 19.33 | 21.80 | 11.98 | 12.37 | 8.06 | 7.79 | 38.27 | 38.43 |
| S.Em. (±) | 1.77 | 2.15 | 0.97 | 0.76 | 0.39 | 0.44 | 0.28 | 0.17 | 0.41 | 0.33 |
| CD at 5% | 5.32 | 6.44 | 2.9 | 2.29 | 1.17 | 1.32 | NS | NS | NS | NS |

per plant and ultimately seed yield per plant. Data on number of seeds per pod also followed the similar trend. Plants under the spacing of 45 × 20 cm being at par with 30 × 20 cm recorded significantly maximum number of seeds per pod. The increasing trend in test weight was also observed with increase in spacing but not up to the level of significance. The superiority of individual plant performance at wider spacing might be attributed to less plant competition for nutrients, moisture, space, solar radiation etc which finally led towards better growth and development of the plants. The results are in close conformity with the findings of Ulemale et al. (2003) and Shastri et al. (2007).

## Effect of topping on yield attributes

The data recorded on effect of topping on different yield attributing characters are depicted in Table 2. It is evident from the result that due to topping practices significant variations were recorded in number of pods per plant, pod weight per plant and seed yield per plant however no significant difference was observed in number of seeds per pod and test weight. Significantly maximum number of pods per plant (65.25), pod weight per plant (22.65 g) and seed yield per plant (13.21 g) was noticed under the influence of topping at 30 days after sowing compared to no topping (54.96, 18.37 and 11.24 g, respectively). The difference between topping at 45 days after sowing and no topping was found non significant except for pod weight per plant during second year where it happened to be significant. The increment in yield attributes under the influence of topping might be attributed to better growth characteristics like dry matter accumulation per plant coupled with profuse branching. The significant difference due to different topping practices was not noticed on number of seeds per pod and test weight but still the increasing trend was observed. The results corroborate with the findings of Lakshmi et al. (1995), Bhattacharjee and Mitra (1999), Kathiresan and Duraisamy (2001) and Marie et al. (2007) who mentioned a significant increase in yield attributes by practice of apical bud topping in different crops.

## Effect of spacing on yield

Different spacing treatments were found to have significant influence on total biomass, seed yield and finally on stalk yield during both the years of investigation (Table 3). In contrast to effect of spacing on growth and yield attributes of individual plant the maximum values of different yield was observed under the influence of closer

**Table 3.** Effect of spacing and topping on biomass, seed and stalk yield of seed crop of sunnhemp.

| Treatments | Biomass yield (q/ha) | | Seed yield (q/ha) | | Stalk yield(q/ha) | |
|---|---|---|---|---|---|---|
| | 2009-2010 | 2010-2011 | 2009-2010 | 2010-2011 | 2009-2010 | 2010-2011 |
| **Spacing** | | | | | | |
| 30 × 10 cm | 93.52 | 91.83 | 14.74 | 15.80 | 78.78 | 76.03 |
| 30 × 20 cm | 74.97 | 75.76 | 12.74 | 13.42 | 62.23 | 62.34 |
| 45 × 10 cm | 81.24 | 82.31 | 13.57 | 14.53 | 67.67 | 67.78 |
| 45 × 20 cm | 67.57 | 67.34 | 11.80 | 12.34 | 55.77 | 55.00 |
| S.Em. (±) | 3.73 | 2.77 | 0.41 | 0.51 | 2.53 | 2.45 |
| CD at 5% | 12.89 | 9.60 | 1.42 | 1.75 | 8.75 | 8.47 |
| **Topping** | | | | | | |
| No topping | 76.91 | 73.97 | 12.47 | 13.28 | 64.44 | 60.69 |
| Topping at 30 DAS | 81.47 | 85.10 | 13.96 | 14.89 | 67.51 | 70.21 |
| Topping at 45 DAS | 79.59 | 78.85 | 13.20 | 13.90 | 66.39 | 64.95 |
| S.Em. (±) | 1.17 | 1.98 | 0.31 | 0.36 | 0.81 | 1.67 |
| CD at 5% | 3.52 | 5.93 | 0.93 | 1.08 | 2.43 | 5.00 |

spacing. The spacing of 30 × 10 cm being statistically on par with that at 45 × 10 cm recorded significantly higher biomass (92.67 q/ha) and seed yield (15.27 q/ha) followed by spacing of 30 × 20 cm. The lowest biomass (67.45 q/ha) and seed yield (12.07 q/ha) was found under the influence of 45 × 20 cm of spacing. The seed yield realized under the spacing of 30 × 10 cm was 24.91 and 28.04% higher compared to spacing of 45 × 20 cm during both the years respectively. The similar trend of spacing was noticed on stalk yield also. This reveals that although at wider spacing seed yield per plant was higher than that at closer spacing but seed yield per hectare was low on account of lower plant density under this spacing. The improvement in yield attributes under wider spacing failed to compensate for lower number of plants per unit area under this spacing. Although at closer spacing plants became sub marginal and produced yield below their potentiality but the cumulative effect of large number of sub marginal plants increased the total yield per hectare. The similar findings were reported by Ulemale et al. (2002) and Shastri et al. (2007) in sunnhemp.

### Effect of topping on yield

Topping practices significantly influenced the total biomass, seed and stalk yield during both the years (Table 3). Topping at 30 days after sowing resulted into significantly higher biomass (83.28 q/ha) and seed yield (14.42 q/ha) compared to no topping (75.44 and 12.87 q/ha, respectively). On an average increase in seed yield was noticed to the tune of 12.04%. The stalk yield also followed similar trend. The difference between no topping and topping at 45 days after sowing was found non significant with respect to biomass, seed and stalk yield. The higher yield under the effect of topping might be attributed to better growth characteristics which resulted in considerable improvement in yield attributing characters like number of pod, pod weight and seed yield per plant and finally reflected into yield. In addition significantly more number of primary and secondary branches carrying more number of pods per plant under topping treatment might have multiplicative effect on seed yield. The significant higher seed yield recorded in topped plants may also be attributed to diversion of photosynthates and metabolites produced by leaves to strong carbohydrate sinks that is, pods, when compared to apical meristem in un-topped plants. The results confirmed the findings of Bhattacharjee and Mitra (1999) and Jagannatham et al. (2008) in jute, Lakshmi et al. (1995) in mesta, Kathiresan and Duraisamy (2001) in sesbania.

## Conclusion

Thus from the present study it may be concluded that sowing of seed crop of sunnhemp at the spacing of 30 × 10 cm coupled with topping at 30 days after sowing is effective in increasing the seed yield of sunnhemp.

## REFERENCES

Bhattacharjee AK, Mitra BN (1999). Jute seed productivity and its quality as influenced by suppression of apical dominance. In Palit et al (eds) Jute and Allied Fibres: Agriculture and Processing, CRIJAF Publication, India, pp. 177-185.

Chittapur BM, Kulkarni SS (2003). Effect of sowing dates on the performance of sunnhemp. J. Maharashtra Agric. Univ. 28(3):331-331.

Jagannatham J, Rajendra KB, Sreelatha T, Rajabapa RV (2008). Effect of topping on seed yield and fibre yield of mesta, *sabdariffa* vs *cannabinus*. In: Palit et al (eds) Abstracts of Papers: International

Symposium on Jute and Allied Fibres Production, Utilization and Marketing held at Kolkata, India. CRIJAF Publication, P. 121.

Kathiresan G, Duraisamy K (2001). Effect of clipping and diammonium phosphate spray on growth and seed yield of dhaincha (*Sesbania aculeata*). Indian J. Agron. 46(3):568-572.

Lakshmi MB, Naidu MV, Reddy DS, Reddy CV (1995). Effect of time of sowing and topping on seed yield of roselle (*Hibiscus sabdariffa*). Indian J. Agron. 40(4):682-685.

Marie AI, Ihsan A, Salih SH (2007). Effect of sowing date, topping and some growth regulators on growth, pod and seed yield of okra (*Abelmoschus esculentus* L.M.). In: African Crop Science Conference Proceedings. 8:473-478.

Obasi MO, Msaakpa TS (2005). Influence of topping, side branch pruning and hill spacing on growth and development of cotton (*Gossypium barbadense* L.) in the Southern Guinea Savanna location of Nigeria. J. Agric. Rural Develop. Trop. Subtrop. 106(2):155-165.

Rajendra KB, Jagannatham J, Sreelatha T, Rajabapa RV (2008). Effect of spacing on seed production of mesta. In: Palit et al (eds) Abstracts of Papers: International Symposium on Jute and Allied Fibres Production, Utilization and Marketing held at Kolkata, India. CRIJAF Publication, P. 122.

Ram H, Singh G (2011). Growth and seed yield of sunnhemp genotypes as influenced by different sowing methods and seed rates. World J. Agric. Sci. 7(1):109-112.

Rotar PP, Joy RJ (1983). 'Tropic Sun' sunn hemp, *Crotalaria juncea* L. Research Extension Series 036. College of Tropical Agriculture and Human Resources, University of Hawaii. P. 1.

Sajjan AS, Shekaragouda M, Badanu VP (2002). Influence of apical pinching and pod picking on growth and seed yield of okra. Karnataka J. Agric. Sci. 15(2):367-372.

Shastri AB, Desai BK, Pujari BT, Halepyati AS, Vasudevan SN (2007). Studies on the effect of plant densities and phosphorus management on growth and seed yield of sunnhemp (*Crotalaria juncea* L.). Karnataka J. Agric. Sci. 20(2):359-360.

Singh F, Kumar R, Kumar P, Pal S (2011). Effect of irrigation, fertility and topping on Indian mustard (*Brassica juncea*). Prog. Agric. 11(2):477- 478.

Tripathi MK, Chaudhary B, Bhandari HR, Harish ER (2012). Effect of varieties, irrigation and nitrogen management on fibre yield of sunnhemp. J. Crop. Weed 8(1):84-85.

Ulemale RB, Giri DG, Shivankar RS, Patil VN (2003). Effect of sowing dates, spacing and phosphate levels on the growth and productivity of sunnhemp. Legume Res. 26(2):121-124.

Ulemale RB, Giri DG, Shivankar RS (2002). Effect of sowing date, row spacing and phosphate level on biomass studies in sunnhemp. J. Maharashtra Agric. Univ. 26(3):323-325.

# *Ex situ* conservation of *Hedychium spicatum* Buch.-Ham. using different types of nursery beds

**Nidhi Lohani[1], Lalit M. Tewari[2], G. C. Joshi[1], Ravi Kumar[1] and Kamal Kishor[2]**

[1]Regional Research Institute of Himalayan Flora, CCRAS, Tarikhet, Ranikhet-263 663, Uttarahand, India.
[2]Department of Botany, D. S. B. Campus, Kumaun University, Nainital-263 201, Uttarakhand, India.

***Hedychium spicatum* Buch.-Ham. belongs to family Zingiberaceae. It is known for its many medicinal uses. Due to great market potential of the plant, it is harvested in uncontrolled way causing the decline of the herb from its natural habitat. Presently, its status in nature is not good. Thus, there is need of its *in situ* as well as *ex situ* conservation and propagation. The present study was conducted on different trials with different organic fertilizers and types of nursery beds to develop a need based agro-technique for mass scale cultivation of the plant in the climatic condition of Ranikhet, Uttarakhand.**

**Key words:***Hedychium spicatum* Buch.-Ham., organic cultivation, conservation, sustainable harvesting.

## INTRODUCTION

*Hedychium spicatum* Buch.-Ham. belongs to family Zingiberaceae, commonly known as "Ginger Lily", found in temperate and sub-temperate zones in Himalayas between 1500 to 2700 m (Naithani, 1985). It is an annual-perennial, erect herb, stem leafy, 5 to 150 cm high. Leaves broadly ovate-lanceolate, 30 to 60 × 10 to 20 cm acuminate, glabrous above, sparsely pubescent beneath. Flowers fragrant, white with an orange-yellow or red base, in dense, terminal, 15 to 25 cm long spikes; floral bracts large, green, 1-flowered. Calyx membranous: three-lobed, ovate, obtuse, shorter than bracts. Corolla tube 5 to 6.5 cm long, much longer than calyx; petals white, linear, spreading; lip white with two-elliptic lobes and orange or yellow base; filaments of stamen-red. Capsules globular: three-valved with an orange-red lining; seeds black with red aril (Gaur, 1999).

*H. spicatum* is highly traded from the Himalayan region under the trade name "Kapur Kachari" or "Sathi". Its rhizome is used in Ayurvedic and Unani medicine. Aromatic rootstock contains essential oil, saccharin, albumin, starch and mucilage. The rhizomes are considered useful as stomachic, carminative and stimulant for the treatment of liver complaints, diarrhea, food poisoning and inflammation. It is useful in the treatment of asthma and bronchitis (Singh, 1983). Rhizome powder is sprinkled as an antiseptic agent and also used in various aches and pains (Thakur et al., 1989). A famous perfume 'abir' is obtained from the rootstocks. The rhizomes are also considered to have insect repelling properties and used for the preservation of cloth in some part of Uttarakhand. The essential oil can be used in perfumes, soaps, hair oils and in cosmetics. Locally rhizomes, are boiled and eaten with salt, powder of roasted rhizome is effective in asthma and respiratory disorders. Seeds are believed to cause abortion. Decoction of rhizome with deodar saw dust is taken for tuberculosis (Gaur, 1999). The rhizome yields an aromatic volatile oil known as 'Kapur Kachri oil' which has p-methoxycinnamate as major chemical constituent (Sarin, 2008).

In recent years, over exploitation of this herb due to demand of raw drugs in pharmaceutical industries has led to habitat destruction. Due to lack of management, this highly valuable species have reached near extinction and require immediate attention towards its *ex situ* and *in situ* conservation.

Here, we are presenting the agro-technique of this highly valuable medicinal plant which may be fruitful in future for farmers and medicinal plant growers. The main objective of our study is *ex situ* conservation of this valuable herb as well as upliftment of the economy of hill people by its cultivation.

## MATERIALS AND METHODS

Germplasm of *H. spicatum* has been collected from the wild sources surrounding Ranikhet before the onset of its dormancy and the trials for this study were conducted at Medicinal Plant Garden of Central Council for Research in Ayurveda and Siddha (CCRAS), Ranikhet (29°38'60 N, 79°25'0 E). The garden is situated at an altitude of 1700 m. It is surrounded by thick pine forest, characterized by *Cedrus*, *Myrica*, *Rhododendron*, *Quercus* and Shrubs like *Berberis*, *Rubus* sp., and *Crataegus* sp. Germplasm of *Polygonatum cirrhifolium* was collected from the wild sources just before the onset of its dormancy. Before plantation, we prepared the land and applied the fertilizer as follows:

The land was dug up or ploughed twice or thrice until a fine tilth was obtained. After those different types of experimental beds namely: beds in plain, beds in slopes and beds with rows and furrows of 1 $m^2$ areas were prepared. Before plantation, these nursery beds were supplemented with different types of organic fertilizers. Three types of organic fertilizers namely: farmyard manure (FYM), forest litter and vermincompost were used to see their effect on the survival, growth and yield. The fertilizers were added in the beds in two doses: first before plantation (at the time of bed preparation) and second after sprouting. In order to study the effect of fertilizers and nature of nursery beds on survival, growth, yield and other related parameters the experiment was designed in randomize block design (RBD) with a total of 16 treatment combinations along with a control set and each treatment was replicated three times.

Plants can be raised by rhizomes as well as by seeds; however, plants raised by seeds can take more time for crop maturity. Hence, rhizomes are best for its propagation (Nautiyal and Nautiyal, 2004; Anonymous, 2008). Therefore, we used rhizome sections for the propagation. In morning hours, these rhizome sections were planted in the different nursery beds. Before plantation, rhizomes were washed with natural spring water of Medicinal Plant Garden and dipped in cow's urine for overnight to prevent them from soil born diseases as suggested in "Varkshayurveda" by Vijyalashmi and Shyam (1993). Rhizome sections were cut into transverse sections of about 4.00 to 5.00 cm long at the intermodal portion with at least two nodes within each section and each rhizome section was planted laterally in different types of nursery beds. A total of 432 rhizome sections were selected for plantation and total 9 rhizome sections were planted in each bed. A row-to-row and plant-to-plant distance of 30 cm was followed for sowing the rhizomes. In all the treatments, irrigation was done at regular intervals through, natural spring water (heavy metal free), depending upon weather condition and moisture requirements of soil. The crop was irrigated once a week during dry months. During the dormancy period, irrigation was done in every 20 days. The beds were kept free from weeds manually.

Data were recorded on both pre-harvest and post-harvest agronomic characters namely: days to sprout, days to flower, number of leaves, leaf length, plant height, fresh weight of rhizomes, dry weight of rhizomes and rate of fresh weight of rhizomes increased. Observations were recorded in 15 days interval. Statistical analysis was carried out to calculate mean values and correlation between different morphological traits. Analysis of variance (ANOVA) and correlation between different morphological traits was calculated by using data analysis tool of Window 2007.

To observe the yield, underground parts (rhizomes) from one replication of each treatment were uprooted at the end of each growing season for three consecutive growth seasons. These rhizomes were properly washed with running water to remove soil particles. Fresh weight was taken after removing rootlets. These rhizomes were then cuts into small slices and kept in the partial shade for drying. After complete drying, dry weight of these rhizomes was taken.

### Intercultural operation

The following fertilizer doses and spacing was used to standardize the agro-technique for commercial cultivation of this herb:

(a) Design: RBD.
(b) Fertilizers: $F_1$, forest litter; $F_2$, vermi compost; $F_3$, FYM; $F_4$, control (quantity of fertilizer: 600 g per bed (60 qut /ha).
(c) Nursery beds: $B_1$, slope; $B_2$, rows; $B_3$, furrows; $B_4$, plain.
(d) No. of treatment combinations: 4 × 4 = 16

| | $F_1$ | $F_2$ | $F_3$ | $F_4$ |
|---|---|---|---|---|
| $B_1$ | $T_1$ | $T_2$ | $T_3$ | $T_4$ |
| $B_2$ | $T_5$ | $T_6$ | $T_7$ | $T_8$ |
| $B_3$ | $T_9$ | $T_{10}$ | $T_{11}$ | $T_{12}$ |
| $B_4$ | $T_{13}$ | $T_{14}$ | $T_{15}$ | $T_{16}$ |

(e) Replication: 3
(f) Spacing: 30 × 30 cm.
(g) Total no. of plots: 48
(h) Plot size: 1 × 1 $m^2$
(i) Number of rhizome cuttings planted in each bed: 9

## RESULTS

Data analyzed for mean values showed that in the first year of cultivation (Table 1), *H. spicatum* takes about 181.56 to 195 days to sprout and 251.89 to 266.44 days to flower from the date of plantation. The percentage of survival was observed 100% in all type of beds. Maximum average number of leaves per plant, 7.56 in the bed $T_2$ (slope + vermin compost) and minimum, 5.37 in the bed $T_{16}$ (plain + control), maximum average leaf length, 33.04 cm in the bed $T_{14}$ (plain + vermi compost) and minimum, 21.49 cm in the bed $T_{12}$ (forrow + control), maximum average height per plant, 76.22 cm in the bed $T_{13}$ (plain + forest litter) and minimum, 39.81 cm in the bed $T_8$ (row + control), maximum fresh weight of rhizomes harvested 113.62 qt/ha in the bed $T_1$ (slope + forest litter) and minimum, 62.98 qt/ha in the bed $T_8$ (row + control), maximum dry weight of rhizomes obtained 24.50 qt/ha in the bed $T_1$ (slope + forest litter) and minimum, 13.90 qt/ha in the bed $T_{12}$ (forrow + control), maximum rate of fresh weight of rhizomes increased 53.05% in the bed $T_1$ (slope + forest litter) and minimum, 12.03% in the bed $T_8$ (row + control) were observed during

**Table 1.** Pre-harvest and post- harvest agronomic characters of *H. spicatum* under different types of nursery beds and fertilizer treatments for first year of cultivation.

| Bed | Treatment | Spacing | % of survival | Pre- harvest agronomic character | | | | | Post- harvest agronomic character | | | |
|---|---|---|---|---|---|---|---|---|---|---|---|---|
| | | | | Days to sprout | Days to flower | Avg. no. of leaves/plant | Avg. leaf length (cm) | Avg. height /plant (cm) | Fresh weight of rhizomes planted (qt/ha) | Fresh weight of rhizomes harvested (qt/ha) | Dry weight of rhizomes obtained (qt/ha) | Rate of fresh weight of rhizomes increased (%) |
| $T_1$ | Slope + Litter | 30 × 30 | 100 | 181.56 ± 3.32 | 251.89 ± 3.69 | 6.96 ± 1.60 | 30.68 ± 10.75 | 61.26 ± 3.49 | 75.12 ± 7.66 | 113.62 ± 5.50 | 24.50 ± 1.18 | 53.05 ± 21.02 |
| $T_2$ | Slope + Vermi | 30 × 30 | 100 | 182.11 ± 4.34 | 252.00 ± 4.53 | 7.56 ± 1.93 | 30.93 ± 7.55 | 68.13 ± 2.92 | 72.83 ± 7.52 | 108.04 ± 4.52 | 23.19 ± 0.95 | 49.94 ± 18.67 |
| $T_3$ | Slope + FYM | 30 × 30 | 100 | 182.00 ± 3.57 | 252.22 ± 3.49 | 5.85 ± 1.66 | 26.17 ± 7.34 | 62.24 ± 3.33 | 68.37 ± 11.52 | 96.30 ± 6.62 | 20.75 ± 1.45 | 45.15 ± 30.27 |
| $T_4$ | Slope + Control | 30 × 30 | 100 | 182.22 ± 3.03 | 252.22 ± 3.77 | 5.56 ± 1.85 | 28.32 ± 8.75 | 53.27 ± 2.27 | 65.42 ± 8.88 | 90.75 ± 4.54 | 19.49 ± 0.97 | 41.33 ± 22.40 |
| $T_5$ | Row + Litter | 30 × 30 | 100 | 181.78 ± 3.31 | 252.00 ± 4.39 | 6.78 ± 1.60 | 27.48 ± 7.35 | 60.89 ± 1.93 | 63.58 ± 9.74 | 74.21 ± 4.64 | 15.87 ± 0.98 | 22.76 ± 19.06 |
| $T_6$ | Row + Vermi | 30 × 30 | 100 | 183.33 ± 2.92 | 254.33 ± 3.24 | 7.37 ± 1.88 | 30.89 ± 9.14 | 70.85 ± 2.48 | 61.73 ± 8.14 | 75.94 ± 3.23 | 16.31 ± 0.76 | 25.08 ± 18.77 |
| $T_7$ | Row + FYM | 30 × 30 | 100 | 188.11 ± 3.98 | 259.56 ± 3.17 | 6.41 ± 2.15 | 26.93 ± 5.97 | 57.73 ± 2.39 | 73.46 ± 9.38 | 85.84 ± 4.72 | 18.44 ± 1.05 | 19.40 ± 18.76 |
| $T_8$ | Row + Control | 30 × 30 | 100 | 187.00 ± 3.04 | 257.56 ± 3.97 | 5.70 ± 1.68 | 22.18 ± 4.50 | 39.81 ± 2.14 | 59.88 ± 8.21 | 62.98 ± 3.28 | 14.24 ± 0.72 | 12.03 ± 12.32 |
| $T_9$ | Furrow + Litter | 30 × 30 | 100 | 194.33 ± 4.24 | 266.00 ± 4.66 | 6.74 ± 1.46 | 29.88 ± 9.16 | 64.50 ± 2.62 | 73.49 ± 6.93 | 102.49 ± 5.42 | 22.00 ± 1.18 | 40.12 ± 9.08 |
| $T_{10}$ | Furrow + Vermi | 30 ×30 | 100 | 187.44 ± 3.32 | 258.89 ± 3.89 | 6.19 ± 1.75 | 29.17 ± 8.57 | 69.81 ± 2.32 | 61.11 ± 8.12 | 83.37 ± 4.65 | 17.91 ± 0.96 | 39.17 ± 24.20 |
| $T_{11}$ | Furrow + FYM | 30 × 30 | 100 | 188.00 ± 4.44 | 259.33 ± 3.20 | 5.93 ± 1.54 | 28.07 ± 7.25 | 60.23 ± 3.23 | 62.95 ± 9.14 | 78.40 ± 3.46 | 16.64 ± 0.73 | 29.24 ± 16.90 |
| $T_{12}$ | Furrow + Control | 30 × 30 | 100 | 195.00 ± 4.24 | 266.44 ± 6.46 | 6.07 ± 1.49 | 21.49 ± 3.69 | 47.37 ± 3.56 | 59.26 ± 7.90 | 64.21 ± 3.69 | 13.90 ± 0.74 | 15.04 ± 10.03 |
| $T_{13}$ | Plane + Litter | 30 × 30 | 100 | 186.00 ± 3.16 | 257.89 ± 4.48 | 6.74 ± 1.43 | 32.14 ± 8.11 | 76.22 ± 3.28 | 67.27 ± 8.47 | 94.47 ± 4.76 | 20.42 ± 1.02 | 42.89 ± 23.29 |
| $T_{14}$ | Plane + Vermi | 30 × 30 | 100 | 187.67 ± 3.08 | 259.78 ± 3.27 | 6.30 ± 1.44 | 33.04 ± 8.86 | 69.58 ± 3.28 | 67.92 ± 7.04 | 95.71 ± 4.42 | 20.66 ± 1.01 | 42.39 ± 17.22 |
| $T_{15}$ | Plane + FYM | 30 × 30 | 100 | 188.00 ± 3.57 | 259.89 ± 3.22 | 6.67 ± 1.57 | 31.55 ± 7.62 | 72.44 ± 3.17 | 75.28 ± 7.47 | 104.95 ± 5.69 | 22.58 ± 1.20 | 40.79 ± 17.77 |
| $T_{16}$ | Plane + Control | 30 × 30 | 100 | 189.11 ± 3.37 | 261.33 ± 3.57 | 5.37 ± 1.84 | 24.13 ± 5.47 | 55.55 ± 3.06 | 71.57 ± 9.95 | 81.49 ± 3.48 | 17.42 ± 0.68 | 18.07 ± 13.35 |
| ANOVA | | | DF | 15 | 15 | 15 | 15 | 15 | 15 | 15 | 15 | 15 |
| | | | MS | 158.91 | 207.07 | 10.77 | 238.21 | 2517.78 | 285.80 | 2052.06 | 91.85 | 1553.17 |
| | | | F-value | 12.30*** | 12.80*** | 3.76*** | 4.00*** | 301.94*** | 3.88*** | 95.65*** | 92.72*** | 4.27*** |

uring first year of cultivation. In the first year of cultivation of *H. spicatum*, ANOVA was found significant for all the pre-harvest and post-harvest agronomic characters namely: days to sprout, days to flower, average number of leaves per plant, average leaf length (cm), average height per plant (cm), fresh weight of rhizomes planted (qt/ha), fresh weight of rhizomes harvested (qt/ha), dry weight of rhizomes obtained (qt/ha) and rate of fresh weight of rhizomes increased (%) with F-value (12.30, 12.80, 3.76, 4.00, 301.94, 3.88, 95.65, 92.72 and 4.27, respectively at $p < 0.001$).

In the second year of cultivation (Table 2), the plant took about 164 to 181.11 days to sprout and 234.22 to 256.78 days to flower from the date of dormancy and 100% survival was recorded in all type of beds. Maximum average number of leaves per plant, 7.89 in the bed $T_2$ (slope + vermicompost) and minimum, 5.61 in the bed $T_{16}$ (plain + control), maximum average leaf length, 33.86 cm in the bed $T_{14}$ (plain + vermi compost) and minimum, 23.11 cm in the bed $T_{12}$ (forrow + control), maximum average height per plant, 77.35 cm in the bed $T_{13}$ (plain + forest litter) and minimum, 40.32 cm in the bed $T_8$ (row + control), maximum fresh weight of rhizomes harvested, 181.23 qt/ha, dry weight of rhizomes obtained 39.12 qt/ha and rate of fresh weight of rhizomes increased 162.62% in the bed $T_1$ (slope + forest litter) and minimum fresh weight of rhizomes harvested 128.37 qt/ha, dry weight of rhizomes obtained 27.58 qt/ha and rate of fresh weight of rhizomes increased 74.86% in the bed $T_8$ (row + control) were observed during second year of cultivation. In the second year of cultivation,

**Table 2.** Pre-harvest and post-harvest agronomic characters of *H. spicatum* under different types of nursery beds and fertilizer treatments for second year of cultivation.

| Bed | Treatment | Spacing | % of survival | Pre-harvest agronomic character | | | | | Post-harvest agronomic character | | | |
|---|---|---|---|---|---|---|---|---|---|---|---|---|
| | | | | Days to sprout | Days to flower | Avg. no. of leaves/plant | Avg. leaf length (cm) | Avg. height /Plant (cm) | Fresh weight of rhizomes planted (qt/ha) | Fresh weight of rhizomes harvested (qt/ha) | Dry weight of rhizomes obtained (qt/ha) | Rate of fresh weight of Rhizomes increased (%) |
| $T_1$ | Slope + Litter | 30 × 30 | 100 | 166.22 ± 3.07 | 236.89 ± 4.23 | 7.44 ± 1.50 | 32.45 ± 8.49 | 62.08 ± 3.43 | 70.32 ± 9.68 | 181.23 ± 8.44 | 39.12 ± 1.84 | 162.62 ± 44.11 |
| $T_2$ | Slope + Vermi | 30 × 30 | 100 | 166.67 ± 4.21 | 237.78 ± 4.15 | 7.89 ± 1.60 | 32.76 ± 7.19 | 69.43 ± 2.84 | 72.32 ± 9.28 | 175.33 ± 7.87 | 37.53 ± 1.64 | 146.94 ± 41.54 |
| $T_3$ | Slope + FYM | 30 × 30 | 100 | 164.00 ± 3.35 | 234.22 ± 4.09 | 6.33 ± 1.41 | 27.18 ± 6.15 | 63.02 ± 2.87 | 71.66 ± 9.50 | 173.57 ± 16.09 | 37.19 ± 3.45 | 145.89 ± 39.03 |
| $T_4$ | Slope + Control | 30 × 30 | 100 | 167.00 ± 3.67 | 238.11 ± 3.59 | 6.00 ± 1.33 | 30.25 ± 9.62 | 53.98 ± 2.10 | 59.52 ± 6.71 | 138.72 ± 10.11 | 29.96 ± 2.22 | 135.69 ± 30.38 |
| $T_5$ | Row + Litter | 30 × 30 | 100 | 169.56 ± 3.54 | 241.44 ± 4.16 | 7.50 ± 1.82 | 29.08 ± 9.07 | 61.62 ± 2.20 | 72.60 ± 9.53 | 144.96 ± 9.78 | 31.25 ± 2.10 | 103.56 ± 34.65 |
| $T_6$ | Row + Vermi | 30 × 30 | 100 | 173.78 ± 3.73 | 248.00 ± 4.39 | 7.72 ± 1.32 | 32.62 ± 9.76 | 71.69 ± 3.34 | 71.35 ± 11.30 | 144.07 ± 10.97 | 30.97 ± 2.36 | 107.28 ± 41.96 |
| $T_7$ | Row + FYM | 30 × 30 | 100 | 172.22 ± 3.80 | 244.56 ± 4.07 | 6.56 ± 1.42 | 28.47 ± 9.31 | 58.58 ± 2.61 | 67.24 ± 7.47 | 135.37 ± 11.74 | 29.08 ± 2.42 | 102.89 ± 22.23 |
| $T_8$ | Row + Control | 30 × 30 | 100 | 174.33 ± 3.61 | 247.89 ± 4.04 | 6.00 ± 1.64 | 23.88 ± 5.89 | 40.32 ± 2.22 | 74.22 ± 7.63 | 128.37 ± 9.99 | 27.58 ± 2.18 | 74.86 ± 24.56 |
| $T_9$ | Furrow+litter | 30 × 30 | 100 | 180.89 ± 4.54 | 253.67 ± 3.94 | 7.17 ± 1.34 | 31.25 ± 8.01 | 65.96 ± 4.68 | 68.46 ± 10.14 | 154.73 ± 10.74 | 33.44 ± 2.31 | 130.62 ± 38.64 |
| $T_{10}$ | Furrow + Vermi | 30 × 30 | 100 | 176.22 ± 2.95 | 250.56 ± 3.81 | 6.50 ± 1.47 | 30.89 ± 8.94 | 70.90 ± 3.44 | 67.72 ± 8.66 | 145.82 ± 13.79 | 31.46 ± 2.94 | 119.39 ± 40.52 |
| $T_{11}$ | Furrow + FYM | 30 × 30 | 100 | 177.33 ± 3.20 | 251.78 ± 4.38 | 6.39 ± 1.50 | 29.94 ± 6.68 | 60.95 ± 2.97 | 63.64 ± 9.63 | 132.87 ± 10.30 | 28.67 ± 2.14 | 113.39 ± 39.96 |
| $T_{12}$ | Furrow + Control | 30 × 30 | 100 | 181.11 ± 4.40 | 254.33 ± 3.39 | 6.44 ± 1.42 | 23.11 ± 5.96 | 48.19 ± 2.78 | 69.39 ± 10.64 | 135.14 ± 14.62 | 29.09 ± 3.03 | 98.24 ± 31.89 |
| $T_{13}$ | Plane + Litter | 30 × 30 | 100 | 175.78 ± 3.31 | 250.78 ± 3.73 | 7.28 ± 1.18 | 33.36 ± 7.55 | 77.35 ± 2.52 | 63.86 ± 8.65 | 153.50 ± 12.49 | 33.13 ± 2.70 | 144.27 ± 35.84 |
| $T_{14}$ | Plane + Vermi | 30 × 30 | 100 | 174.78 ± 4.21 | 249.00 ± 5.07 | 6.56 ± 1.50 | 33.86 ± 7.98 | 70.36 ± 2.81 | 65.79 ± 10.12 | 153.51 ± 11.68 | 32.88 ± 2.51 | 137.62 ± 36.41 |
| $T_{15}$ | Plane + FYM | 30 × 30 | 100 | 180.11 ± 4.59 | 254.67 ± 4.03 | 6.94 ± 1.21 | 33.00 ± 8.34 | 72.98 ± 2.71 | 67.92 ± 10.09 | 155.43 ± 11.71 | 33.42 ± 2.57 | 134.94 ± 50.47 |
| $T_{16}$ | Plane + Control | 30 × 30 | 100 | 180.33 ± 4.64 | 256.78 ± 3.80 | 5.61 ± 1.69 | 25.27 ± 6.41 | 56.55 ± 3.47 | 67.00 ± 9.12 | 131.47 ± 12.24 | 28.25 ± 2.64 | 99.11 ± 29.50 |
| ANOVA | | | DF | 15 | 15 | 15 | 15 | 15 | 15 | 15 | 15 | 15 |
| | | | MS | 288.48 | 464.635 | 8.07 | 234.88 | 1715.66 | 134.26 | 2386.79 | 109.27 | 4994.29 |
| | | | F-value | 19.56*** | 28.04*** | 3.74*** | 3.73*** | 190.48*** | 1.54NS | 17.74 *** | 17.76*** | 3.64*** |

ANOVA was found significant for days to sprout, days to flower, average number of leaves per plant, average leaf length (cm), average height per plant (cm), fresh weight of rhizomes harvested (qt/ha), dry weight of rhizomes obtained (qt/ha) and rate of fresh weight of rhizomes increased (%) with F value (19.56, 28.04, 3.74, 3.73, 190.48, 17.74, 17.76 and 3.64, respectively at $p < 0.001$), wherein found non-significant for fresh weight of rhizomes planted (qt/ha) with F-value (1.54 at $p > 0.05$).

In the third year of cultivation (Table 3), the plant took about 160 to 183 days to sprout and 229 to 259.67 days to flower from the date of dormancy and 100% survival was recorded in all type of beds. Maximum average number of leaves per plant 8.56 in the bed $T_6$ (row + vermicompost) and minimum 6.22 in the bed $T_{16}$ (plain + control), maximum average leaf length 34.49 cm in the bed $T_{15}$ (plain + FYM) and minimum 24.04 cm in the bed $T_{12}$ (forrow + control), maximum average height per plant 78.73 cm in the bed $T_{13}$ (plain + forest litter) and minimum 41.22 cm in the bed $T_8$ (row + control), maximum fresh weight of rhizomes harvested 225.48 qt/ha, dry weight of rhizomes obtained 48.72 qt/ha and rate of fresh weight of rhizomes increased 232.34% in the bed $T_1$ (slope + forest litter) and minimum fresh weight of rhizomes harvested 158.62 qt/ha, dry weight of rhizomes obtained 34.06 qt/ha and rate of fresh weight of rhizomes increased 123.85% in the bed $T_8$ (row + control) were observed during second year of cultivation. In the third year of cultivation,

**Table 3.** The pre-harvest and post-harvest agronomic characters of *H. spicatum* under different types of nursery beds and fertilizer treatments for third year of cultivation.

| | | | | Pre-harvest agronomic character | | | | | Post-harvest agronomic character | | | |
|---|---|---|---|---|---|---|---|---|---|---|---|---|
| Bed | Treatment | Spacing | % of survival | Days to sprout | Days to flower | Avg. no. of leaves/plant | Avg. leaf length (cm) | Avg. height /plant (cm) | Fresh weight of rhizomes planted (qt/ha) | Fresh weight of rhizomes harvested (qt/ha) | Dry weight of rhizomes obtained (qt/ha) | Rate of fresh weight of rhizomes increased (%) |
| $T_1$ | Slope + Litter | 30 × 30 | 100 | 160.00 ± 3.12 | 229.00 ± 3.50 | 7.89 ± 1.45 | 32.83 ± 4.79 | 62.86 ± 4.36 | 68.87 ± 9.37 | 225.48 ± 10.45 | 48.72 ± 2.25 | 232.34 ± 43.24 |
| $T_2$ | Slope + Vermi | 30 × 30 | 100 | 163.00 ± 2.78 | 232.89 ± 3.69 | 8.44 ± 1.81 | 33.31 ± 5.41 | 69.98 ± 2.61 | 69.58 ± 11.00 | 217.33 ± 10.24 | 46.74 ± 2.27 | 220.79 ± 62.14 |
| $T_3$ | Slope + FYM | 30 × 30 | 100 | 165.00 ± 2.69 | 236.00 ± 4.24 | 7.33 ± 1.00 | 27.48 ± 6.22 | 64.42 ± 2.16 | 69.90 ± 8.61 | 214.17 ± 19.48 | 46.03 ± 3.99 | 210.43 ± 45.20 |
| $T_4$ | Slope + Control | 30 × 30 | 100 | 163.22 ± 3.99 | 232.00 ± 4.85 | 6.89 ± 1.27 | 31.08 ± 5.98 | 55.13 ± 1.28 | 68.38 ± 6.87 | 189.86 ± 14.19 | 40.74 ± 3.07 | 180.51 ± 39.67 |
| $T_5$ | Row + Litter | 30 × 30 | 100 | 167.11 ± 3.22 | 238.89 ± 4.57 | 8.11 ± 1.05 | 29.48 ± 6.68 | 62.20 ± 2.58 | 72.57 ± 8.41 | 180.56 ± 12.44 | 38.98 ± 2.78 | 152.20 ± 38.14 |
| $T_6$ | Row + Vermi | 30 × 30 | 100 | 169.22 ± 4.35 | 242.22 ± 3.99 | 8.56 ± 1.13 | 33.53 ± 5.21 | 72.00 ± 2.93 | 70.80 ± 8.98 | 177.55 ± 13.23 | 38.20 ± 2.88 | 153.77 ± 33.03 |
| $T_7$ | Row + FYM | 30 × 30 | 100 | 174.00 ± 3.77 | 246.89 ± 3.86 | 7.33 ± 1.41 | 28.74 ± 5.59 | 59.73 ± 1.98 | 67.82 ± 8.07 | 166.63 ± 13.09 | 35.89 ± 2.73 | 148.71 ± 34.94 |
| $T_8$ | Row + Control | 30 × 30 | 100 | 172.33 ± 3.91 | 244.56 ± 4.10 | 6.89 ± 1.17 | 24.16 ± 6.39 | 41.22 ± 2.61 | 72.48 ± 12.00 | 158.62 ± 12.46 | 34.06 ± 2.61 | 123.85 ± 37.77 |
| $T_9$ | Furrow + litter | 30 × 30 | 100 | 176.89 ± 3.98 | 249.89 ± 4.43 | 7.78 ± 1.64 | 32.04 ± 4.60 | 67.13 ± 3.21 | 71.14 ± 10.27 | 190.85 ± 13.65 | 41.16 ± 2.93 | 172.29 ± 35.52 |
| $T_{10}$ | Furrow + Vermi | 30 × 30 | 100 | 174.78 ± 3.31 | 248.78 ± 3.99 | 7.44 ± 1.51 | 30.99 ± 9.84 | 71.62 ± 3.43 | 68.77 ± 10.61 | 181.64 ± 17.30 | 39.11 ± 3.84 | 169.59 ± 46.86 |
| $T_{11}$ | Furrow + FYM | 30 × 30 | 100 | 174.33 ± 3.46 | 247.56 ± 3.40 | 7.11 ± 1.69 | 30.57 ± 6.86 | 62.37 ± 2.51 | 64.93 ± 7.98 | 167.91 ± 14.08 | 36.26 ± 2.96 | 160.64 ± 26.12 |
| $T_{12}$ | Furrow + Control | 30 × 30 | 100 | 176.67 ± 3.08 | 248.56 ± 4.61 | 7.33 ± 1.58 | 24.04 ± 5.86 | 48.83 ± 2.65 | 72.87 ± 10.50 | 168.02 ± 17.93 | 36.08 ± 3.85 | 133.71 ± 32.68 |
| $T_{13}$ | Plane + Litter | 30 × 30 | 100 | 179.44 ± 3.71 | 254.44 ± 4.30 | 8.00 ± 1.58 | 33.62 ± 4.59 | 78.73 ± 2.89 | 64.39 ± 8.65 | 190.12 ± 15.64 | 40.90 ± 3.49 | 199.00 ± 39.11 |
| $T_{14}$ | Plane + Vermi | 30 × 30 | 100 | 177.11 ± 3.10 | 251.78 ± 3.96 | 7.22 ± 1.20 | 34.34 ± 10.59 | 72.54 ± 2.78 | 67.85 ± 8.63 | 192.42 ± 14.42 | 41.43 ± 3.11 | 188.78 ± 48.80 |
| $T_{15}$ | Plane + FYM | 30 × 30 | 100 | 175.89 ± 3.55 | 250.67 ± 4.00 | 7.89 ± 1.05 | 34.49 ± 5.77 | 74.20 ± 2.20 | 61.89 ± 6.39 | 171.98 ± 12.45 | 37.09 ± 2.70 | 179.88 ± 28.63 |
| $T_{16}$ | Plane + Control | 30 × 30 | 100 | 183.00 ± 4.64 | 259.67 ± 3.97 | 6.22 ± 1.09 | 25.88 ± 5.95 | 58.16 ± 3.60 | 71.08 ± 9.64 | 164.51 ± 15.26 | 35.32 ± 3.26 | 136.65 ± 49.37 |
| | | DF | | 15 | 15 | 15 | 15 | 15 | 15 | 15 | 15 | 15 |
| ANOVA | | MS | | 397.97 | 683.27 | 3.41 | 109.43 | 869.09 | 86.72 | 3577.72 | 167.31 | 9042.56 |
| | | F-value | | 31.01*** | 40.47*** | 1.80* | 2.61** | 109.15*** | 1.02NS | 17.37*** | 17.55*** | 5.38*** |

Significant levels (* = P<0.05; ** = P<0.01; ***P<0.001); NS, non significant.

ANOVA was found significant for days to sprout, days to flower, average height per plant (cm), fresh weight of rhizomes harvested (qt/ha), dry weight of rhizomes obtained (qt/ha) and rate of fresh weight of rhizomes increased (%) with F-value (31.01, 40.47, 109.15, 17.37, 17.55 and 5.38, respectively at $p < 0.001$); average number of leaves per plant with F-value (1.80 at $p < 0.05$); average leaf length (cm) with F-value (2.61 at $p < 0.01$), whereas found non-significant for fresh weight of rhizomes planted (qt/ha) with F-value (1.02 at $p > 0.05$).

## Correlation in morphological traits

In the first year of cultivation of *H. spicatum* (Table 4), days to sprout showed significant positive correlation with days to flower (r = 0.99 at p = 0.05). Average number of leaves per plant was significantly and positively correlated with average leaf length (cm) and average height/plant (cm) (r = 0.63 and 0.61, respectively at p = 0.05). Average leaf length (cm) showed significant positive correlation with average height/plant (cm), fresh weight of rhizomes harvested (qt/ha), dry weight of rhizomes obtained (qt/ha) and rate of fresh weight of rhizomes increased (%) (r = 0.88, 0.71, 0.70 and 0.75, respectively at p = 0.05). Average height/plant (cm) showed significant positive

**Table 4.** Correlation between different morphological traits of *H. spicatum* for first year of cultivation.

| Parameter | Days to sprout | Days to flower | Avg. no. of leaves/plant | Avg. leaf length (cm) | Avg. height/ plant (cm) | Fresh weight of rhizomes planted (qt/ha) | Fresh weight of rhizomes harvested (qt/ha) | Dry weight of rhizomes obtained (qt/ha) | Rate of fresh weight of rhizomes increased (%) |
|---|---|---|---|---|---|---|---|---|---|
| Days to sprout | 1.00 | | | | | | | | |
| Days to flower | 0.99* | 1.00 | | | | | | | |
| Avg. no. of leaves/plant | -0.27 | -0.26 | 1.00 | | | | | | |
| Avg. leaf length (cm) | -0.33 | -0.27 | 0.63* | 1.00 | | | | | |
| Avg. height/plant (cm) | -0.21 | -0.14 | 0.61* | 0.88* | 1.00 | | | | |
| Fresh weight of rhizomes planted (qt/ha) | -0.07 | -0.05 | 0.30 | 0.44 | 0.36 | 1.00 | | | |
| Fresh weight of rhizomes harvested (qt/ha) | -0.28 | -0.26 | 0.44 | 0.71* | 0.59* | 0.84* | 1.00 | | |
| Dry weight of rhizomes obtained (qt/ha) | -0.29 | -0.26 | 0.43 | 0.70* | 0.57* | 0.83* | 1.00* | 1.00 | |
| Rate of fresh weight of rhizomes increased (%) | -0.41 | -0.40 | 0.39 | 0.75* | 0.64* | 0.51* | 0.90* | 0.90* | 1.00 |

DF=14; p= 0.05; r= 0.497; * = P<0.05.

**Table 5.** Correlation between different morphological traits of *H. spicatum* for second year of cultivation.

| Parameter | Days to sprout | Days to flower | Avg. no. of leaves/plant | Avg. leaf lengt (cm) | Avg. height/ plant (cm) | Fresh weight of rhizomes planted (qt/ha) | Fresh weight of rhizomes harvested (qt/ha) | Dry weight of rhizomes obtained (qt/ha) | Rate of fresh weight of rhizomes increased (%) |
|---|---|---|---|---|---|---|---|---|---|
| Days to sprout | 1.00 | | | | | | | | |
| Days to flower | 0.99* | 1.00 | | | | | | | |
| Avg. no. of leaves/plant | -0.25 | -0.26 | 1.00 | | | | | | |
| Avg. leaf length (cm) | -0.17 | -0.14 | 0.64* | 1.00 | | | | | |
| Avg. height/plant (cm) | 0.004 | 0.05 | 0.60* | 0.86* | 1.00 | | | | |
| Fresh weight of rhizomes planted (qt/ha) | -0.19 | -0.21 | 0.37 | -0.27 | -0.17 | 1.00 | | | |
| Fresh weight of rhizomes harvested (qt/ha) | -0.55* | -0.57* | 0.59* | 0.54* | 0.52* | 0.27 | 1.00 | | |
| Dry weight of rhizomes obtained (qt/ha) | -0.55* | -0.57* | 0.59* | 0.55* | 0.52* | 0.26 | 1.00* | 1.00 | |
| Rate of fresh weight of rhizomes increased (%) | -0.44 | -0.45 | 0.42 | 0.71* | 0.62* | -0.25 | 0.86* | 0.87* | 1.00 |

DF=14; p = 0.05; r = 0.497; * = P<0.05.

correlation with fresh weight of rhizomes harvested (qt/ha), dry weight of rhizomes obtained (qt/ha) and rate of fresh weight of rhizomes increased (%) (r = 0.59, 0.57 and 0.64, respectively at p = 0.05). Fresh weight of rhizomes planted (qt/ha) showed significant positive correlation with fresh weight of rhizomes harvested (qt/ha), dry weight of rhizomes obtained (qt/ha) and rate of fresh weight of rhizomes increased (%) (r = 0.84, 0.83 and 0.51, respectively at p = 0.05). Fresh weight of rhizomes harvested (qt/ha) showed perfect positive significant correlation with dry weight of rhizomes obtained (qt/ha) (r = 1.00 at p = 0.05) and it was significantly and positively correlated with rate of fresh weight of rhizomes increased (%) (r = 0.90 at p = 0.05) and dry weight of rhizomes obtained (qt/ha) was significantly and positively correlated with rate of fresh weight of rhizomes increased (%) (r = 0.90 at p = 0.05). In the second year of cultivation (Table 5), days to

**Table 6.** Correlation between different morphological traits of *H. spicatum* for third year of cultivation.

| Parameter | Days to sprout | Days to flower | Avg. no. of leaves/plant | Avg. leaf length (cm) | Avg. height /plant (cm) | Fresh weight of rhizomes planted (qt/ha) | Fresh weight of rhizomes harvested (qt/ha) | Dry weight of rhizomes obtained (qt/ha) | Rate of fresh weight of rhizomes increased (%) |
|---|---|---|---|---|---|---|---|---|---|
| Days to sprout | 1.00 | | | | | | | | |
| Days to flower | 0.99* | 1.00 | | | | | | | |
| Avg. no. of leaves/ plant | -0.39 | -0.36 | 1.00 | | | | | | |
| Avg. leaf length (cm) | -0.17 | -0.13 | 0.62* | 1.00 | | | | | |
| Avg. height/ plant (cm) | 0.10 | 0.16 | 0.58* | 0.85* | 1.00 | | | | |
| Fresh weight of rhizomes planted (qt/ha) | -0.18 | -0.21 | -0.10 | -0.61* | -0.58* | 1.00 | | | |
| Fresh weight of rhizomes harvested (qt/ha) | -0.67* | -0.65* | 0.43 | 0.46 | 0.39 | -0.03 | 1.00 | | |
| Dry weight of rhizomes obtained (qt/ha) | -0.67* | -0.65* | 0.44 | 0.47 | 0.40 | -0.04 | 1.00* | 1.00 | |
| Rate of fresh weight of rhizomes increased (%) | -0.54* | -0.52* | 0.44 | 0.64* | 0.57* | -0.38 | 0.93* | 0.93* | 1.00 |

DF=14; p = 0.05; r = 0.497; * = P<0.05.

sprout showed significant positive correlation with days to flower (r = 0.99 at p = 0.05). Days to flower showed significant but negative correlation with fresh weight of rhizomes harvested (qt/ha) and dry weight of rhizomes obtained (qt/ha) (r = -0.57 and -0.57, respectively at p = 0.05). Average number of leaves per plant was significantly and positively correlated with average leaf length (cm), average height/plant (cm), fresh weight of rhizomes harvested (qt/ha) and dry weight of rhizomes obtained (qt/ha) (r = 0.64, 0.60, 0.59 and 0.59, respectively at p = 0.05). Average leaf length (cm) showed significant positive correlation with average height/plant (cm), fresh weight of rhizomes harvested (qt/ha), dry weight of rhizomes obtained (qt/ha) and rate of fresh weight of rhizomes increased (%) (r = 0.86, 0.54, 0.55 and 0.71, respectively at p = 0.05). Average height/plant (cm) showed significant positive correlation with fresh weight of rhizomes harvested (qt/ha), dry weight of rhizomes obtained (qt/ha) and rate of fresh weight of rhizomes increased (%) (r = 0.52, 0.52 and 0.62, respectively at p = 0.05). Fresh weight of rhizomes harvested showed perfect positive significant correlation with dry weight of rhizomes obtained (qt/ha) (r = 1.00 at 0.05) and showed significant positive correlation with rate of fresh weight of rhizomes increased (%) (r = 0.86 at p = 0.05) and dry weight of rhizomes obtained (qt/ha) was significantly and positively correlated with rate of fresh weight of rhizomes increased (%) (r = 0.87 at p = 0.05).

In the third year of cultivation (Table 6), days to sprout showed significant positive correlation with days to flower (r = 0.99 at p = 0.05) but it showed significant negative correlation with fresh weight of rhizomes harvested (qt/ha), dry weight of rhizomes obtained (qt/ha) and rate of fresh weight of rhizomes increased (%) (r = -0.67, -0.67 and -0.54, respectively at p = 0.05). Average number of leaves per plant showed significant positive correlation with average leaf length (cm) and average height/plant (cm) (r = 0.62 and 0.58, respectively at p = 0.05). Average leaf length (cm) was significantly and positively correlated with average height/plant (cm) and rate of fresh weight of rhizomes increased (%) (r = 0.85 and 0.64, respectively at p = 0.05) but showed negative significant correlation with fresh weight of rhizomes planted (qt/ha) (r = -0.61 at p = 0.05). Average height/plant (cm) showed significant positive correlation with rate of fresh weight of rhizomes increased (%) (r = 0.57 at p = 0.05) but showed significant negative correlation with fresh weight of rhizomes planted (qt/ha) (r = -0.58 at p = 0.05). Fresh weight of rhizomes harvested showed perfect significant positive correlation with dry weight of rhizomes obtained (qt/ha) (r = 1.00 at p = 0.05) and it was significantly and positively correlated with rate of fresh weight of rhizomes increased (%) (r = 0.93 at p = 0.05) and dry weight of rhizomes obtained (qt/ha) showed significant positive correlation with rate of fresh weight of rhizomes increased (%) (r = 0.93 at p = 0.05) (Figure 1).

## DISCUSSION

The implementation of plant species conservation involves two broad approaches: (i) *in situ*- protection of species within habitats (where the protected area networks play a crucial role) and (ii) *ex situ* - use of botanical gardens, arboreta and *in vitro* methods including high-tech cryopreservation.

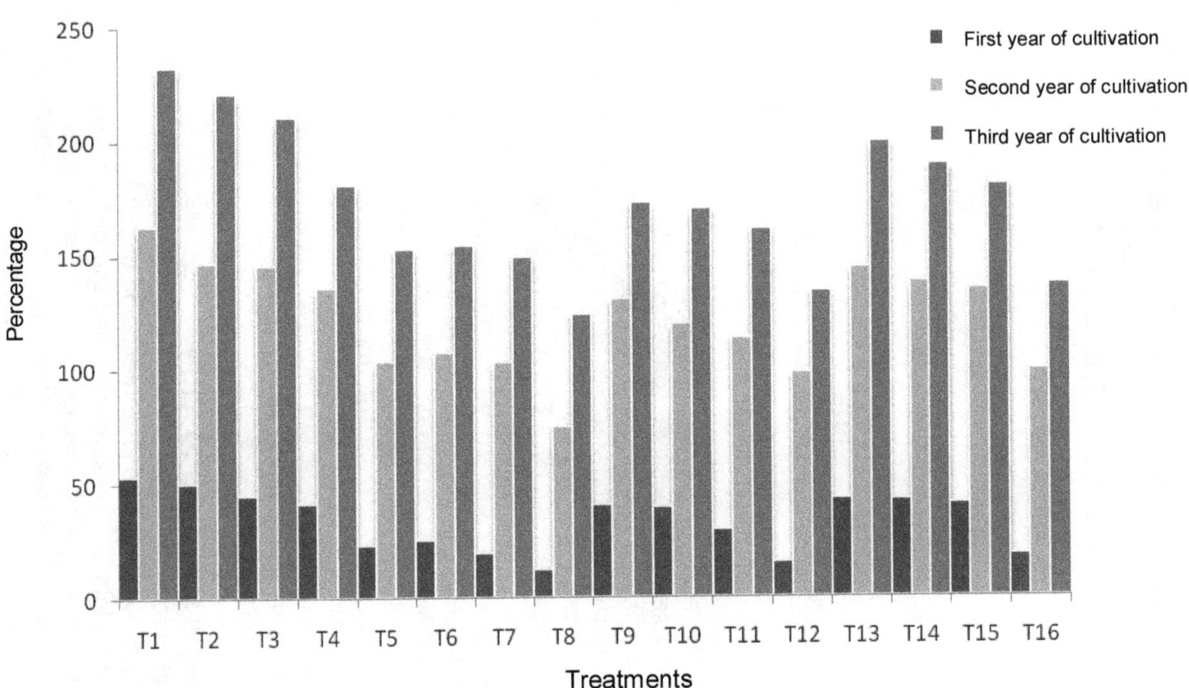


**Figure 1.** Showing rate of fresh weight of rhizomes increased under experimental cultivation of *H. spicatum* in different types of treatments for first, second and third year.

*Ex situ* cultivation of threatened plant taxa, particularly those in high demand for trade, has been seen as a practical step, not only indirectly supporting *in situ* conservation (by diverting attention from *in situ* harvesting) but also in reaching a sustainable supply of raw material.

Economic yield was found lower under control beds as compared to litter, FYM and vermicompost in all type of beds prepared in slopes, rows, furrows and plain. In control condition, minerals are not enough for proper growth and development of plants. Chauhan and Bhatt (2000) concluded that addition of 5 to 10 tons/ha of FYM resolve nutritional problems of various hill crops and deteriorating physical condition of soil. Chauhan and Nautiyal (2005) found that economic yield increased with addition of manure (FYM) in all treatments and altitudes compared to control. Application of FYM and leaf litter in cultivated field is traditional practices in Kumaun Himalaya for the better yield and production. Plants grown under different treatments showed much higher economic yield as compared to control beds. However, yield was higher at $T_1$ (slope + forest litter) as compared to other beds, this difference may be due to low level of mineral nutrient in alpine soil. Several workers (Patidar and Mali, 2001; Saharan et al., 2001; Kasera and Sharan, 2002) also found increase in biomass production with the use of FYM in different crops, while in our study highest incensement in biomass is obtained with the use of leaf litter. Higher economic yield in manure as well as litter beds could become possible only due to availability of essential mineral nutrients, which were needed for growth and development. Addition of manure, litter and vermicompost increased moisture content of soil and retained it for quite some time. It also improved physical, chemical and microbial properties of soil and thereby its productivity. Increased soil fertility enhanced vegetative growth of plants and additional food got stored in underground rhizomes, which improved economic yield of plants. Similar observations were earlier made by Ramamurthy et al. (1998), Phirke (2001) and Sharma (2002) and also supported addition of bio-fertilizer for improvement of soil quality.

Spacing treatment generally does not have any significant effect on economic yield. Generally, spacing showed effect on biomass when population density increased at per unit area and plants compete for space, moisture, mineral nutrient or sunlight. But during short period of 3 years, spacing could not show significant difference in economic yield production due to low density during 3-year growth; therefore, we used same spacing during all three growth seasons. However, spacing will certainly affect economic yield in latter stages of development due to increase in above ground and below ground surface area per plant in per unit area. Such study will require long time for fruitful result due to long maturation time. Maturity stage is 7 to 8 years for high altitude species and maximum gain can be obtained by fully matured plants (Rai et al., 2000; Nautiyal et al.,

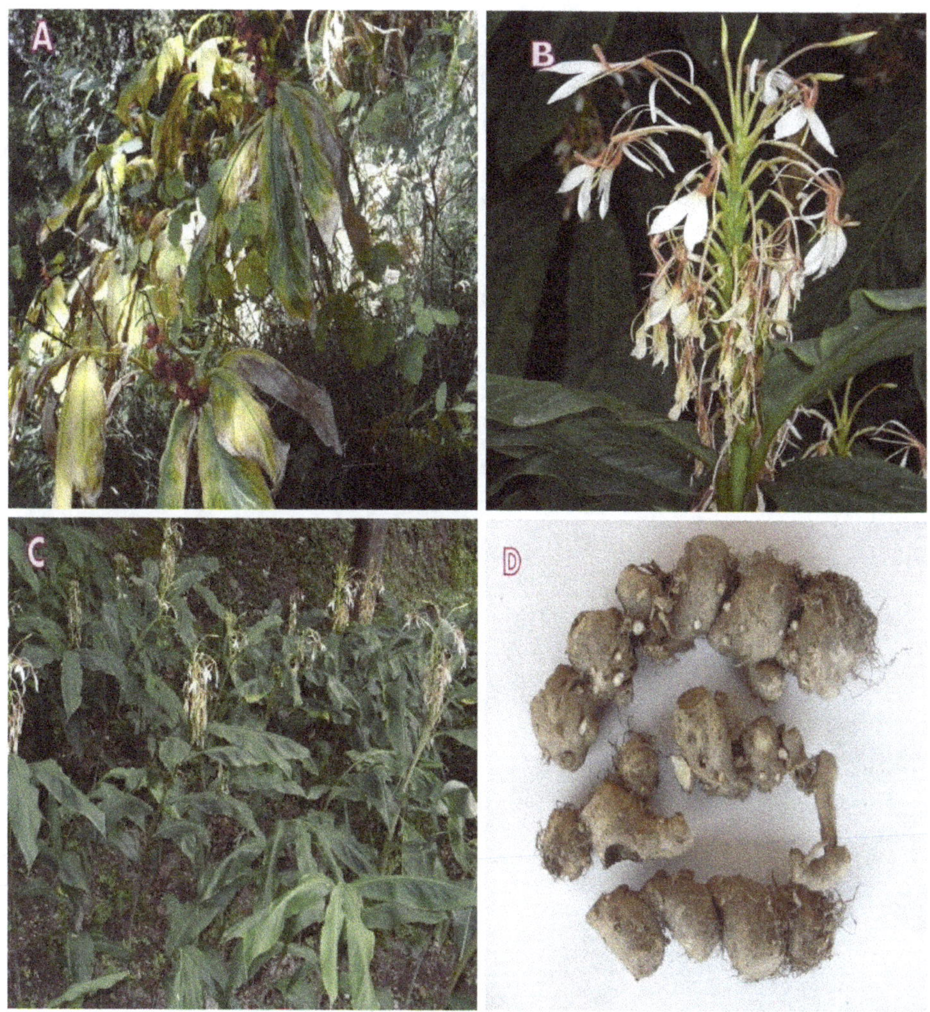


**Plate 1.** Different stages of *H. spicatum*: (A) In wild. (B) Flowering. (C) In experimental beds. (D) Harvested rhizomes.

2002b) (Plate 1).

## Conclusion

From the present study, it may be concluded that the maximum economic yield of rhizomes is obtained in the beds prepared in slope in comparison to beds prepared in plain, rows and furrows. However, best results are obtained in the bed prepared in slope and supplemented with forest litter. It means that the plantation in slopes supplemented with forest litter will results maximum economic yield. The observations are based on 3 years of study. This study on the cultivation technology of *H. spicatum* may help in sustainable utilization, fulfillment of pharmaceutical demand and maintenance of its population in wild.

## ACKNOWLEDGEMENT

The authors are thankful to Director General, C.C.R.A.S., New Delhi for providing necessary facilities for this work.

## REFERENCES

Anonymous (2008). Agro-techniques of selected medicinal plants. National medicinal plant Board, New Delhi. 1.

Chauhan RS, Nautiyal MC (2005). Commercial viability of cultivation of an endangered medicinal herb *Nardostachys jatamansi* at three different agro climatic zones. Curr. Sci. 89:1481-1488.

Chauhan VS, Bhatt JC (2000). Agriculture in Uttarakhand: From subsistence towards self-sufficiency. In- *Uttarakhand statehood: Dimensions of Development*. (Editors- M. C. Sati and S. P. Sati). Indus Pub. Co. New Delhi. pp. 168-180.

Gaur RD (1999). Flora of District Garhwal North West Himalaya. Transmedia Srinagar (Garhwal). P. 715.

Kasera PK, Sharan P (2002). Economics of *Evolvus alsinoides* (Sankhpusphi) from Indian Thar Desert. Annals of Forestry, 10 (1):167-171.

Naithani BD (1985). Flora of Chamoli. Vol. II. Botanical Survey of India, Howarth.

Nautiyal MC, Nautiyal BP (2004). Agrotechniques of High Altitude Medicinal and aromatic Plants. Bishen Singh Mahendra Pal Singh, Dehradun.

Nautiyal MC, Vinay P, Nautiyal BP (2002b). Cultivation technique of some high altitude medicinal herb. Ann. For. 1: 62-67.

Patidar M, Mali AL (2001). Integrated nutrient management in sorghum (*Sorghum bicolour*) and its residual effect on wheat (*Triticum aestivum*). Ind. J. Agric. Sci. 71:587-590.

Phirke NV (2001). Biotech inputs for improving the yield of Banana. D. Phil. Thesis, North Maharastra Univ. Jalgawn. pp. 1-154.

Rai LK, Prasad P, Sharma E (2000). Conservation threats to some important medicinal plants of the Sikkim Himalaya. Biol. Conserv. 93:27-33.

Ramamurthy V, Sharma RK, Kothari RM (1998). Microbial conservation of Ligno-Cellulosic waste into soil conditioner. In-Advances in Biotechnology, (Editor- A. Pandey). Educational Pub. New Delhi. pp. 433-438.

Saharan P, Shkula JK, Kesera PK, Chawan DD (2001). Effect of different nutritional treatments on growth and biomass of Sankhpushpi. Ind. J. Bot. Soc. 80:127-131.

Sarin YK (2008). Principal Crud Herbal Drugs of India- An Illustrated Guide to Important, Largely Used and Traded Medicinal Raw Materials of Plant Origin. Bishen Singh Mahendra Pal Singh, Dehradun. pp. 176-177.

Sharma R (2002). Aushdhiya Avyam Sugandhiya Paudhon Ki Krishi Technique (Agro-technique of Medicinal and Aromatic Plants). Daya Publishing House Delhi.

Singh RS (1983). Vanausadhi Nidarshika (Ayurvedic Pharmacopeia). Uttar Pradesh Hindi Sansthan, Lucknow.

Thakur RS, Puri HS, Husain A (1989). Major Medicinal Plants of India. CIMAP, Lucknow.

Vijyalashmi SS, Shyam S (1993). Varkshayurveda-An Introduction to Indian Plant Science, LSPSS Monograph No. 9. Lok Swasthya Parampara Samvardhan Samithi, South Madras- 600086.

# 21

# Genetic diversity in soybean genotypes under water stress and normal condition using factor analysis and cluster analysis

**Sarkaut Salimi**

Islamic Azad University, Mariwan Branch, Mariwan, Iran.

**To study the relationships between morphological characters of soybean plant an experiment was conducted in randomized complete blocks design (RCBD) in two replications under drought stress condition at Agricultural College of Guilan University in 2008. Result of analysis of variance showed that there was significant difference among the studied soybean genotypes in the majority of traits. The result of factor analysis in under water stress condition showed that 7 independent factors for characters to explain 86.4% variation of all data and under normal condition 5 independent factors for characters to explain 82% variation of all data. A similarity factor was constructed using UPGMA method for morphological characters varieties were classified into 8 groups for water stress condition and 4 groups for normal condition. Classifying the results of the cluster analysis identified Hamilton genotype suitable for sown in water stress condition and majority genotypes suitable for sown in normal condition.**

**Key words:** Cluster analysis, factor analysis, genetic diversity, normal condition, soybean, water stress.

## INTRODUCTION

Soybean seed is a major source of high-quality protein and oil for human consumption (Katerji et al., 2001). The unique chemical composition of soybean has made it one of the most valuable agronomic crops worldwide (Thomas et al., 2003). Its protein has great potential as a major source of dietary protein. The oil produced from soybean is highly digestible and contains no cholesterol (Essa and Al-ani, 2001). Genetic diversity analysis reveals genetic backgrounds and relationships of germplasm, and also provides strategies to establish, unitize, and manage crop core collections (Brown-Guedira et al., 2000; Roussel et al., 2004). Soybean genetic diversity and relationships can be assessed by the differences in morphological and agronomic traits, pedigree information, geographic origins, isozymes and DNA markers (Dong et al., 2004; Guan et al., 2010; Wang et al., 2010). The importance of genetic diversity in plant breeding is obvious from the results obtained in different crops (Ghafoor et al., 2001; Smart, 1990; Upadhyaya, 2003; Upadhyaya et al., 2002). Water stress is considered one of the most important factors limiting plant performance and yield worldwide (Boyer, 1982). Water stress during reproductive development often decreases the seed size in soybean (Kadhem et al., 1985; Momen et al., 1979; Sionet and Kramer, 1977). Waterlogging is defi ned as prolonged soil saturation with water at least 20% higher than the field capacity (Aggarwal et al., 2006). Increasing soil water deficiency correlated with reduction in dry matter accumulation (Lopez et al., 1996; Lazcano-Ferrat and Lovatt, 1999; Grieu et al., 2001). The best option for crop production, yield improvement and yield stability under water stress conditions is to develop water tolerant crop varieties. One of the main goals in breeding programs is selection of the

**Table 1.** Used genotypes.

| S/N | Genotypes | Origin (country) | S/N | Genotypes | Origin (country) |
|---|---|---|---|---|---|
| 1 | Line 33 | Iran | 11 | Zane | Iran |
| 2 | Hill | Iran | 12 | Hack | Iran |
| 3 | Union | Iran | 13 | Dw2 | Iran |
| 4 | Bp | Iran | 14 | Line 32 | Iran |
| 5 | Hamilton | Iran | 15 | Clark | Iran |
| 6 | Streslland | Iran | 16 | Gorgan 3 | Iran |
| 7 | Williams | Iran | 17 | Talar | Iran |
| 8 | Tnh 56 | Iran | 18 | Century | Iran |
| 9 | Dpx | Iran | 19 | Line 17 | Iran |
| 10 | Williams 82 | Iran | | | |

**Table 2.** Analysis of variance (RCBD) for studied traits

| S.O.V | df | MS | | | | | | | | |
|---|---|---|---|---|---|---|---|---|---|---|
| | | Plant height | Number of pod | Leaf area | Number of Seed on Pod | Number of seed on plant | 100 grain weight | Day to 50% flowering | Day to maturity | Grain yield |
| Replication | 2 | 0/013$^{ns}$ | 0/71$^{ns}$ | 0/02$^{ns}$ | 0/012$^{ns}$ | 0/87$^{ns}$ | 1/90$^{ns}$ | 1/28$^{ns}$ | 7/6$^{*}$ | 0/05$^{ns}$ |
| Genotype | 18 | 0/031$^{ns}$ | 12/53$^{ns}$ | 1/22$^{ns}$ | 0/029$^{ns}$ | 12/70$^{ns}$ | 5/37$^{**}$ | 170/21** | 41/06** | 2/77* |
| Error | 18 | 0/032 | 5/84 | 0/97 | 0/03 | 6/25 | 10/31 | 0/34 | 1/38 | 1/19 |
| %CV | | 8/64 | 22/38 | 13/98 | 8/49 | 16/72 | 7/96 | 2/42 | 2/77 | 17/83 |

*,** Significantly different at 5 and 1% probability level respectively; ns, non significant.

best genotypes under water stress conditions. Insufficient water, especially during emergence, flowering and pod-filling stages lower the yield of soybean (Abayomi, 2008). Optimally supplying water during different growth stages such as the early of growth season, flowering, pod set, and grain–filling improves crop yield and quality; and plants need more water from flowering to onset of grain–filling than any other times (Speeht et al., 2001). Narjesi et al. (2007) reported that 5 independent factors for characters 30 soybean genotypes to explain 80.2% variation of all data. The first factor alone 22.54 of the data changes can be justified and called the phonological properties. (Arshad et al., 2006) showed that cluster analysis the genotypes divided in 3 groups. The first group was involved 14 genotypes, second group 33 genotypes and third group involved 11 genotypes. The objectives of this study are to evaluate the genetic diversity of soybean cultivar using factor analysis and cluster analysis, to analyze and characterize population structure within soybean cultivars and compare effect traits on grain yield under water stress and normal condition.

## MATERIALS AND METHODS

The present study was conducted to evaluate the effects of water stress and normal condition on soybean genotypes. To study the relationships between morphological characters of soybean plant an experiment was conducted in randomized complete blocks design (RCBD) with two replication sat water stress conditions and normal condition at Agricultural College of Guilan University, Iran during 2008. The material consisted of 19 soybean genotypes (Table 1). The seeds sown in the spring season and genotypes were grown in two row plots, each plot included four ridges, and each ridge was 3.5 m in length and 50 cm apart. Agronomic characteristics were including plant height, number of pod, leaf area, number of seed on pod, number of seed on plant, 100 grain weight, day to 50% flowering, day to maturity and grain yield. Data were recorded on 5 competitive plants of each plot and grain yield (kg/ha ) was calculated for the entire plot. Data were statistically analyzed using ANOVA appropriate for RCBD with SAS ver. 9.1 and factor analysis and cluster analysis using SPSS 16 software's.

## RESULTS AND DISCUSSION

Result of analysis of variance (Table 2) showed that there were significant differences among the studied soybean genotypes for yield and component yield traits. This illustrates the high potential of these genotypes to use the genetically source for breeding purposes.

### Factor analysis

Factor analysis based principal component analysis and after varimax rotation under water stress condition

**Table 3.** Factor analysis of studied traits under water stress condition.

| Parameter | PC1 | PC2 | PC3 | PC4 | PC5 |
|---|---|---|---|---|---|
| Eigen values | 6/155 | 3/506 | 2/842 | 2/011 | 1/952 |
| Cumulative Eigen values | 6/155 | 9/661 | 12/503 | 14/514 | 16/466 |
| Proportion of variance | 26/176 | 15/309 | 14/365 | 13/489 | 12/991 |
| Cumulative variance | 26/176 | 41/485 | 55/850 | 69/339 | 82/331 |
| **Traits** | | | | | |
| Plant height | 0/681 | 0/365 | -0/403 | 0/215 | 0/216 |
| Number of pod | -0/437 | 0/014 | 0/210 | 0/735 | 0/190 |
| Leaf area | 0/038 | 0/825 | -0/115 | 0/241 | 00/145 |
| Number of seed per pod | -0/228 | -0/275 | 0/327 | 0/014 | 0/678 |
| Number of seed per plant | 0/392 | -0/138 | -0/193 | 0/364 | 0/128 |
| 100 grain weight | 0/876 | -0/774 | -0/246 | 0/227 | -0/169 |
| Day to 50% flowering | 0/868 | 0/061 | 0/097 | 0/2 | -0/241 |
| Day to maturity | 0/109 | 0/19 | 0/058 | 0/868 | -0/309 |
| grain yield | 0/203 | 0/254 | 0/101 | 0/804 | -0/344 |
| Oil content | 0/034 | 0/88 | -0/221 | 0/267 | 0/105 |
| Protein content | 0/346 | 0/486 | 0/442 | 0/045 | 0/132 |

**Table 4.** Factor analysis of studied traits under normal (non stress) condition.

| Parameter | PC1 | PC2 | PC3 | PC4 | PC5 | PC6 | PC7 |
|---|---|---|---|---|---|---|---|
| Eigen values | 6/155 | 3/246 | 2/407 | 1/608 | 1/456 | 1/294 | 1/128 |
| Cumulative Eigen values | 6/155 | 9/401 | 11/808 | 13/416 | 14/872 | 16/166 | 17/294 |
| Proportion of variance | 15/685 | 14/891 | 14/406 | 11/346 | 11/312 | 10/023 | 8/833 |
| Cumulative variance | 15/685 | 30/549 | 44/955 | 56/300 | 67/612 | 77/635 | 86/468 |
| **Traits** | | | | | | | |
| plant height | 0/109 | -0/037 | 0/331 | -0/241 | 0/744 | -0/107 | 0/068 |
| Number of pod | 0/114 | 0/257 | 0/151 | -0/123 | -0/180 | 0/839 | 0/116 |
| Leaf area | 0/049 | -0/157 | -0/017 | 0/194 | 0/793 | 0/120 | 0/097 |
| Number of seed per pod | -0/271 | 0/046 | -0/009 | -0/740 | -0/087 | -0/224 | 0/041 |
| Number of seed per plant | 0/383 | 0/595 | -0/048 | -0/137 | 0/195 | 0/138 | 0/756 |
| 100 grain weight | 0/009 | 0/154 | 0/330 | 0/155 | 0/386 | 0/337 | 0/716 |
| Day to 50% flowering | 0/920 | -0/335 | 0/369 | 0/261 | 0/361 | 0/372 | 0/304 |
| Day to maturity | 0/919 | 0/060 | 0/206 | 0/107 | 0/050 | 0/134 | 0/042 |
| grain yield | 0/123 | 0/049 | 0/213 | 0/177 | 0/080 | 0/061 | 0/034 |
| Oil content | -0/092 | 0/023 | -0/002 | 0/010 | -0/128 | 0/030 | -0/028 |
| Protein content | 0/059 | 0/964 | 0/793 | 0/034 | 0/157 | -0/003 | 0/128 |

showed (Table 3) that 7 independent factors for characters to explain 86.4% variation of all data. The first factor because of high day to 50% flowering and day to maturity alone 15.68% of the data changes can be justified and called the phonological factor properties. The second factor because high number of seed per plant and protein content was called quality factor and alone 14.89% of the data changed can be justified and total 86.4% variation of all data under water stress condition. Factor analysis under normal (non stress) condition showed (Table 4) that 5 independent factors for characters to explain 82% variation of all data. The first factor because of high plant height, 100 grain weight and Day to 50% flowering alone 26.17% of the data changes can be justified and called yield and yield component factor properties. The second factor because high oil content was called quality factor and alone 15.3% of the data changed can be justified and total 82% variation of all data under normal condition (non stress) condition. Figures 1 and 2 showed projection of the agro-morphological and seed quality traits on the planes defined in water stress and normal condition by principal

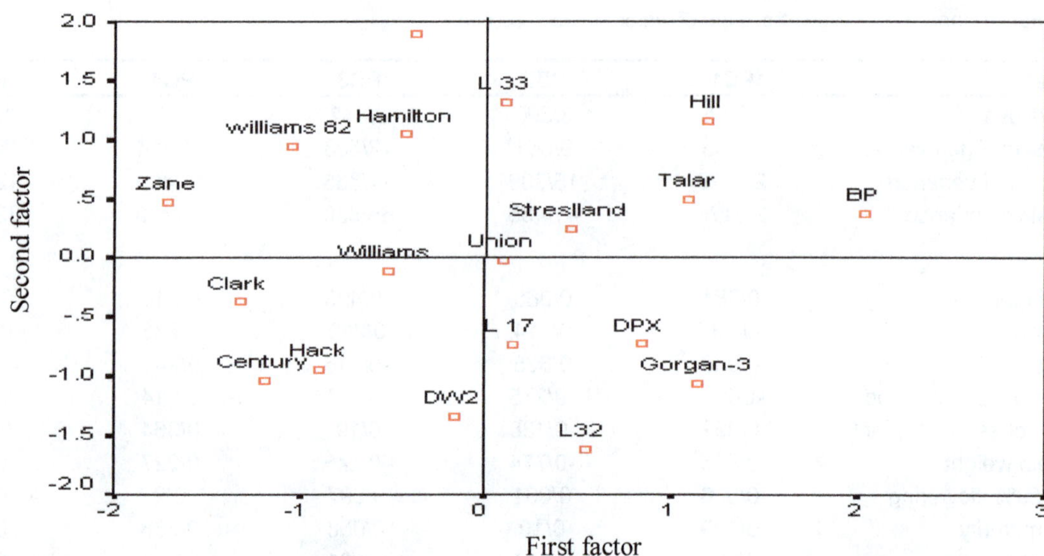


**Figure 1.** Plot of graph first factor (phonological properties) and second factor (quality properties) in soybean.

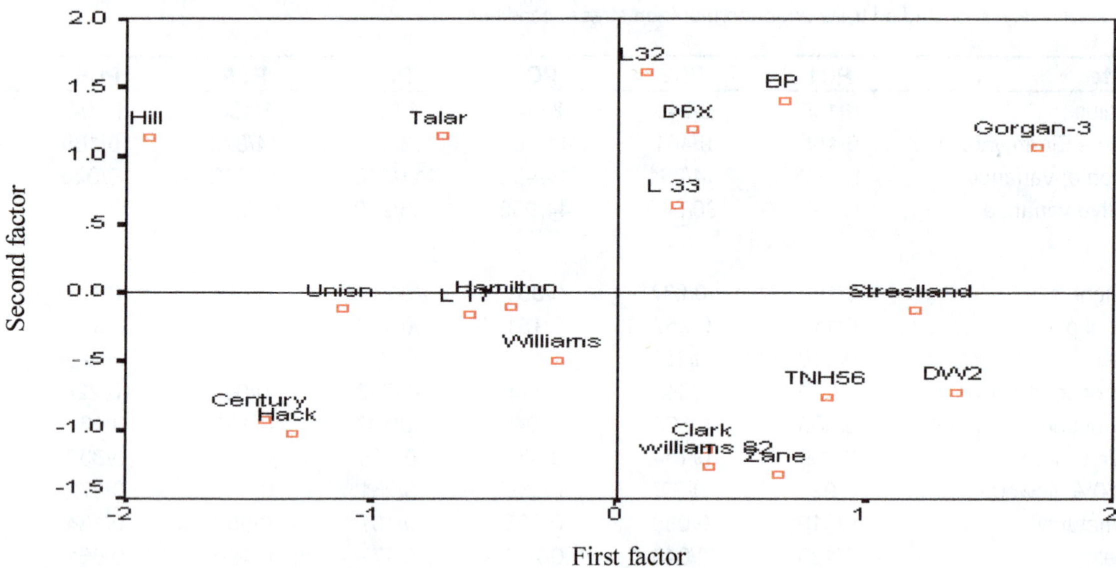


**Figure 2.** Plot of graph first factor (yield and yield component properties) and second factor (quality properties) in soybean.

components. In water stress condition first factor determined according to high day to 50% flowering (0/92) and day to maturity (0/919) and called phonological factor properties and second factor determined according to high number of seed per plant (0/595) and protein content (0/964) was called quality factor (Figure 1) and in normal condition first factor determined according to high plant height (0/681), number of seed per plant (0/868) and grain yield (0/876) called yield and yield component factor properties and second factor determined according to high oil content (0/88) and 100 grain weight (-0/774) was called quality factor (Figure 2). Principal component analysis is useful as it gives information about the groups where certain traits are more important allowing the breeders to conduct specific breeding programs. The results of present studies are in line with those of Narjesi et al. (2007). Principal component analysis is useful as it gives information about the groups where certain traits are more important allowing the breeders to conduct specific

**Table 5.** Means value and variance of 7 cluster of studied traits under water stress condition.

| Traits | Cluster I | | Cluster II | | Cluster III | | Cluster IV | | Cluster V | | Cluster VI | | Cluster VII | | Cluster VIII | |
|---|---|---|---|---|---|---|---|---|---|---|---|---|---|---|---|---|
| | Mean | Variance | Mean | Variance | Mean | Variance | Mean | Mean | Variance | Variance | Mean | Variance | Mean | Variance | Mean | Variance |
| plant height | 99 | 32/16 | 87/62 | 318/7 | 93 | 84/5 | 95/75 | 70 | 690/75 | 1/12 | 85/66 | 72/3 | 101/75 | 21/12 | 70 | 690/75 |
| Number of pod | 139/7 | 988/07 | 130/72 | 1028/18 | 209/23 | 4387/5 | 121 | 171/53 | 136/1 | 551/12 | 132/8 | 2263/93 | 112/5 | 5325/1 | 171/53 | 136/1 |
| Leaf area | 55/71 | 40/98 | 50/17 | 64/68 | 55/55 | 6/91 | 60/22 | 41/95 | 213/08 | 121/13 | 53/83 | 76/27 | 48/84 | 111/52 | 41/95 | 213/08 |
| Number of seed on pod | 2/07 | /008 | 2/03 | 0/038 | 1/93 | 0/00 | 2/13 | 2/19 | 0/001 | 0/002 | 2/12 | 0/004 | 2/2 | 0/008 | 2/19 | 0/001 |
| Number of seed per plant | 17/55 | 6/29 | 17/28 | 7/78 | 19/09 | 16/84 | 17/28 | 17/67 | 3/7 | 0/01 | 16/19 | 6/62 | 19/55 | 6/26 | 17/67 | 3/7 |
| 100 grain weight | 43/76 | 84/52 | 38/31 | 137/32 | 53/98 | 8/16 | 39/18 | 46/71 | 3/43 | 66/52 | 36/56 | 25/18 | 44/49 | 139/36 | 46/71 | 3/43 |
| Day to 50% flowering | 255/2 | 879/56 | 222/27 | 1797/6 | 296/45 | 4077/04 | 228/05 | 286 | 571/59 | 3353/8 | 233/46 | 2012/7 | 221/2 | 1897/2 | 286 | 571/59 |
| Day to maturity | 68 | 40/66 | 64/5 | 23 | 71/5 | 0/5 | 62/5 | 66 | 21 | 12/5 | 67 | 36 | 67 | 0/00 | 66 | 21 |
| grain yield | 141/7 | 128/25 | 134/5 | 144/33 | 148 | 8 | 134/5 | 138 | 112/3 | 24/5 | 137/67 | 156/33 | 138 | 0/00 | 138 | 112/3 |
| Oil content | 14/13 | 3/55 | 15/006 | 1/75 | 14/83 | 7/5 | 12/59 | 16/52 | 3/64 | 0/55 | 15/49 | 9/96 | 14/26 | 0/77 | 16/52 | 3/64 |
| Protein content | 80/71 | 116 | 85/68 | 57/06 | 84/73 | 244/74 | 71/89 | | | 17/91 | 88/48 | 324/8 | 81/43 | 25/17 | 118/7 | 94/36 |

**Table 6.** Mean value and variance of 4 cluster of studied traits under normal (non stress) condition.

| Traits | Cluster I | | Cluster II | | Cluster III | | Cluster IV | |
|---|---|---|---|---|---|---|---|---|
| | Mean | Variance | Mean | Variance | Mean | Variance | Mean | Variance |
| Plant height | 95/83 | 203/37 | 93 | 249 | 83 | 0/5 | 57/5 | 690/75 |
| Number of pod | 121/25 | 492/59 | 134/22 | 895/68 | 169/6 | 4493/5 | 164 | 136/1 |
| Leaf area | 58/13 | 136/91 | 51/73 | 119/43 | 79/25 | 300/61 | 34/42 | 213/08 |
| Number of seed on pod | 2/06 | 0/005 | 2/16 | 0/02 | 2/09 | 0/019 | 2/19 | 0/001 |
| Number of seed per plant | 17/76 | 5/41 | 18/81 | 7/24 | 16/88 | 0/89 | 15/59 | 3/7 |
| 100 grain weight | 54/48 | 99/47 | 42/73 | 150/22 | 28/66 | 0/13 | 12/13 | 3/43 |
| Day to 50% flowering | 307/12 | 1042/51 | 235/56 | 3328/41 | 161/7 | 92/48 | 81/3 | 571/5 |
| Day to maturity | 66/33 | 22 | 68/14 | 22/47 | 65/5 | 40/5 | 61 | 21 |
| Grain yield | 139 | 74/25 | 141/28 | 87/57 | 135/5 | 220/5 | 125 | 112/3 |
| Oil content | 14/47 | 5/07 | 15/2 | 1/63 | 13/92 | 6/66 | 13/28 | 3/64 |
| Protein content | 82/65 | 165/4 | 86/79 | 53/25 | 79/48 | 217/2 | 75/85 | 118/7 |

breeding programs.

## Cluster analysis

Cluster analysis based on nine agro-morphological and two seed quality traits during 2008, divided 19 genotypes of glycin max into 8 clusters in water stress condition and 4 cluster in normal condition. Mean values and variance for each of 8 and 4 clusters are presented in (Tables 5 and 6). In water stress condition Cluster I consisted of 4 genotypes (6, 9, 10 and 14) and these genotypes were medium early in flowering, medium early in maturity, tall, medium number of seed per plant, medium grain yield, medium oil content and medium protein contents (Figure 3). Cluster II comprised of 4 genotypes (2, 11, 12 and

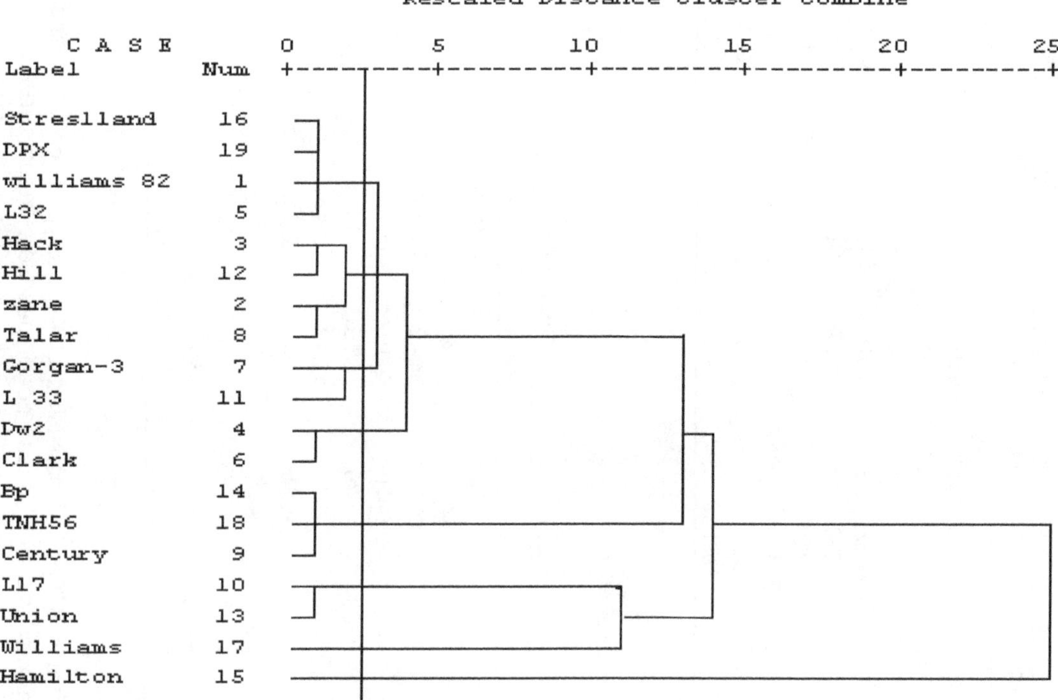


**Figure 3.** Dendogram of cluster analysis of soybean genotypes classified according to all the traits studied in water stress condition.

17) and these genotypes were early in flowering, early in maturity, short, low number of seed per plant, low grain yield, medium oil contents and medium protein contents. There were 2 genotypes (1 and 16) in Cluster III and these genotypes were late in flowering, late in maturity, medium height, high number of seed per plant, high grain yield, medium oil contents and medium protein contents. Cluster IV consisted of 2 genotypes (13 and 15) and these genotype were early in flowering, early in maturity, tall, low number of seed per plant, low grain yield and low oil contents and very low protein contents. There were 3 genotypes (4, 8 and 18) in cluster V and these genotypes were medium in flowering, medium in maturity, short, low number of per plant, low grain yield, high oil contents and high protein contents. Cluster VI had 2 genotypes (3 and 19) and these genotypes were characterized by medium in flowering, medium in maturity, tall, low number of seed per plant, high grain yield, medium oil contents and medium protein contents. Cluster VII comprised of Williams genotype characterized as medium in flowering, late in maturity, medium very tall, medium number of seed per plant, medium grain yield, medium oil contents and low protein contents. Cluster VIII comprised of Hamilton genotype characterized as medium in flowering, medium in maturity, very short, high number of seed per plant, high grain yield, high oil contents and high protein contents (Figure 4). In normal (non stress) condition Cluster I consisted of 9 genotypes (1, 6, 7, 8, 10, 14, 15 and 16) and these genotypes were medium in flowering, medium in maturity, tall, very high number of seed per plant, very high grain yield, medium oil content and medium protein contents. Cluster II comprised of 7 genotypes (3, 4, 5, 9, 11, 17 and 19) and these genotypes were medium in flowering, medium in maturity, medium height, very low number of seed per plant, medium grain yield, medium oil contents and medium protein contents. There were 2 genotypes (2 and 18) in cluster III and these genotypes were medium in flowering, medium in maturity, short, very low number of seed per plant, very low grain yield, medium oil contents and medium protein contents. Cluster IV consisted Hack genotype and this genotype were early in flowering, early in maturity, very short, very low number of seed per plant, very low grain yield and medium oil contents and medium protein contents. Classifying the results of the cluster analysis identified Hamilton genotype suitable for sown in water stress condition and majority genotypes suitable for sown in normal (non stress) condition and Hamilton genotype in water stress condition which confirm the results of the compared means yield. This genotype could be used as source of germplasm for breeding for water tolerance. The results of cluster analysis suggested that there is variation among the genotypes for different agro-morphological and seed quality traits. Genotypes with greater similarity for agro-morphological and seed quality traits were placed in the same cluster. Results of present studies are in agreement with those of Chang et al. (1998).

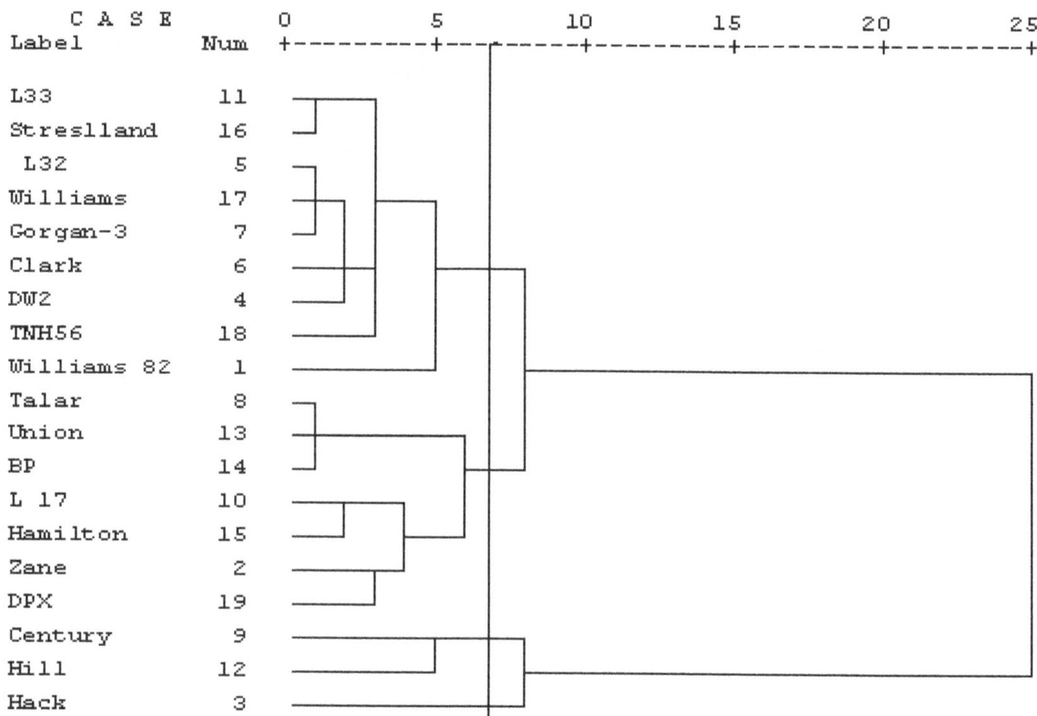


**Figure 4.** Dendogram of cluster analysis of soybean genotypes classified according to all the traits studied in normal (non stress) condition.

## Conclusions

Result of analysis of variance showed that there was significant difference among the studied soybean genotypes in the majority of traits. Principal component analysis is useful as it gives information about the groups where certain traits are more important allowing the breeders to conduct specific breeding programs. Classifying the results of the cluster analysis identified Hamilton genotype suitable for sown in water stress condition and majority genotypes suitable for sown in normal condition. Identifying a genetic structure within soybean genotypes is useful for establishing strategies for sampling and managing germplasm. Crosses between the Hamilton genotype with other genotypes could be used to create genotype resistant under stress condition.

## REFERENCES

Abayomi YA (2008). Comparative Growth and Grain Yield Responses of Early and Late Soybean Maturity Groups to Induced Soil Moisture Stress at Different Growth Stages. World J. Agric. Sci. 4(1):71-78.

Aggarwal PK, Kalra N, Chander S, Pathak H (2006). InfoCrop: a dynamic simulation model for the assessment of crop yields, losses due to pests, and environmental impact of agroecosystems in tropical environments. I. Model description; Agric. Syst. 89:1–25.

Arshad M, Naazar A, Ghafoor A (2006). Character correlation and path coefficient in soybean [Glacin max (L.)] Merill. Pak. J. Bot. 38(1):121-130.

Boyer JS (1982). Plant productivity. Environ. Sci. 218: 43-448.

Brown-Guedira GL, Thompson JA, Nelson RL, Warburton ML (2000). Evaluation of genetic diversity of soybean introductions and north American ancestors using RAPD and SSR markers. Crop Sci. 40:815–823.

Chang RZ, Sun JY, Qiu LJ (1998). The development of soybean germplasm in China. Crops 3:7–9.

Dong YS, Zhao LM, Liu B, Wang ZW, Jin ZQ, Sun H (2004). The genetic diversity of cultivated soybean grown in China. Theor. Appl. Genet. 108:931-936.

Essa TA, Al-ani DH (2001). Effect of salt stress on the performance of six soybean genotypes. Pak. J. Biol. Sci. 4:175-177.

Ghafoor A, Sharif A, Ahmad Z, Zahid MA, Rabbani MA (2001). Genetic diversity in blackgram (Vigna mungo L. Hepper). Field Crops Res. 69:183-190.

Grieu P, Lucero DW, Ardiani R, Ehleringer JR (2001). The mean depth of soil water uptake by two temperate grassland species over time subjected to mild water deficit and competitive association. Plant Soil. 230:179-209.

Guan R, Chang R, Li Y, Wang L, Liu Z, Qiu L (2010). Genetic diversity comparison between Chinese and Japanese soybeans (Glycine max (L.) Merr.) revealed by nuclear SSRs. Genet. Resour. Crop. Evol. 57:229-242.

Kadhem FA, Specht JE, Williams JH (1985). Soybean irrigation serially timed during stages R1 to R6. II. Yield component responses. Agron. J. 77:299-304.

Katerji N, Van Hoorn JW, Hamdy A, Mastrorilli M, Oweis T, Erskine W (2001). Response of two varieties of lentil to soil salinity. Agric. Water Manag. 47:179-190.

Lazcano-Ferrat I, Lovatt CJ (1999). Relationship between relative water content, nitrogen pools, and growth of Phaseolus vulgaris L. and P. acutifolius A. Gray during water deficit. Crop Sci. 39:467-475.

Lopez FB, Johansen C, Chauhan YS (1996). Effects of timing of drought stress on phenology, yield and yield components of shortduration pigeon pea. J. Agron. Crop Sci. 177:311-320.

Momen NN, Carlson RE, Shaw RH, Arjmand O (1979). Moisture stress effects on the yield components of two soybean cultivars. Agron J. 71:86-90.

Narjesi V, Khaneghah HZ, Zali EE (2007). Assesment of genetical relationship in few important agronomic characters with grain yield in soybean by multivariate ststistical analysis. Agric. Nat. Res. Sci. 41(11):227-234.

Roussel V, Koenig J, Bechert M, Balfouriter F (2004). Molecular diversity in French bread wheat accessions related to temporal trends and breeding programmes. Theor. Appl. Genet. 108:920–930.

Sionet N, Kramer PJ (1977). Effect of water stress during different stages of growth of soybean. Agron. J. 69:274-278.

Smart J (1990). Evolution of genetic resources in grain legumes. Cambridge University Press. Cambridge, pp.140-175.

Speeht JE, Chase K, Macrander M, Greaf GL, Chung J, Markwell TP, German JHO, Lark KG (2001). Crop Sci. 40:493-509.

Thomas JMG, Boote KJ, Allen LH, Gallo-Meagher M, Davis J (2003). Seed physiology and metabolism: Elevated temperature and carbon dioxide effects on soybean seed composition and transcript abundance. Crop Sci. 43:1548-1557.

Upadhyaya HD (2003). Phenotypic diversity in groundnut (Arachis hypogaea L.) core collection assessed by morphological and agronomical evaluations. Genet. Res. Crop Evol. 50:539-550.

Upadhyaya HD, Ortiz R, Bramel PJ, Singh S (2002). Phenotypic diversity for morphological and agronomic characteristics in chickpea core collection. Euphytica 123(3):333-342.

Wang M, Li RZ, Yang WM, Du WJ (2010). Assessing the genetic diversity of cultivars and wild soybeans using SSR markers. Afr. J. Biotechnol. 9:4857-4866.

# Response of Nutripellet placement on Marigold yield and its components

**Muthukrishnan R., Arulmozhiselvan K. and Jawaharlal M.**

Department of Soil Science and Agricultural Chemistry, Tamil Nadu Agricultural University, Coimbatore – 641 003, Tamil Nadu, India.

**Nutripellet Pack has 3 parts viz., top bioinoculant mixture, central manure pellet and bottom fertilizer pellet. On the top, bioinoculants mixture responsible for N2 fixation, P and Zn solubilization and biocontrol agents are placed as a powder or granules. Highly decomposed manure having C:N ratio below 30:1 enriched with P, micronutrients and pesticide/ fungicide is pelleted with pelleting device and placed at the centre. At the bottom, a mixture of NPK fertilizers made in pellet form and encapsulated in polymer paper (bio degradable) pouch is placed. The nutrients in fertilizer pellet are in amount equal to the yield target of the crop. To study the Nutripellet (Nutripellet is a packet in tubular form having several inputs viz., fertilizers, manures; for preparing manure pellet, enriched vermicompost was used, fungicide and bioinoculants) application of nutrients on the yield of marigold (*Tagetes erecta* L.), a field experiment was conducted with following treatments viz., T1, Control; $T_2$ and T3, 100 and 75% NPK surface/broadcast application of fertiliser, $T_4$ and $T_5$, 100 and 75% NPK- Nutripellet (Diammonium phosphate (DAP) as P source) ;$T_6$ and $T_7$, 100 and 75% NPK- Nutripellet (Single super phosphate (SSP) as P source). The results of the experiment indicated the 100% NPK Nutripellet recorded the highest flower yield of 45.5 t $ha^{-1}$ which was 98.8% higher than surface broadcast. On an average, Nutripellet with DAP recorded flower weight, petal-calyx ratio, and number of petals per flower. The advantages of root zone placement of Nutripellet were one time placement at transplanting, no top dressing of fertilizers and slow release of nutrients expected throughout the crop period.**

**Key words:** Deep placement, fertilizer, flower yield.

## INTRODUCTION

African marigold (*Tagetes erecta* L.) is an important traditional flower crop and constitutes one of the five most commonly cultivated and used flowers in urban and rural India. They are extensively used for making garlands, beautification, religious offerings, social functions and other purposes namely pigment and oil extraction besides therapeutic uses. Apart from these uses, marigold is widely grown in gardens and pots for display purpose. It is a highly suitable bedding plant and also ideal for newly planted shrubberies to provide colour and to fill up the space. It has a great economic potential in loose flower trade. It gains popularity among garden lovers and loose flower dealers on account of its free flowering habit, short duration, attractive colour, shape and keeping quality. The crop also finds industrial application in several areas like preparation of natural dyes and essential oils. It is used as mosquito and nematode repellents. It is also used as a feed additive for poultry industry (Bose and Yadav, 1993).

Nutripellet technique implements the benefits of integrated nutrient management and deep placement in the root zone of crops. On placement in soil, eachNutripellet is expected to give continuous nutrient support to the crop in the rhizosphere region.

Its promising effects have been successfully evaluated under surface irrigation and drip irrigation (Radhika, 2008). Nutripellet is a packet in tubular form having several inputs viz., fertilizers, manures, pesticides and bioinoculants. By this composition, it is possible to place NPK fertilizers, manure and bioinoculants just below the plant or by the side of plant at the time of transplanting, so that nutrients are efficiently utilized by crop in the active root zone. This technique was first developed and tested for rice in pot study with $^{15}N$ tracer using Nutriseed Pack having seed and fertilizer. The results showed 57.1% of fertilizer N recovery, which exceeded two folds of recovery noted for surface broadcast, with a grain yield increase to the tune of 81.8% over conventional surface broadcast (Asha and Arulmozhiselvan, 2006).

Deivanai (2005) and Arulmozhiselvan et al. (2009) recorded remarkable yield increase in rice with the Nutriseed Pack technique. Vengatesan (2007) has conducted experiment with Nutriseed Packs consisting seed, manure and fertilizer. Pellets of Nutriseed Pack placement increased maize yield to the tune of 56% over surface application of straight fertilizers. With this background, attempt was made with different levels of nutrients in Nutripellet under surface irrigation condition for marigold yield and its components.

## MATERIALS AND METHODS

### Preparation of Nutripellets

The fertilizer nutrient dose generally adopted for carnation in many parts of India is 90:75:75 kg NPK/ha. In this study, the fertilizers (in the form of urea, single super phosphate (SSP) or diammonium phosphate (DAP) and muriate of potash) needed to supply 100 or 75% of N, P and K as per treatment for a single marigold plant were taken and placed in the pelleting device and 30 mm long fertilizer pellets were formed. Then the pellets were encapsulated by placing in a degradable polyester coated paper pouch and sealing with hot flat wire. For preparing manure pellet, enriched vermicompost was used. For this purpose 10 kg of vermicompost was enriched with 1 kg of single super phosphate and incubated for 30 days with adequate moisture. At the end of the period, the enriched manure was pelleted in the pelleting machine. Each manure pellet weighed about 3 g. Bioinoculants viz., Azophos (mixture of *Azospirillum* and Phosphobacteria), *Trichoderma* were also mixed with vermicompost and kept in powder form.

Each Nutripellet was constructed by combining fertilizer and manure pellets on a 10 x 10 cm newspaper by placing one over other and then wrapping as a roll. First the encapsulated fertilizer pellet was placed, coinciding to the bottom edge. Over the fertilizer pellet, the manure pellet was placed. Then, one end of paper was flipped over the pellets and then folded as a roll. In the top cavity, about 0.5 g of bioinoculant mixture was added. Finally, the top of Nutripellet was closed with adhesive, then air dried and stored in cartons. The roll wrap which contained fertilizer pellet at bottom, manure pellet in the middle and bioinoculant on top, in total, is called as Nutripellet.

### Field experiment

In order to evaluate the effect of levels of nutrients in Nutri-Packs, a field experiment was conducted with the test crop of marigold under surface irrigation. The field experiment was conducted in Farmer field, Puliampatti, Coimbatore during 2012 -13. The field was well ploughed, leveled and raised beds (20 cm) and channel space (35 cm) were formed alternatively. The field was divided into plots of 40 square meters in randomized block design. The experimental field experiences semiarid climate with dry summer extending from March to August. The mean annual rainfall is 893 mm, out of which 39.8% is distributed during South West Monsoon, 42% during North East Monsoon, 2.1% during winter and 16.1% during summer. The daily maximum and minimum temperature ranges at 33.5 and 25.3°C during South West Monsoon, 30.9 and 21.1°C during North East Monsoon, 30.9 and 20.8°C during winter and 36.4 and 24.7°C during summer, respectively.

Sowing of African marigold seeds in protrays was done at 1 seed per cell. The seeds placed in the trays were covered with cocopeat and the trays were kept one above the other and covered with a polythene sheet till germination. After four days, the germinated protrays were individually placed on the raised beds inside the shade net. Watering was done using rose can every day (twice/day) until seed germination. 19:19:19 + 0.5% ferrous sulphate and 0.5% manganese sulphate solution were drenched using rose can at 15 days after sowing. Twenty days old seedlings were transplanted in the mainfield during evening hours. Planting was done at a spacing of 60 x 45 cm. Gap filling was done at $5^{th}$ day after transplanting. The treatments details are as follow T1, Control; T2, 100% NPK - surface application of fertilizers; $T_3$–75%, NPK - surface application of fertilizers; $T_4$ and $T_5$, 100 and 75% NPK- Nutripellet (DAP as P source) ;$T_6$ and $T_7$ – 100 and 75% NPK- Nutripellet (SSP as P source). Recommended cultural practices (ploughing, irrigation, weeding and plant protection measures) were followed throughout the growing period. At harvest yield parameters (petal calyx ratio, number of harvest or picking, flower diameter, individual flower weight, number of flowers per plant and yield were recorded.

### Statistical analysis

The growth, yield parameters and yield obtained in the study were subjected to statistical scrutiny by mean values with randomized block design.

## RESULTS AND DISCUSSION

### Initial characteristic of physico-chemical properties of soil

The soil of the experimental site was neutral in reaction, nonsaline and low in organic carbon (0.38%), with cation exchange capacity (CEC) of 17.7 cmol ($p^+$) $kg^{-1}$ soil. The nutrient status of soil was low in available N (235 kg $ha^{-1}$), P (8.96 kg $ha^{-1}$) and high in K (727.1 kg $ha^{-1}$). The status of available micronutrients was deficient, indicating DTPA-extractable content of 1.88 ppm for iron, 1.95 ppm for manganese, 0.59 ppm for zinc and 0.17 ppm for Copper.

### Yield and yield attributes at harvest

Nutrition of marigold crop by surface broadcast and Nutripellet placement tried in different levels has influenced yield (Table 1) and yield attributes (Table 2) of

**Table 1.** Yield (kg $ha^{-1}$) at Harvest under Deep placement of fertilizer (Nutripellet).

| Treatment | | Flower yield per plant (kg) | Flower yield per hectare (tonnes) | % yield increase over control |
|---|---|---|---|---|
| T1 | Control (No fertilizers) | 0.361 | 22.9 | - |
| T2 | Broadcast application - 100% NPK | 0.602 | 38.2 | 66.8 |
| T3 | Broadcast application - 75 % NPK | 0.532 | 33.8 | 47.4 |
| T4 | Nutripellet with DAP - 100% NPK | 0.717 | 45.5 | 98.8 |
| T5 | Nutripellet with DAP - 75 % NPK | 0.593 | 37.6 | 64.4 |
| T6 | Nutripellet with SSP - 100% NPK | 0.624 | 39.6 | 73.0 |
| T7 | Nutripellet with SSP - 75 % NPK | 0.529 | 33.6 | 46.6 |
| SEd | | 0.004 | 0.25 | - |
| CD (5%) | | 0.009 | 0.54 | - |

**Table 2.** Yield attributes at harvest under deep placement fertilizer (Nutripellet).

| Treatments | | Petal - calyx ratio | Number of harvest (No. of picking) | Flower diameter (cm) | Number of flowers per plant |
|---|---|---|---|---|---|
| T1 | Control (No fertilizers) | 49.14 | 5.32 | 5.94 | 41.04 |
| T2 | Broadcast application - 100% NPK | 48.65 | 7.95 | 6.35 | 47.45 |
| T3 | Broadcast application - 75 % NPK | 48.21 | 7.87 | 6.14 | 44.72 |
| T4 | Nutripellet with DAP - 100% NPK | 47.56 | 8.29 | 7.13 | 49.91 |
| T5 | Nutripellet with DAP - 75 % NPK | 47.71 | 7.98 | 6.94 | 45.68 |
| T6 | Nutripellet with SSP - 100% NPK | 47.60 | 8.16 | 7.04 | 47.06 |
| T7 | Nutripellet with SSP - 75 % NPK | 47.93 | 7.94 | 6.89 | 43.54 |
| SEd | | 0.021 | 0.036 | 0.017 | 0.103 |
| Critical Difference (0.05) | | 0.046 | 0.080 | 0.037 | 0.224 |

marigold by exhibiting significant variations in yield and yield attribute parameters. The highest fresh flower yield (45.5 t $ha^{-1}$) was achieved with Nutripellet Pack containing DAP as P source, followed by Nutripellet Pack containing SSP as P source. Comparing DAP and SSP sources, DAP is easily soluble and slow release of nutrients from Nutripellet. Under broadcast, the yield recorded (38.2 t $ha^{-1}$) was moderate (Figure 1), and under control, very low yield (22.9 t $ha^{-1}$) was recorded. The yield attained was higher under Nutripellet placement when compared to conventional surface broadcast of fertilizers due to the placement of Nutripellet pack brought out that added nutrients in slow state of release in Nutripellet Pack was sufficient to reach the maximum attainable yield of the marigold. The high flower yield recorded under Nutripellet Pack placement was 98.8% higher than surface broadcast (Farmers practice).

Chauhan et al. (2005) in marigold cv. Pusa Narangi Gainda, the application of vermicompost at 1000 g per square meter recorded higher yield of flowers (1757.76 g / square meter) compared to application of vermicompost at 500 g per square meter (1429.00 g/$m^2$). Swapna (2010) recorded surface broadcast application of 100% NPK along with humic acid 0.2% as foliar spray recorded the highest value (41.50, 41. 05 and 40.77) in the first season, second season and pooled mean, respectively, for marigold cv. Among the treatments and control (conventional method of fertilizer application and irrigation), control exhibited lowest flower yield (31.35, 30.38 and 31.12) in the first season, second season and pooled mean, respectively.

Nutripellet placement and surface broadcast application influenced the yield attributes (Table 2) are equally by recording petal calyx ratio from 47.56 - 49.14. The effect of nutrition was much pronounced in Nutripellet placement application showing the highest flower diameter (7.13 cm), number of flowers per plant (49.91), number of harvest or picking of flowers (8.29) and individual flower weight (Figure 2) (14.37 g). The yield attributes were poor under control.

The promising effect of Nutripellet pack would be attributed to the controlled release of fertilizers which were precisely placed within 5 cm distance from soil surface in the root zone of the crop. The fertilizer pellet positioned nearer to manure pellet of Nutripellet pack in combination would have remained as nutrient pile allowing mainly radial movement of N, P and K nutrients within the soil in root zone, as the pellet was fully covered by polycoat paper as an encapsulation, having diffusion area all around. The high status of available nutrients in

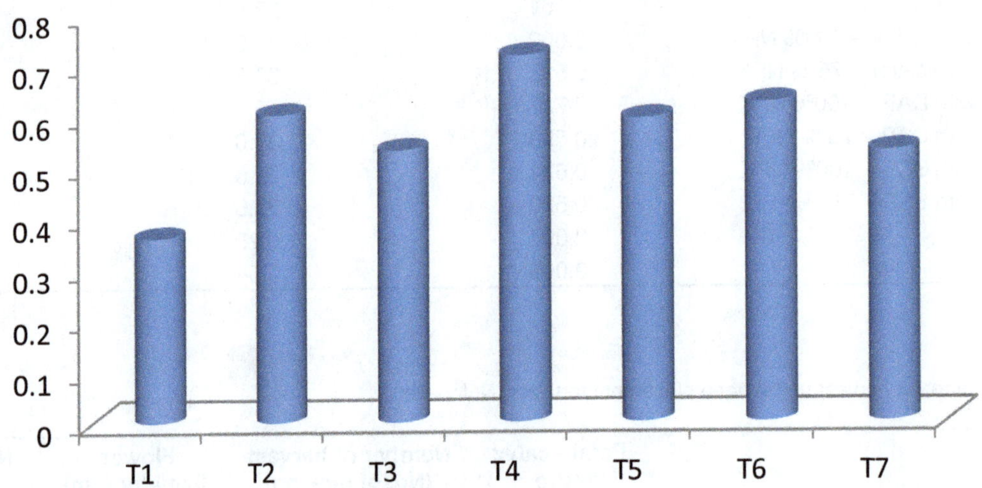


**Figure 1.** Effect of surface broadcast and Nutripellet placement on flower yield per plant.

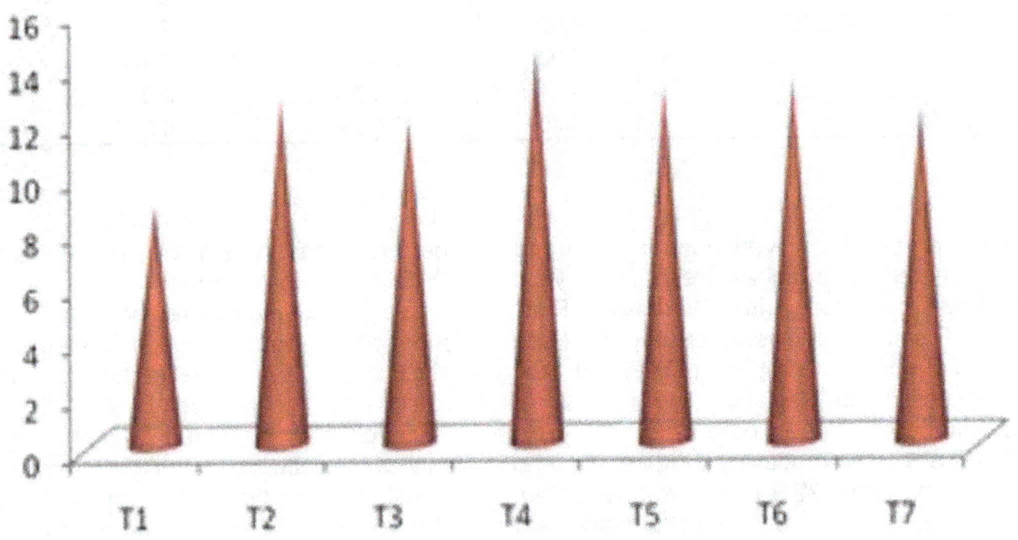


**Figure 2.** Effect of surface broadcast and Nutripellet placement on individual flower weight.

deep placement throughout period of crop could be the evidence of the controlled release phenomenon. With higher nutrient availability and high nutrient uptake resulted under deep placement might have influenced yield and yield attributes of marigold.

## Conclusion

Marigold crop requires continuous supply of nutrients since its vegetative as well as flowering stage. Effort made in the present study to maintain the constant nutrient support by Nutripellet placement was successful in achieving desired flower production with appreciable yield and quality. Among the two methods, the placement of Nutripellet was much convenient as it was one time placement at the time of transplanting and there after only irrigation was monitored. In surface broadcast method was done to distribute the nutrients erratic, low nutrient efficiency. Hence, Nutripellet placement could be preferred for marigold, to save labour and reduce the fertilizer cost and increase yield and yield parameters of marigold.

**Abbreviations: SSP,** Single super phosphate; **DAP,** diammonium phosphate; **CEC,** cation exchange capacity.

**REFERENCES**


Arulmozhiselvan K, Vengatesan R, Deivanai M (2009). Nutriseed Holder Technique for increasing nutrient use efficiency and yield under wetland and upland situations in rice and maize. Crop Research and Research on Crops. 10(3):473-480.

Asha VS, Arulmozhiselvan K (2006). $^{15}$N Tracer technique for studying efficiency of deep placed fertilizer through Nutriseed holder in direct seeded rice. J. Nuclear Agric. Boil. 35(1):1-14.

Bose TK, Yadav LP (1993). Commercial flowers, P. 713.

Chauhan S, Singh CN, Singh AK (2005). Effect of Vermicompost and pinching on growth and flowering in marigold cv. Pusa narangi gainda. Prog. Hortic. 37(2):419- 422.

Deivanai M (2005). Dynamics of deep placed fertilizer nutrients in soil column under controlled irrigation for direct seeded rice. M.Sc. (Ag.) Thesis, Tamil Nadu Agricultural University, Coimbatore.

Panse VG, Sukhatme BV (1967). Statistical Methods for Agricultural Workers. *ICAR* Publication, New Delhi, pp. 100-109 and 152-161.

Radhika K (2008). Standardization of Nutriseed Holder technique for enhancing yield of maize under surface and drip irrigation methods. Ph.D. (Ag.) Thesis. Tamil Nadu Agricultural University, Coimbatore.

Swapna C (2010). Investigation on production system efficiency of precision farming in comparison with conventional system in marigold (*Tagetes erecta* L.). Ph.D. (Hort.) Thesis. Tamil Nadu Agricultural University, Coimbatore.

Vengatesan R (2007). Nutriseed Holder Technique for Enhancing Yield of Maize Under Surface, Micro Sprinkler And Drip Irrigation Methods. M.Sc. (Ag.) Thesis. Tamil Nadu Agricultural University, Coimbatore.

# Molecular weights and tanning properties of tannin fractions from the *Acacia mangium* bark

**Bo Teng, Tao Zhang, Ying Gong and Wuyong Chen**

National Engineering Laboratory for Clean Technology of Leather Manufacture, Institute of Life Sciences, Sichuan University, Chengdu, 610065, China.

**The bark of *Acacia mangium* was smashed and extracted by acetone solution, and the extract was then degreased by petroleum ether. The degreased solution was extracted by diethyl ether and ethyl acetate successively. In this way, the tannin of *A. mangium* was divided as diethyl ether fraction, ethyl acetate fraction, and water fraction, respectively. The molecular weight of these tannins was measured by gel permeation chromatography (GPC), and the particle sizes of tannin were examined by Zetasizer ZS instrument. The results showed that the ether fraction consisted of the smallest size molecule with average molecular weight of 415 Da, possessing weaker tanning ability but fast penetration rate. The molecular weight of ethyl acetate fraction was 1788 Da, and it showed a significant tanning ability. For the water fraction, the molecular weight was 2808 Da with better tanning ability, and the biggest particle size was shown in the tannin. The thermal stability of the hide powder and cowhide tanned by these tannins was as followed: Water fraction > ethyl acetate fraction > diethyl ether fraction; the penetration was: diethyl ether fraction > ethyl acetate fraction > water fraction. These results could provide a valuable reference for the use of the *A. mangium* tannin.**

**Key words:** *Acacia mangium*, extraction, molecular weights, tanning ability tannage.

## INTRODUCTION

*Acacia mangium* belongs to a Mimosaceae family. It is an evergreen arbor, originally planted in the Queensland Australia, southwestern New Guinea, and eastern Indonesia (Alamsyah et al., 2007). Nowadays *A. mangium* has already been widely planted in Asia as one of the most popular fast-growing trees (Tsai, 1988). As the other kind of the condensed type (Pasch et al., 2001), *A. manguim* tannin (AMT) is a condensed tannin consisting of profisetinidin, prorobintinidin, and prodelphinidin units connected through carbon-carbon bonds (Zhang et al., 2010), which contribute to the formation of oligomers and polymers with molecular weights ranging from 500 to 3000 Da. Yeoh Beng Hoong used MALDI- TOF and CP-MAS 13CNMR to study the structure of AMT, and, for the first time, suggested that the AMT consists of 'angular' and 'twice-angular' polymer structure with more than 7 flavonoid units (Figure 1).

The tanning ability and colloid chemical properties of AMT have been well investigated (Teng et al., 2010). However, there are almost no studies on the relation between molecular weight and tanning ability of these tannins. For a better understanding of the correlation between tanning ability of tannins and their molecular weight, some different polarity organic solvents were used to extract tannin fractions having different molecular weight and polarity, providing a valuable evidence for

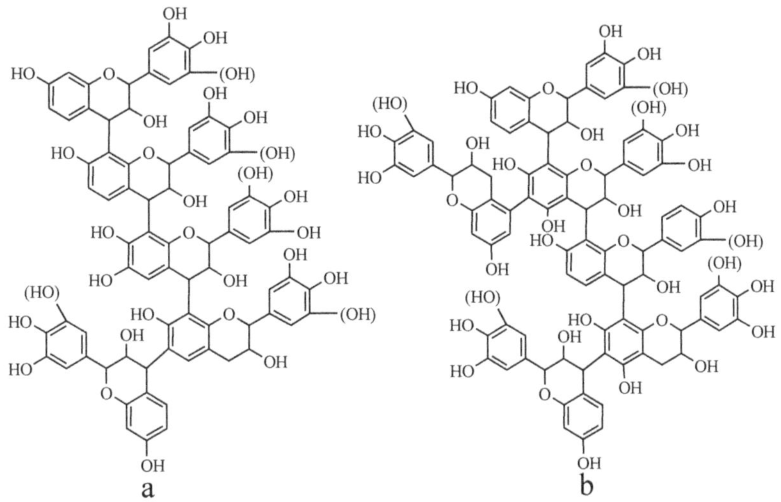

**Figure 1.** Angular' (a) and 'Twice Angular' (b) Structure of *A. mangium* tannin (Hoong et al., 2010).

preparation and application of AMT in the future.

## MATERIALS AND METHODS

*A. mangium* bark (5 years of its tree age) with 36.75% (Liang et al., 2009) tannin content was collected from Baise Tree Farm in Guangxi, China. The hide powder was purchased from Nanjin Forestry and Chemistry Research Institute. The pickle hide was collected from Zhengda Tannery Chengdu. In addition, other reagents were research grade.

### Preparation of tannins

The bark was air-dried and smashed into particles of 1.0 to 2.0 mm for extraction. 3 L Acetone/water solution (7:3, v/v) was used to extract 1 kg bark at 20°C. Acetone/water extract was changed every 48 h and then was stored in dark place. Again, the acetone water extraction was degreased by petroleum ether, and then extracted by diethyl ether and ethyl acetate successively. After pressure distillation was reduced and freezing drying, 223.18 g dry extractions were obtained including 145.20 g water fraction (65% for total extraction), 74.33 g ethyl acetate fraction (33% for total extraction), and 3.56 g diethyl ether fraction (2% for total extraction). The detailed operations are shown in Figure 2.

### Molecular weight distribution and particle size of tannins

One milligram of the tannin was dissolved in 1 ml tetrahydrofuran, and was subjected to gel permeation chromatography (GPC) analysis. A 150-C ALC/GPC instrument (Waters) equipped with a differential refractive index detector, and a combination of μ-Styrage GPC column (10, 50, 100 and 1000 nm) was used to measure the molecular weight of the tannins. Tetrahydrofuran (THF) as the mobile phase was pumped into the column at the flow-rate of 1.0 ml/min, 150 bars at 20°C (Cadahfa et al., 1996). A molar mass standard curve was obtained using monodispersed polystyrene as a standard sample, and the standard curve equation was Y=11.451-0.3767X ($R^2$=0.9975), which showed a good linear correlation. The diethyl ether fraction was used as a control for testing the accuracy of the GPC experiment for 5 times. Standard deviation of the data was less than 5.17, and relative standard deviation was below 0.90%, indicating a highly accurate GPC test (Cheng, 1993). Two gram of tannin was dissolved in acetone/water (7:3, v: v) solution, and the solution was constant to 10 ml by adding acetone water (7:3, v: v) solution. Then the mixture was filtered with a 450 nm Millipore filter and transferred into the sample cell. Particle sizes of these tannins were tested using a Zetasizer Nano ZS-series instrument (Malvern) at 25°C.

### Sorptive ability and binding capacity with hide powder

1.50 g tannin was dissolved in 30 ml acetone/water solution (7:3, v: v). 5 ml of acetone water solution was dried at 120°C and weighed ($M_0$); another 25 ml was employed as a tanning agent. After 2.00 g of dry hide powder was soaked with 25 ml distilled water in a conical flask for 4 h and 25 ml tanning agent was added and kept in a CHZ-82 shaker incubator (Jintan Fuhua) at a rotation speed of 120 r/min (20°C) rotation for 24 h. Then the mixture was filtered with Büchner funnel and 10 ml filtrate was dried at 120°C and weighed ($M_1$). Meanwhile the tanned hide powder (A) was collected. Sorptive ability was calculated based on the following equation:

$$\text{Sorptive ability} = 5(M_0\text{-}M_1)/2.00 \times 100\% \quad (1)$$

The tanned hide powder (A) was washed with 1 L distilled water every 8 h, and 1% gelatin-sodium chloride solution was used to react with filtrate. After 72 h, there was no white sediment formed, which meant no reaction between gelatin-sodium chloride solution with filtrate, the washed hide powder was transferred into a oven dried at 50°C and weighed (B), and also reserved for thermal stability test. Binding ability was counted at a basis of the following equation:

$$\text{Binding ability} = (B\text{-}2.00)/2.00 \times 100\% \quad (2)$$

### Thermal stability of tanned collagen and penetration rate in leather

The thermal properties of tanned hide powder were measured in a

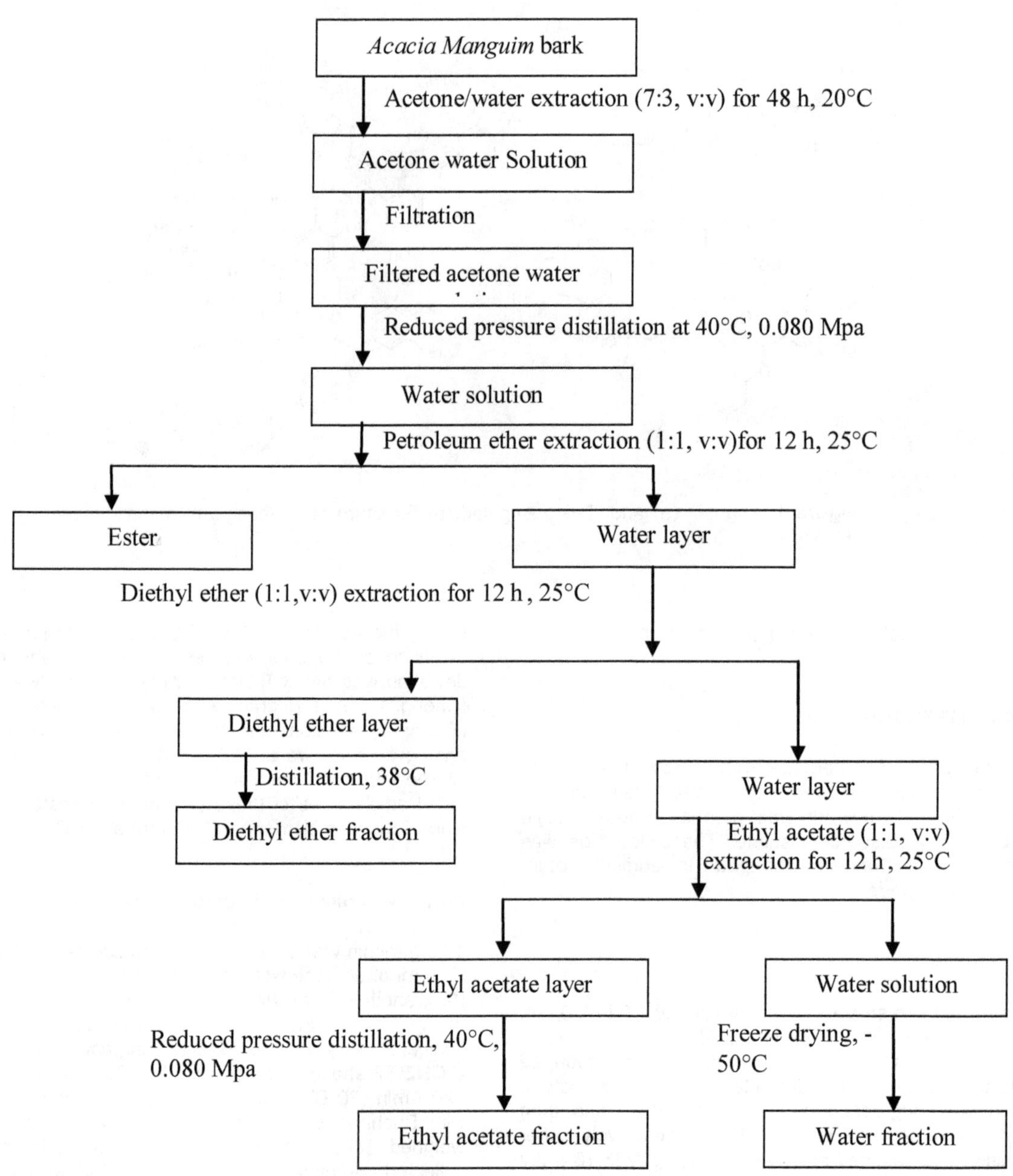


**Figure 2.** Schematic of the extraction process.

differential scanning colorimeter DSC-2C instrument (PERKIN-ELmyer). 4 mg of tanned hide powder (B) was used. The experiments were carried out under a nitrogen atmosphere at a rate of 5°C/min, heating from 20 to 120°C, and indium was used as a control (Komanowsky, 1991). Marched skin samples (10 g average mass) were cut from pickled hide with pH 5.0. These hides were tanned by the three fractions under the same conditions. The samples were treated with 2.00 g tannins (or without tannins as blank sample) dissolved in 10 ml water acetone (7:3, v:v) solution in tanning vessel stirred at 25°C. Penetration was observed every 8 h using an eyepiece micrometer on a microscope. The penetration depth at the grain side ($D_1$) and flesh side ($D_2$) were recorded. The penetration rate was calculated as following equation:

Penetration rate = $(D_1 + D_2)$/ Leather thickness × 100%

## RESULTS AND DISCUSSION

### Molecular weight distributions

Table 1 shows that the $\overline{Mn}$, average polymerization degree and particle size of tannins had a quite large differences. For the diethyl ether fraction, the $\overline{Mn}$ had a relatively small molecular mass (415 Da). On the contrary, the molecular weight of the water fraction was the highest at 2808 Da. This indicated that the AMT can well be separated through some different organic solvents with different polarity. Also, there should exist

**Table 1.** Molecular weight and particle size of tannins.

| Fraction | Mn(Da) | Average polymerization degree* | Particle size (nm) |
|---|---|---|---|
| Diethyl ether fraction | 415 | 1.4 | 64.11 |
| Ethyl acetate fraction | 1788 | 6.2 | 128.7 |
| Water fraction | 2808 | 9.7 | 282.9 |

*Date was calculated based on catechin unit (290Da).

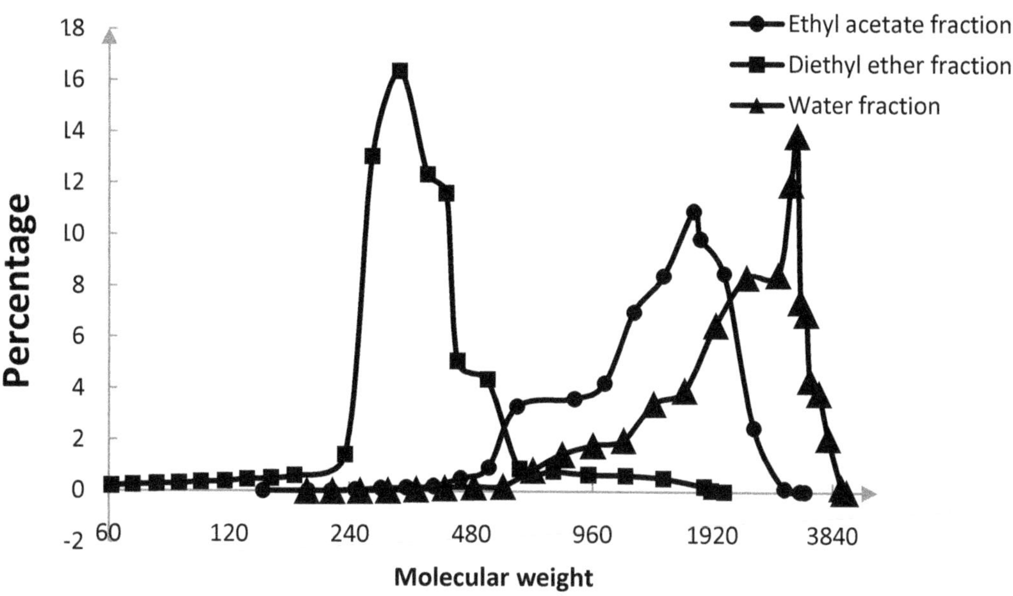


**Figure 3.** Molecular weight ($\overline{Mn}$) distribution of the tannins.

some relations between the molecular weight of tannin and the tanning capacity. This will be discussed thus. The $\overline{Mn}$ distribution of the tannins is shown in Figure 3. There was a narrow distribution ranging from 200 to 600 Da for the diethyl ether fraction. By contrast, for the ethyl acetate fraction and water fraction, there was a wider distribution ranging from 500 to 2500 Da and 800 to 3800 Da respectively. It is indicated that there was some overlap between these fractions.

## Tanning ability

### *Sorptive ability and binding ability*

As a first step in determining the tanning ability with different fractions, the sorptive ability and binding ability of these tannins was tested. The sorptive ability is defined as an ability of a certain amount of tannin to be bound with hide powder; the definition of binding ability is an ability of a tannin to be irreversibly combined with collagen. Results of these tests could be used for judging the tanning ability of different fractions. Table 2 shows that the water fraction had the best properties in terms of binding and sorption. On the contrary, the diethyl ether fraction showed minimum values for both properties. This was the same as the $\overline{Mn}$ sequence. A positive correlation between the $\overline{Mn}$ and the binding ability and the sorptive ability was observed. That is to say that the higher the $\overline{M}$ n of the tannin, the stronger tanning ability it will present.

### *Penetration and thermal stability*

Figure 4 shows the penetration time with the penetration rate of the different tannins. For the water fraction, penetration time was the longest, with up to 96 h required for its biggest particle size in the solution. Conversely, diethyl ether fraction with the smallest particle size penetrated quickly, and only 36 h was required for complete penetration. Otherwise, for ethyl acetate fraction with medium particle size, its penetration time was 56 h between diethyl ether fraction and water fraction. A positive correlation was observed between penetration rate and particle size, as determined by $\overline{Mn}$. Thermal stability of leather can be characterized by its

**Table 2.** Sorptive ability and binding ability of tannins.

| Ability | Diethyl ether fraction | Ethyl acetate fraction | Water fraction |
|---|---|---|---|
| Sorptive ability (%) | 32 | 63 | 72 |
| Binding ability (%) | 11 | 47 | 66 |

**Table 3.** Thermal mutation temperature of tanned collagen.

| Thermal (°C) | Diethyl ether fraction | Ethyl acetate fraction | Water fraction | Untreated collagen |
|---|---|---|---|---|
| Td | 90.4 | 96.5 | 101.7 | 82.8 |
| Ts | 69.5 | 76.7 | 79.2 | 59.8 |

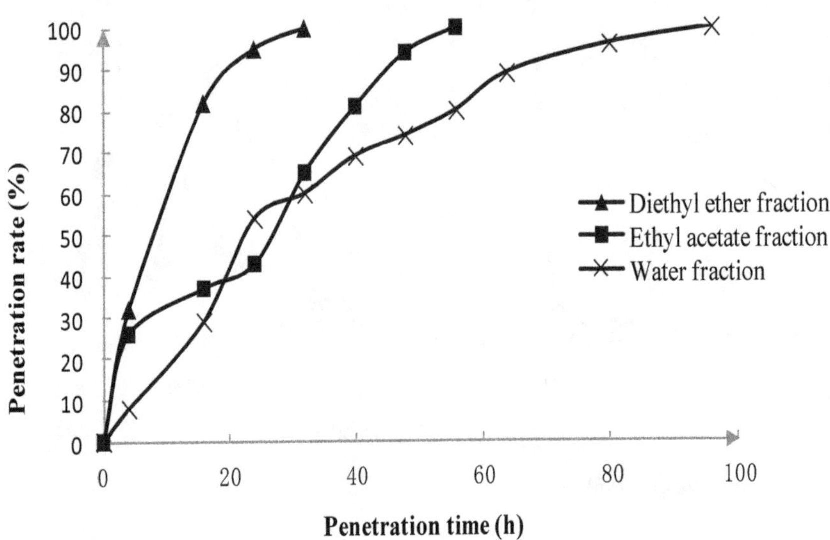


**Figure 4.** The penetration rate of the tannins.

shrinkage temperature (Ts) and thermal denaturating temperature (Td). White (1958) suggested that the molecular weight of vegetable tannin with tanning ability should range from 500 to 3000 Da. Vegetable tannin with molecular weight higher than 3000 Da could hardly penetrate. On the other hand, tannin with weight lower than 500 Da could not combine with collagen (White, 1954). The Td and Ts of collagen treated by water fraction was the highest, while the collagen tanned with diethyl ether fraction presented the minimum value of Td and Ts (Table 3), indicating that the Td and Ts of tanned collagen increased with the $\overline{Mn}$ (Table 1). So, it has been concluded that the tanning ability of the tannin was determined by $\overline{Mn}$ of the tannins.

## Conclusion

In summary, the AMT could well be separated to different fractions through some different organic solvents with different polarity and the separated fractions have different properties.

1. There is a positive correlation between the $\overline{Mn}$ and the tanning ability. In general, the higher the $\overline{Mn}$ of the tannin, the stronger tanning ability it will present.
2. The ether fraction consisted of tannin species having $\overline{Mn}$ of 415 Da with the smallest size, possessing weaker tanning ability but the fastest penetration rate.
3. The molecular weight of ethyl acetate fraction was 1788 Da, and it exhibited a significant tanning ability.
4. For water fraction, the molecular weight was 2808 Da and it had the biggest particle size, so it was judged to have the best tanning ability.

## ACKNOWLEDGEMENT

Authors thank the Baise Tree Farm for providing plant materials.

## REFERENCES

Alamsyah EM, Nan LC, Yamada M, Taki K, Yoshida H. (2007). Bondability of tropical fast-growing tree species. III: Curing behavior of resorcinol formaldehyde resin adhesive at room temperature and effects of extractives of *Acacia mangium* wood on bonding. J. Wood Sci. 53:208-213.

Cadahfa E, Conde M, Garcfa C, Fernfindez B (1996). Gel permeation chromatographic study of the molecular weight distribution of tannins in the wood, bark and leaves of *Eucalyptus* spp, Chromatographia 42:95-100.

Cheng ZY (1993). Application of Gel Permeation Chromatography. China Petrochemical Press Inc. Beijing, pp. 289-290.

Hoong YB, Pizzi A, Tahir PM, Pasch H (2010). Characterization of *Acacia mangium* polyflavonoid tannins by MALDI-TOF mass spectrometry and CP-MAS 13C NMR. Eur. Polym. J. 46:1268-1277.

Komanowsky M (1991). Thermal stability of hide and leather at different moisture contents. JALCA 86:269-280.

Pasch H, Pizzi A, Rode K (2001). MALDI-TOF mass spectrometry of polyflavonoid tannin. Polymer 42:7531-7539.

Liang FX, Yan XZ, Tan XM, Huang JX, Shi B (2009). Property of Acacia Mangium Tannin Extract and its Applications on Leather Processing. China Leather 15:11-19.

Teng B, Gong Y, Chen WY (2010). Colloid chemistry and tanning properties of *Acacia manguim* tannin extracts. China Leather 39:18-25.

White T (1958). Chemistry and Technology of Leather, Reinhold Publishing Corporation, New York 1:133-135.

Zhang LL, Chen JH, Wang YM, Wu DM, Xu M (2010). Phenolic extracts from *Acacia mangium* bark and their antioxidant activities. Molecules 15:3567-3577.

Tsai LM (1988). Studies on *Acacica mangium* in kemasul forest. Malaysia. i .Biomass and productivity. J. Trop. Ecol. 4:293-302

# Mathematical model for nursery raising inside greenhouse

**S. H. Sengar[1] and S. Kothari[2]**

[1]Department of Electrical and Other Energy Sources, College of Agricultural Engineering and Technology, DBSKKV, Dapoli, Dist: Ratnagiri-415712, India.

[2]College of Technology and Engineering, Department of Renewable Energy Sources, Maharana Pratap University of Agriculture and Technology, Udaipur-313001, India.

**Farmers require initially high input to raise nursery seedling/cuttings. Farmers' further activity related to cultivation depends on healthy and available seedlings. Once a farmer has prepared the land and cannot get good seedling, he becomes weak for some season. Studying of environment is a must for planning nursery raising according to thermal profile which can be obtained in greenhouse. Keeping this in mind, mathematical model was developed and evaluated for predicting thermal environment inside the 80 km$^2$ arch shaped greenhouse. The maximum increase in greenhouse air temperature was 13.92°C for solar radiation of 500 W/m$^2$ with 0°C ambient temperature. Experimental and calculated values of greenhouse temperature were almost the same with variation of 2 to 3°C. The theoretical values obtained from model were in reasonably good agreement with the experimental results. Therefore, the mathematical model could be used to predict temperature conditions inside the greenhouse for a variety of climatic parameters. Selection of plants' nursery inside the greenhouse was determined according to predicted thermal environment. As par the thermal profile found inside the greenhouse, plumery plants' nursery was selected and its sprouting, survival percentage and economic were calculated for the farmers' awareness.**

**Key words:** Arch shaped greenhouse, mathematical model, nursery, economics.

## INTRODUCTION

Farmers are looking towards the technology, which is economical and less laborious in present situation. Solution to such scenario of plants' nursery growing inside the greenhouse is best technology for year round production in water scarcity areas. For successful plantation programme, cuttings/seed must be raised first in nursery. Production of healthy seedlings is important where the planting stock is raised seed or cuttings and is maintained for some months (Thakur and Thakur, 1993). Cultivation of nursery also improves the overall growth of plant substantially in terms of height compared to outside condition. A study is therefore undertaken to find out the thermal environment inside the greenhouse by validation of thermal model for selection of suitable nursery in order to increase the germination and survival percentage of plants for higher benefits to farmers. It is also felt that developed model should be more versatile so that it can be used under any climatic conditions, all months and at any location. Accordingly, modified arch shaped greenhouse was selected for cultivation of nursery to perform better where cooling is required (Amita and Tiwari, 2002). An arch shaped greenhouse was designed covering a soil area of 13.4 m × 6.0 m, that is, 80 m$^2$ as shown in Figure 1. Orientation is in East-west direction.

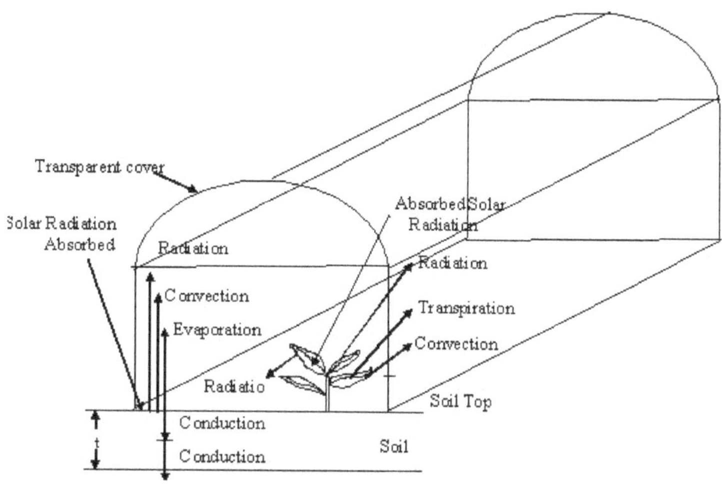


**Figure 1.** Energy transfer mechanism in greenhouse.

The greenhouse was covered by ultra violet stabilized low-density ethylene sheet of 200-micron thickness.

## METHODOLOGY

Here, various energy balance equation was validated for the prediction of thermal environment inside the arch shaped greenhouse. Effect of solar radiation and ambient temperature on greenhouse air temperature were also predicted through model. As par the predicted values for particular month, types of plants were selected for growing nursery inside the greenhouse. Predicted values were compared with observed values for finding the actual performance of sprouting percentage and survival percentage of plants. On the basis of growth parameter economics of this technology was calculated.

### Energy balance analysis

By considering the number of complexities of the heat and mass transfer mechanisms occurring in greenhouse (Garg, 1987; Kaushick, 1988) (Figure 1), modeling the greenhouse as a single component is too cumbersome. Therefore, it is more rational approach to divide a greenhouse into separate components and model them independently. Mathematical models have been evaluated to predict the hourly variation in greenhouse environment, that is, temperature of cover, enclosed air, plants and soil surface and relative humidity of enclosed air for operating condition separately (Cooper and Fuller, 1983). The energy balance equations for finite difference technique and result for solving different components of the system were obtained by using computer programme prepared in Microsoft Office Excel. The energy transfer mechanism inside the greenhouse for greenhouse, plant and soil is shown in Figures 1 to 3.

The expression for temperature of cover, enclosed air, plants and soil surface and relative humidity of enclosed air can be written as

$$T_{co} = \frac{1}{D_4}[T_{st}A_4 + T_{gh}B_4 + T_pC_4 + E_4 + A_{co}(h_{coa}/1005)h_{sg}(W_{gh}-W \quad (1)$$

Where, $A_4 = A_{st}\, h_{rsco;}$ $B_4 = A_{co}h_{coa}$ ;$C_4 = A_p\, h_{rpco}$ $D_4 = A_{co}h_{coa} + A_c\, h_{rcco} + A_p\, h_{rpco} + A_{co}h_{coo} + A_{co}\, h_{rcosky}$, , and $E_4 = I_{co}\, A_{co}\, \alpha_{co} + T_a\, A_{co}h_{coo} + T_{sky}\, A_{co}\, h_{rcosky}$.

$$T_{gh} = \frac{1}{B_3}[T_{st}A_3 + T_cC_3 + T_{co}\, D_3 + E_3] \quad (2)$$

Where, $A_3 = h_{sa}\, A_s$ , $B_3 = h_{coa}\, A_{co} + h_{pa}\, A_p\, L_i + h_{sa}\, A_s + LVC_{pc}\, \rho_a + m_v\, C_{pa} + m_h\, C_{pa}$; $C_3 = h_{pa}\, A_p\, L_i$ , $D_3 = h_{coa}\, A_{co}$ , $E_3 = T_a(LVC_{pc}\, \rho_a + m_v\, C_{pa}) + T_h\, (m_h\, C_{pa})$

$$T_p = \frac{1}{C_2}[T_{st}A_2 + T_{gh}\, B_2 + E_2 - k_tL_iA_p(W_p - W_{gh})h_{sg}] \quad (3)$$

Where, $A_2 = h_{rps}\, A_p$ , $B_2 = h_{pa}A_pL_i$; $C_2 = h_{pa}A_pL_i + h_{rps}A_p + h_{rpsky}A_p$ , $E_2 = I_p\alpha_pA_p + h_{rpsky}A_pT_{sky}$.

$$T_{st} = \frac{1}{A_1}[T_{gh}\, B_1 + T_p\, C_1 + E_1 - h_{ds}\, h_{sg}\, (W_{st} - W_{gh})\, A_s] \quad (4)$$

Where, $A_1 = h_{sa}\, A_s + h_{rsp}\, A_s + \frac{kA_s}{t}$ , $B_1 = h_{sa}\, A_s$ , $C_1 = h_{rsc}\, A_s$, and $E_1 = \alpha_s\, Is\, A_s - \frac{kA_s}{t}\, T_{sb}$.

### Humidity ratio

For determining $W_{co}$, $W_p$, $W_{st}$ and $W_{gh}$ saturation conditions can be assumed at the cover, leaf and soil surfaces. From psychrometric relations, various humidity terms can be written as follows.

$$W_{co} = 0.622 \times \frac{P_{sco}}{P - P_{sco}} \quad (5)$$

$$W_p = 0.622 \times \frac{P_{sp}}{P - P_{sp}} \quad (6)$$

$$W_{st} = 0.622 \times \frac{P_{sst}}{P - P_{sst}} \quad (7)$$

The mass balance equation for the greenhouse air could be written

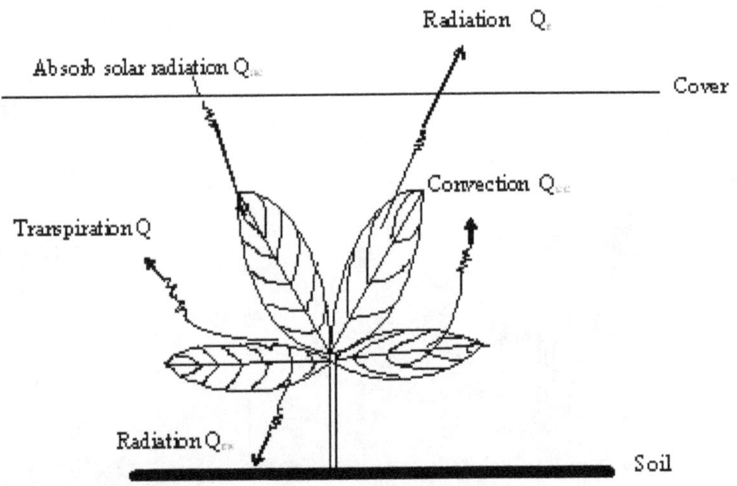


**Figure 2.** Energy transfer mechanism for plant.

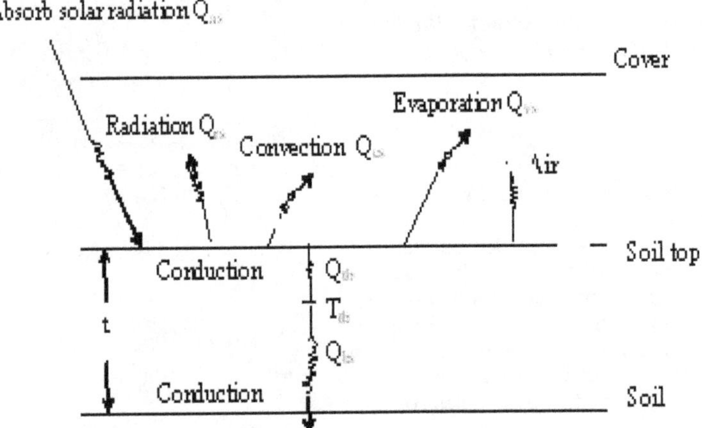


**Figure 3.** Energy transfer mechanism for soil.

as:

$$W_{gh} = 0.622 \times \phi \times \frac{P_{sgh}}{P - P_{sgh}} \tag{8}$$

Where $\phi$ is the relative humidity inside the greenhouse. The comparisons of different expressions developed by various researchers for calculating saturation vapour pressure corresponding to different temperatures were made with the steam table (Tiwari and Goyal, 1998). According to this expression, vapour pressure at saturation at cover, plant, soil and greenhouse air temperature could be given by:

$$P_{sco} = 6894.76 \exp [51.59 - \frac{6834.3}{T_{co} + 273.15} - 5.17 \ln(T_{co} + 273.15) \tag{9}$$

$$P_{sp} = 6894.76 \exp [51.59 - \frac{6834.3}{T_{sp} + 273.15} - 5.17 \ln(T_{sp} + 273.15)] \tag{10}$$

$$P_{sst} = 6894.76 \exp [51.59 - \frac{6834.3}{T_{st} + 273.15} - 5.17 \ln(T_{st} + 273.15)] \tag{11}$$

$$P_{sgh} = 6894.76 \exp [51.59 - \frac{6834.3}{T_{gh} + 273.15} - 5.17 \ln(T_{gh} + 273.15)] \tag{12}$$

The actual vapour pressure in the greenhouse was determined by HORTITRANS model (Jolliet, 1994).

$$P_{gh} = \frac{\upsilon a I + h_t P_{sgh} + h_v P_{so}}{h_t + h_c + h_v} \tag{13}$$

The relative humidity inside the greenhouse can be determined by

$$\phi = \frac{P_{gh}}{P_{sgh}} \times 100 \tag{14}$$

$$a = c_1 \ln(1 + c_2 L_i^{c_3}) \tag{15}$$

$$h_t = c_4 L_i (1 - c_5 e^{(-1/c_6)}) \tag{16}$$

$$h_v = \rho_a C_{pa} \frac{q}{A} \quad (17)$$

### Material for nursery bed preparation

Out of total 80 $m^2$ floor area, 55 $m^2$ area is used for plant seedling and 25 $m^2$ area is left for movement in the greenhouse carrying out agricultural operations. In 55 $m^2$ area of greenhouse, 9700 seedling could be raised with 0.075 × 0.075 m spacing in 20 pits. Each pit of size 2.75 × 1 m was filled with locally available garden soil, sand and vermicompost in 1:1:1 ratio. No chemical was used to control soil properties because moderate temperature was predicted inside the greenhouse. The hard-wood stem cuttings of about 20 to 25 cm (8 to 9 inch) long were prepared from one year old mature shoots. This was done by giving a slant cut at the basal portion about 1 cm below a bud and another round cut was made at the top 3 cm away from the bud. The cuttings were about 10 to 12 mm thickness. All cuttings were treated with rootex (Toky and Srinivasu, 1994) for 30 s and used for propagation on nursery beds inside the greenhouse.

### Growth parameter

The different growth parameters such as number of cutting planted, sprouted, transferred, survival percentage, number of days required for sprouting, rooting and transplanting, number of leaves per seedling /cutting, height of plant and length of the longest root per seedling were observed.

### Parameter of economic consideration

For the success and commercialization of this technology, different economic indicators were calculated for economic analysis of the arch shaped greenhouse in this study (Kothari and Panwar, 2004).

### Net present worth

The mathematical statement for net present worth (NPW) can be written as:

$$\text{NPW} = \sum_{t=1}^{t=n} \frac{B_t - C_t}{(1+i)^t} \quad (18)$$

Where, Ct = Cost in each year, $B_t$ = Benefit in each year, t = 1, 2, 3.....n and i = discount rate.

### Benefit cost ratio

The mathematical benefit-cost ratio can be expressed as:

$$\text{Benefit-cost ratio} = \frac{\sum_{t=1}^{t=n} \frac{B_t}{(1+i)^t}}{\sum_{t=1}^{t=n} \frac{C_t}{(1+i)^t}} \quad (19)$$

### Payback period

It shows the length of time between cumulative net cash outflow recovered in the form of yearly net cash inflows.

## RESULTS AND DISCUSSION

Application of thermal model and values used in calculation are in Table 1. The maximum increase in greenhouse air temperature was 13.92°C for a solar radiation of 500 W/$m^2$ with 0°C ambient temperature (Figure 4). The above results indicate that under cold and sunny climate we could cultivate those crops inside the greenhouse which cannot be grown outside the greenhouse at low temperature.

It is clear from Figure 5 that difference in greenhouse air temperature is more with increasing solar radiation and decreasing ambient temperature while this difference is less with increasing solar radiation and ambient temperature (Cooper and Fuller, 1983). Cover temperature, soil temperature, plant temperature and greenhouse temperature are shown in Table 2.

### Greenhouse air temperature

The increase in greenhouse air temperature above ambient temperature is up to 15°C more than ambient temperature during sunshine hours.

### Cover temperature

Cover temperatures were found as lowest from 1 to 8 a.m. and increased with increasing ambient temperature and solar radiation up to 13 to 14 p.m.; and later on, decreased with decreasing ambient temperature and solar radiation.

### Soil temperature

Trend of temperature changes for soil was also the same as cover temperature but slightly more than cover temperature.

### Crop temperature

Changes were observed in the temperature of crop as an intermediate stage of cover temperature and soil temperature.

Experimental and calculated values of greenhouse temperature were almost the same with variation of 2 to 3°C (Figure 6). It may be inferred from these results that the theoretical values were in reasonably good agreement with the experimental results. Therefore, the mathematical model could be used to predict temperature conditions inside the greenhouse for a variety of climatic parameters.

**Table 1.** Specification and properties used for modeling.

| Parameter | Values | Unit |
|---|---|---|
| Length of greenhouse (L) | 13.4 | meter |
| Thickness of polythene ($Th_{pe}$) | 200 | micron |
| Absorptivity of cover ($\alpha_{co}$) | 0.25 | Dimensionless |
| Transmitivity of cover ($\tau_{co}$) | 0.75 | Dimensionless |
| Density of polythene ($\rho_{pe}$) | 1150 | kg/m$^3$ |
| Specific heat of cover ($C_{co}$) | 2302 | J/kg °C |
| Emissivity of cover ($\varepsilon_{co}$) | 0.9 | Dimensionless |
| Specific heat of air (Ca) | 1005 | J/kg °C |
| Thermal conductivity of air (Ka) | 0.028 | W/m$^2$ °C |
| Density of air ($\rho_a$) | 1.2 | Kg/m$^3$ |
| Specific heat of plant ($C_p$) | 3190 | J/kg °C |
| Emissivity of plant ($\varepsilon_p$) | 0.5 | Dimensionless |
| Specific heat of soil (Cs) | 2300 | J/kg °C |
| Density of soil ($\rho_s$) | 1250 | Kg/m$^3$ |
| Absorptivity of soil ($\alpha_s$) | 0.80 | Dimensionless |
| Emissivity of soil ($\varepsilon_s$) | 0.05 | Dimensionless |
| Wind velocity (W) | 5 | Km/h |
| Area of ventilation ($A_{vent}$) | 0.09 | m$^2$ |
| Stefan-Boltsman constant ($\sigma$) | 5.67 × 10$^{-8}$ | W/m$^2$ K$^4$ |
| Prandtl Number (Pr) | 0.7 | Dimensionless |
| Atmospheric pressure ($P_{atm}$) | 101.325 | Kg/m$^2$ |

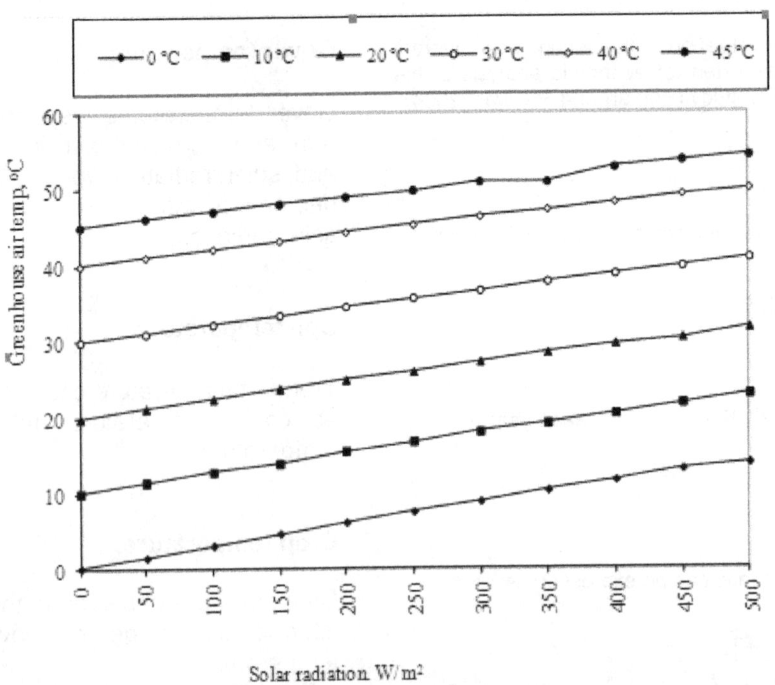


**Figure 4.** Greenhouse air temperature for different solar radiation and ambient temperature (°C).

## Economic analysis of nursery

Based on the predicted environment inside the arch shaped greenhouse, the cuttings of Champa (*Plumeri* sp.) were selected for nursery raising. The different growth parameters observed are shown in Table 3. It

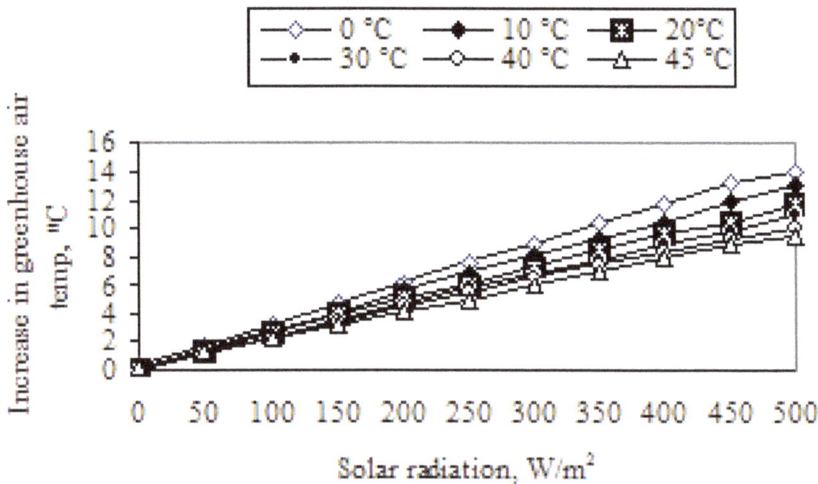


**Figure 5.** Increase in greenhouse air temperature with solar radiation.

**Table 2.** Hourly variation in greenhouse cover, floor, crop and greenhouse temperatures for the month of March at Udaipur.

| Hour of the day | Ambient temperature (°C) | Solar radiation (W/m$^2$) | Cover temperature (°C) | Soil temperature (°C) | Crop temperature (°C) | Greenhouse temperature (°C) |
|---|---|---|---|---|---|---|
| 1 | 15.5 | 0 | 20.4 | 20.65 | 20.53 | 20.53 |
| 2 | 14.9 | 0 | 20.7 | 20.94 | 20.82 | 20.8 |
| 3 | 14.4 | 0 | 20.11 | 20.35 | 20.23 | 20.21 |
| 4 | 13.7 | 0 | 19.61 | 19.86 | 19.73 | 19.7 |
| 5 | 13.3 | 0 | 18.92 | 19.17 | 19.04 | 19.02 |
| 6 | 12.8 | 1 | 18.52 | 18.78 | 18.65 | 18.63 |
| 7 | 17 | 65 | 18.06 | 18.34 | 18.2 | 18.29 |
| 8 | 18 | 259 | 19.47 | 21.15 | 20.36 | 20.57 |
| 9 | 20.3 | 473 | 24.89 | 29.64 | 27.43 | 27.55 |
| 10 | 22.8 | 653 | 31.76 | 39.04 | 35.63 | 35.58 |
| 11 | 24.9 | 778 | 37.02 | 45.96 | 41.74 | 41.58 |
| 12 | 26.3 | 842 | 40.74 | 50.58 | 45.9 | 45.66 |
| 13 | 27.2 | 844 | 42.72 | 52.85 | 48.01 | 47.72 |
| 14 | 33.2 | 769 | 43.21 | 53.08 | 48.36 | 48.02 |
| 15 | 28.1 | 635 | 42.24 | 51.2 | 46.92 | 46.58 |
| 16 | 27.9 | 458 | 40.1 | 47.59 | 44.03 | 43.68 |
| 17 | 27.6 | 249 | 36.74 | 42.2 | 39.62 | 39.31 |
| 18 | 26.4 | 64 | 33.01 | 36.05 | 34.62 | 34.39 |
| 19 | 24.5 | 1 | 29.35 | 30.46 | 29.95 | 29.8 |
| 20 | 22.9 | 0 | 26.49 | 26.74 | 26.63 | 26.56 |
| 21 | 21.8 | 0 | 24.87 | 25.1 | 24.99 | 24.94 |
| 22 | 21 | 0 | 23.78 | 24.01 | 23.9 | 23.86 |
| 23 | 20.2 | 0 | 22.98 | 23.22 | 23.1 | 23.07 |
| 24 | 18.4 | 0 | 22.19 | 22.42 | 22.31 | 22.23 |

shows that the whole process from propagation to transplanting is completed within two months inside the greenhouse. Table 4 shows the details of income and expenditure for *Plumeri* sp. nursery under greenhouse conditions. Selling price for Champa Rs 7/- per plant is based on average yearly price. Total benefit from the 80

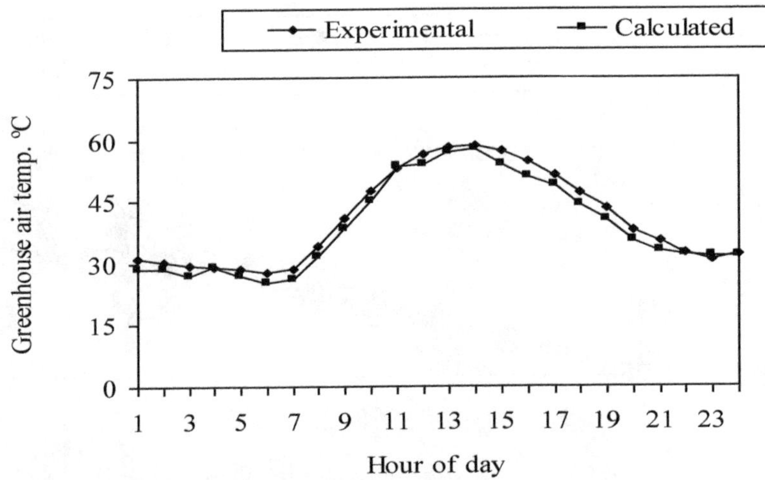


**Figure 6.** Comparison of theoretical and experimental values of greenhouse air temperature for the month of May at Udaipur.

**Table 3.** Overall growth parameter of *Plumeri sp.* plant inside the greenhouse.

| Type of plants | | *Plumeri* sp. |
|---|---|---|
| No. of days required for | Sprouting | 20 |
| | Rooting | 40 |
| Number of leaves after days | 30 | 7 |
| | 60 | 15 |
| Longest root after 60 days (cm) | Primary root | 10 |
| | Primary root | 4 |
| Days required for transplanting | | 60 |

**Table 4.** Details of income and expenditure for different crops under greenhouse conditions.

| S/N | Particulars/ crops | Plant |
|---|---|---|
| 1 | Nursery plants | *Plumeri* sp. |
| 2 | No. of sprouted plants out of 9700 | 9215 |
| 3 | No. of survival plants out of 9700 | 8642 |
| 4 | Total Revenue(Rs) | 60494 |
| 5 | Common Cost for labour (Rs) | 12400 |
| 6 | Cost of cuttings(Rs) | 2000 |
| 7 | Cost of cultivation (Rs) | 14400 |
| 8 | Initial investment | 100000 |
| 9 | Cost of plastic every five year (Rs) | 4000 |
| 10 | Cost of electricity (Rs) | 1200 |
| 11 | Total operation and maintenance cost(Rs) | |
| | Every year | 15600 |
| | Every 5$^{th}$ year | 19600 |

m$^2$ greenhouse was Rs. 44,894/- for three months, which could be sufficient for recovering the cost of greenhouse within 3.7 years; whereas net present worth and benefit cost ratio was Rs.3,54,822/- and 3.80 respectively (Table 5).

**Table 5.** Economic indicators for different plants in greenhouse conditions.

| S/N | Economic Indicators | *Plumeri* sp. |
|---|---|---|
| 1 | NPW (Rs) | 354822 |
| 2 | B/C ratio | 3.80 |
| 3 | Payback period (years) | 3.7 |

## Conclusion

1. This model can be useful to predict the thermal environment inside the greenhouse for selection of plant nursery for different months, season and any location.
2. Difference of greenhouse air temperature is more with increasing solar radiation and decreasing ambient temperature while this difference is less with increasing solar radiation and ambient temperature.

## ACKNOWLEDGEMENT

Authors are totally thankful to Ministry of Non-conventional Energy Sources for providing the financial assistance to carry out the research work. They are also thankful to the Department of Renewable Energy Sources, College of Technology and Engineering, Udaipur for providing all sorts of required facilities for the study.

**Nomenclature:** **A,** Area, $m^2$; **$A_s$, $A_{co}$, $A_p$,** soil, cover and plant area; **C,** specific heat, J/kg °C; **$C_1$-$C_6$,** coefficient of equation; **$C_{pa}$ $C_{pp}$,** specific heats of greenhouse air and plant; **h,** heat transfer coefficient, W/ $m^2$ °C; **$h_{ds}$.** mass transfer coefficient for floor; **$h_{rpco}$, $h_{rsco}$, $h_{rcosky}$.** Radiative heat transfer between plant to cover, soil to cover and cover to sky; **$h_{rpsky}$.** Radiative heat transfer between plant and sky; **$h_{pa}$, $h_{coa}$, $h_{coo}$.** convective heat transfer coefficients between crop and greenhouse air, cover and greenhouse air, cover and ambient; **$h_{sa}$.** convective heat transfer between floor and greenhouse air; **$h_{sg}$,** latent heat of evaporation, kJ/kg; **$h_t$,** coefficient of heat transfer for transpiration; **$h_v$,** coefficient of heat transfer for ventilation; **$I_s$, $I_p$, $I_{co}$,** solar radiation on soil, plant and cover; **k,** thermal conductivity of floor; **L,** length of greenhouse; **$L_i$,** leaf area index; **$m_h$,** mass flow rate out of cooling/ dehumidifying device; **$m_v$,** Mass flow rate due to natural or forced venting with ambient air; **Nu,** Nusalt number; **P,** vapour pressure of outside air; **$P_{gh}$,** vapour pressure inside greenhouse; **$P_{sco}$, $P_{dst}$, $P_{sc}$,** saturation vapour pressure at cover, floor and crop temperature; **$Q_{ap}$, $Q_{aco}$, $Q_{as}$,** energy absorbed by crop, cover and floor from solar radiation; **$Q_{bs}$,** energy transferred by conduction; **$Q_c$,** Energy transfer by condensation; **$Q_{cp}$, $Q_{cs}$, $Q_{cps}$,** Energy transfer by convection between plant, soil and plant and soil; **$Q_{cco}$, $Q_{ccs}$,** energy transfer by convection between cover and cover and soil; **$Q_h$, $Q_i$,,** energy transfer due to cooling device and infiltration; **$Q_{pci}$, $Q_{pc}$, $Q_{ps}$,** energy transfer by convection between plant and infiltration, plant and soil and greenhouse air; **$Q_r$, $Q_{rs}$,** energy transfer by radiation between plant and ambient and between soil and plant; **$Q_{rpco}$, $Q_{rsco}$,** energy transfer by radiation between plant and cover and between soil and cover; **$Q_{rcosky}$, $Q_{rpsky}$,** energy transfer by radiation between cover and skay and plant and sky; **$Q_t$,** energy transfer by plant transpiration; **$Q_{tb}$,** energy transfer by conduction between top surface layer and main mass of soil; **$Q_{ve}$,** energy transfer due to ventilation with ambient air; **$Q_{vs}$,** energy transfer due to evaporation from soil; **R,** radius; **Re,** Reynolds number; **Tp,** plant temperature; **T,** thickness of soil; **Ttb, Tst,** temperature of toplayer of soil and temperature of top layer and soil sink; **$\alpha_p$ , $\alpha_{co}$, $\alpha_s$** plant, cover and floor absorbance; **$\varepsilon_p$, , $\varepsilon_{st}$** plant and floor emittance; **$\sigma$,** Stefan bolts man constant; **$\tau$,** transmittance.

### REFERENCES


Amita G, Tiwari GN (2002). Performance Evaluation of Greenhouse for Different Climatic Zones of India. J. Solar Energy Soc. India 12(1):45-57

Thakur ML, Thakur RK (1993). Forest protection in arid zones, problem and research priorities". In aforestation of arid lands (Edited by Dwivedi A.P. and G.N. Gupta). Scientific Publisher, Jodhpur, pp. 511-21.

Garg HP (1987). Advances in solar Energy Technology Vol. 3: Heating, Agricultural and Photovoltac Application of Solar Energy'. D Reidel Publishing Co, Holland.

Kaushick SC (1988). Thermal Control in Solar Passive Building'. Geo-Environ Academic Press, IBT Publisher, New Delhi.

Cooper PI, Fuller RJ (1983) A Transient Model of the Interaction Between Crop, Environment and Greenhouse Structure for Predicting Crop Yield and Energy Consumption'. J. Agric. Eng. Res. 28:401-417.

Tiwari GN, Goyal RK (1998). Greenhouse Technology. Narosa Publishing House New Delhi, India

Jolliet O (1994). A Model for Predicting and Optimizing Humidity and Transpiration in Greenhouse, J. Agric. Eng. Res. 57:23-37.

Toky OP, Srinivasu V (1994). Response of sodium bicarbonate on survival, seedling growth and plant nurseries of four multipurpose arid trees. Annal Arid Zone 18(3):115-119.

Kothari S, Panwar NL (2004). Economic Evaluation of Greenhouse for cultivation of babchi, Indian Farming 54(6):16-18.

# Achievement of half whiteness of botanicals by single step bleaching process for dry flower making

**Manickam Visalakshi, Murugiah Jawaharlal and Manickam Kannan**

Department of Floriculture and Landscaping, Horticultural College and Research institute, Tamil Nadu Agricultural University, Coimbatore-3, Tamil Nadu, India.

**Studies were conducted during 2010 to 2013 to identify the ideal bleaching agents for achieving half whiteness of different parts of plant used in dry flower making. Dried pods of *Acacia auriculiformis, Sesamum indicum, Gossypium hirsutum, Pongamia glabra*, and cones of *Pinus spp* were given bleaching treatment in a single step. Different chemicals viz., sodium hydroxide (10 %), sodium chlorite (10 and 20%), sodium hypochlorite (30%), hydrogen peroxide (10, 20 and 30%), hydrochloric acid (5%) and sodium silicate (1%) were used in 10 different combinations. The time taken for achieving required half whiteness (Yellow White Group – 158 – A) was observed as per RHS colour chart and the percentage of damage was calculated. Strength of whiteness was assessed and evaluated using RHS colour chart. Whiteness index was measured using Mini scan colour measurement system. Botanicals were scored for quality parameter *viz.*, shape retention. The results indicated that, the treatment involving 20% sodium chlorite + 5% hydrochloric acid (cold water) proved superior with minimum bleaching time of 6 h and lowest rate of damage with highest whiteness index and maximum score for shape retention.**

**Key words:** Dry flower making, botanicals, bleaching agents, half whiteness, RHS colour chart.

## INTRODUCTION

In the present era of eco consciousness, use of natural products like dry flowers and their parts has become the premier choice of the consumers in their life styles for interior decoration. The dry flowers are gaining popularity amongst the floriculturists and buyers as it is an inexpensive, everlasting and eco friendly product available throughout the year (Muthukumaran, 2009). Some dry flower botanicals are too rustic in colour and this interferes with desired colour in dyeing. Otherwise aesthetically attractive plant material which is inherently colored by unwanted pigment and brown lignin can be bleached. Much of the preserved ornamental plant materials are bleached at some stage during preserving process and recolouring with dyes. So before colouring, the materials should be bleached which enhances the colour absorption during treatment with dyes. Bleaching ornamental plant materials lightens the colour of the material and provides a striking contrast and enhances the absorption of dyes (Palisoc and Camalate, 2002). In the process, the appearance and the value of the product are improved.

Different bleaching chemicals can be used but the nature and concentration of the chemical will have influence on the quality of the produce. Oxidative (hypochlorite, chlorite, and peroxide) or reductive (sulfite and borohydride) bleaches were used on ornamental plant materials. The former tend to break down coloured compounds whereas the latter tend to modify them into colourless compounds (Joyce, 1998; Arulmurugan et al., 2007). Literatures were available on bleaching action of different bleaching chemicals. Sodium chlorite was found to be the most commonly used bleach for plant foliage owing to its selective mode of action on lignin without damaging the fiber. Also this bleach was reported to remove the entire colour from cellulose based material with minimal damage to the cellulose even during prolonged contact (Masschelein, 1979; Dubois and Joyce, 2005).

Fabian (2001) reported that, sodium chlorite is a very cheap oxidant and has been extensively used in water treatment and as a bleaching agent in paper and textile industries. Peroxides are important industrial bleaching agent for cellulosic products (Hassan, 2003; Khristova et al., 2003; Shatalov and Pereira, 2005; Samanta and Deepali, 2007). Also bleaching with hydrogen peroxide is energy efficient, eco friendly, less expensive, and economical. The addition of magnesium sulphate and sodium silicate to the medium is required to stabilize the peroxide in alkaline conditions. Yogita (2000) conducted studies to standardize bleaching and dyeing technology of hybrid tea roses and *Aerva sp* and found that, sodium chlorite was the most ideal chemical for bleaching roses under distilled water medium. In the same way, Lourdusamy et al. (2002) found that, sodium chlorite was the most effective bleaching agent for gomphrena flowers at a concentration of 10%.

Others factors affecting the whiteness of a bleached material are quality of water and concentration of hydrogen peroxide used (Naresh and Deepak, 2006; Ziaie et al., 2008). However, no standard comprehensive information on effective bleaching related to desired whiteness with less damage is available. Milder bleaches and standardization of bleaching chemical combinations in single step will help to preserve the structure and to prevent damage. The present study was undertaken with an objective to optimize bleaching chemicals to attain half whiteness of botanicals suitable for dry flower making in a single step.

## MATERIALS AND METHODS

Experiments were conducted in completely randomized block design with 3 replications to identify the ideal bleaching agents for achieving half whiteness of different plant parts used in dry flower making. Dried pods of *Acacia auriculiformis, Sesamum indicum, Gossypium hirsutum, P. glabra* and cones of *Pinus spp* were collected and given bleaching treatment in a single step. The required solutions were prepared using distilled water. In case of hot water treatment, the distilled water was boiled to 100°C. 250 g of botanicals were soaked in 1 L of different treatment combination solution and kept at room temperature for further observations. Treatment combinations were $T_1$ - 10% sodium hydroxide, 1% sodium silicate, 10% hydrogen peroxide, $T_2$ - 10% sodium hydroxide, 1% sodium silicate, 20% hydrogen peroxide, $T_3$ - 10% sodium hydroxide, 1% sodium silicate, 30% hydrogen peroxide, $T_4$ - 10% sodium chlorite + 5% hydrochloric acid (hot water), $T_5$ - 20% sodium chlorite + 5% hydrochloric acid (hot water), $T_6$ - 10% sodium chlorite + 5% hydrochloric acid (cold water) ,$T_7$ - 20% sodium chlorite + 5% hydrochloric acid (cold water), $T_8$ - 30% sodium hypochlorite + 20% hydrogen peroxide, $T_9$ - 30% calcium hypochlorite + 20% hydrogen peroxide, $T_{10}$ - Control (5% sodium hydroxide, 1% sodium silicate , 10% hydrogen peroxide) .

The required half whiteness was fixed as Yellow White Group – 158 – A as per RHS colour chart. The time taken for achieving required half whiteness was observed at periodical intervals of 3, 6 and 9 h. The pods were thoroughly rinsed in water and shade dried. The strength of whiteness was assessed and evaluated using RHS colour chart. The whiteness index was measured using Mini scan colour measurement system model MS (Hunter Associates laboratory, Reston VA) L values which represent the lightness of sample were recorded before and after bleaching treatments and whiteness index was calculated. L value of 100 represents white while L value of 0 represents black.

The percentage of damage was calculated by assessing initial weight and final weight of dried botanicals. Sensory scores on visual observation were done on quality parameter viz., shape retention. A panel of judges comprising 12 members from all the age groups related to dry flower industry judged the samples by feel and visual method and scored based on the score chart (0-0.4 – very low, 0.5 – 1.4 – low, 1.5-2.4 – medium, 2.5-3.4 – high, 3.5 – 4.0 – very high). In the experiment the data recorded were statistically analysed for standard error of deviation and critical mean difference for significance using single factor analysis of variance (ANOVA) using AGRES (7.01, Pascal International software, USA). When ANOVA showed significant effects, mean separation was done using LSD test ($p < 0.05$).

## RESULTS AND DISCUSSION

All the parameters studied in the experiment were significantly (SED., Cd 0.5%) effected by bleaching application. The results indicated in Table 1 that $T_7$ proved superior with minimum bleaching time of 6 h for dried pods of *A. auriculiformis, S. indicum, G. hirsutum, P. glabra* and cones of *Pinus spp* with required half whiteness (HW). $T_4$, $T_5$ and $T_6$ are on par with $T_7$ for required whiteness but with less whiteness index values (Table 2). Sharon (2006) reported that, bleaching process works best on woodier dried materials. The botanicals treated under treatment $T_7$ scored highest whiteness index values (Table 2) for Botanical 1 (53.77), Botanical 3 (52.99), Botanical 4 (46.34), and Botanical 5 (47.67) respectively with minimum percentage of damage, except Botanical 2, followed by Treatment $T_5$ [Botanical 1 (52.25), Botanical 3 (50.56), Botanical 4 (45.16) and Botanical 5 (45.27) respectively].

The minimum percentage of damage (Figure 1) was observed in $T_7$ for Botanical 1 (4.64%), Botanical 3 (2.74%) Botanical 4 (0.78%), and Botanical 5 (3.74%) respectively except Botanical 2 followed by $T_6$ [Botanical 1 (7.88%), Botanical 3 (5.40%), Botanical 4 (3.30 %) and Botanical 5 (4.92%) respectively]. Sensory scores on shape retention were highest in Treatment $T_7$ with maximum scores for Botanical 1 (3.36) Botanical 2 (3.57), Botanical 3 (3.43), Botanical 4 (3.07) and Botanical 5 (3.43) respectively. Scores on shape retention for all botanicals was minimum in Treatments $T_4$ and $T_5$ except *pinus spp*. Since in these treatments hot water was used, more percentage of damage and minimum scores for shape retention was recorded under this treatment.

Exceptionally the cones of *pinus spp* scored high values of whiteness index (60.54) with maximum score for

**Table 1.** Effect of different bleaching agents on time taken for complete bleaching of botanicals to get half whiteness.

| Treatment | *A. auriculiformis* pods (Botanical 1) | | *Pine spp* cones (Botanical 2) | | *S. indicum* pods (Botanical 3) | | *G. hirsutum* pods (Botanical 4) | | *Pongamia pinnata* pods (Botanical 5) | |
|---|---|---|---|---|---|---|---|---|---|---|
| | Time taken (h) | Strength of whiteness | Time taken (h) | Strength of whiteness | Time taken (h) | Strength of whiteness | Time taken (h) | Strength of whiteness | Time taken (h) | Strength of whiteness |
| $T_1$ | 9 | BL | 9 | UB | 9 | BL | 9 | YL | 9 | YL |
| $T_2$ | 9 | BL | 9 | YL | 9 | BL | 9 | YL | 9 | YL |
| $T_3$ | 9 | BL | 9 | HW | 9 | BL | 9 | HW | 9 | YL |
| $T_4$ | 6 | HW | 6 | YL | 6 | HW | 6 | YL | 6 | HW |
| $T_5$ | 6 | HW | 6 | HW | 6 | HW | 6 | HW | 6 | HW |
| $T_6$ | 6 | HW | 6 | YL | 6 | HW | 6 | YL | 6 | HW |
| $T_7$ | 6 | HW | 6 | HW | 6 | HW | 6 | HW | 6 | HW |
| $T_8$ | 9 | UB | 9 | UB | 9 | UB | 9 | UB | 9 | UB |
| $T_9$ | 9 | UB | 9 | UB | 9 | UB | 9 | UB | 9 | UB |
| $T_{10}$ | 9 | BL | 9 | UB | 9 | BL | 9 | YL | 9 | UB |

HB; Half white, YL; yellowing, UB; unbleached, BL; blackening.

**Table 2.** Effect of different bleaching agents on L value and whiteness index of botanicals.

| Treatment | *A. auriculiformis* pods (Botanical 1) | | | *Pine spp* cones (Botanical 2) | | | *S. indicum* pods (Botanical 3) | | | *G. hirsutum* pods (Botanical 4) | | | *Pongamia pinnata* pods (Botanical 5) | | |
|---|---|---|---|---|---|---|---|---|---|---|---|---|---|---|---|
| | L values | | Whiteness index | L values | | Whiteness index | L values | | Whiteness index | L values | | Whiteness index | L values | | Whiteness index |
| | Before bleaching | After bleaching | | Before bleaching | After bleaching | | Before bleaching | After bleaching | | Before bleaching | After bleaching | | Before bleaching | After bleaching | |
| $T_1$ | 18.35 | 12.4 | -5.95 | 14.53 | 18.53 | 4 | 20.23 | 15.56 | -4.67 | 25.36 | 52.31 | 26.95 | 25.36 | 57.36 | 32 |
| $T_2$ | 19.26 | 11.45 | -7.81 | 15.62 | 56.78 | 41.16 | 22.34 | 16.43 | -5.91 | 24.38 | 53.42 | 29.04 | 26.45 | 54.36 | 27.91 |
| $T_3$ | 20.36 | 12.93 | -7.43 | 16.47 | 69.43 | 52.96 | 24.53 | 15.47 | -9.06 | 27.29 | 70.23 | 42.94 | 25.88 | 56.12 | 30.24 |
| $T_4$ | 21.15 | 70.54 | 49.39 | 15.34 | 64.11 | 48.77 | 22.81 | 71.68 | 48.87 | 25.47 | 51.49 | 26.02 | 26.85 | 71.23 | 44.38 |
| $T_5$ | 19.89 | 72.13 | 52.25 | 14.88 | 75.42 | 60.54 | 21.67 | 72.23 | 50.56 | 26.71 | 71.86 | 45.16 | 27.67 | 72.94 | 45.27 |
| $T_6$ | 20.69 | 71.45 | 50.76 | 15.92 | 58.46 | 42.54 | 22.46 | 72.01 | 49.55 | 24.73 | 53.18 | 28.45 | 26.32 | 72.4 | 46.08 |
| $T_7$ | 18.76 | 72.53 | 53.77 | 15.34 | 72.53 | 57.19 | 20.46 | 73.45 | 52.99 | 25.81 | 72.15 | 46.34 | 25.78 | 73.45 | 47.67 |
| $T_8$ | 18.13 | 23.12 | 4.99 | 16.44 | 17.53 | 1.09 | 22.35 | 19.04 | -3.31 | 24.37 | 25.34 | 0.97 | 26.07 | 27.53 | 1.46 |
| $T_9$ | 19.43 | 22.75 | 3.32 | 16.23 | 18.59 | 2.36 | 21.49 | 19.45 | -2.04 | 25.61 | 26.32 | 0.71 | 26.45 | 27.21 | 0.76 |
| $T_{10}$ | 18.62 | 9.32 | -9.30 | 14.84 | 17.34 | 2.5 | 21.36 | 15.01 | -6.35 | 24.32 | 45.23 | 20.91 | 25.72 | 26.14 | 0.42 |
| SEd | - | - | 0.46 | - | - | 0.56 | - | - | 0.46 | - | - | 0.43 | | - | 0.47 |
| CD (5%) | - | - | 1.03 | - | - | 1.25 | - | - | 1.02 | - | - | 0.97 | | - | 1.05 |

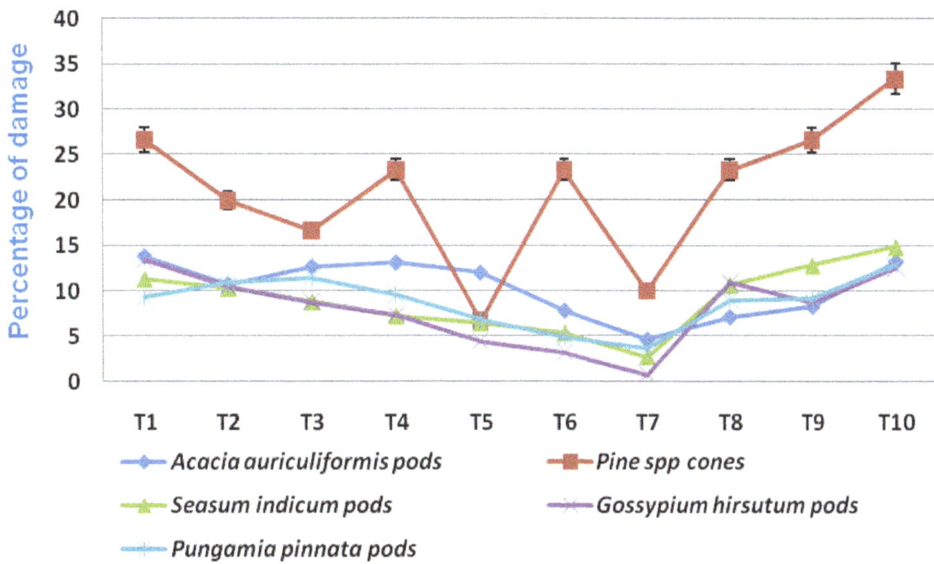


**Figure 1.** Percentage of damage Of Botanicals under different bleaching treatments.

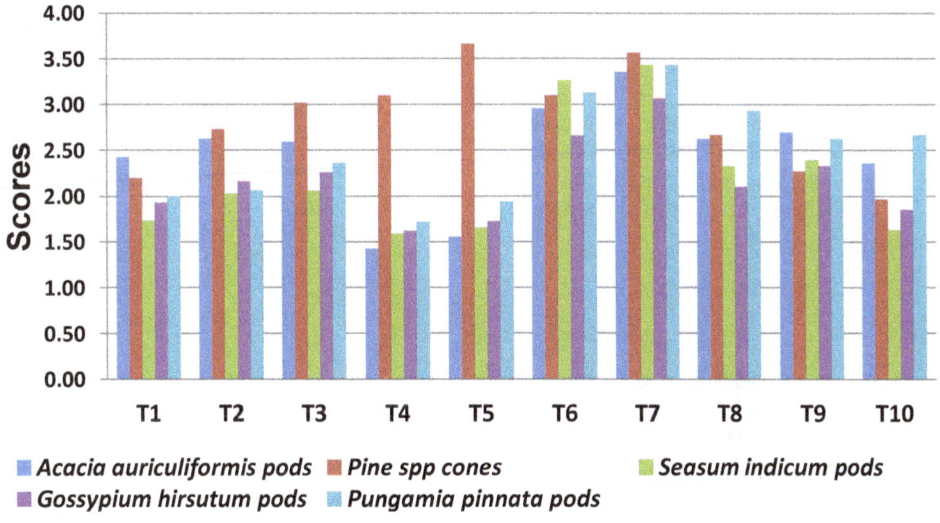


**Figure 2.** Scoring on shape of botanicals under different bleaching treatments.

shape retention (3.67) and minimum damage of 6.70% under Treatment $T_5$. It might be due to effective bleaching of hard materials like cones of *pinus spp* hot water treatment is suitable.

The factors affecting the whiteness of a bleached material are temperature of water and concentration of chemicals used (Naresh and Deepak, 2006). This was in accordance with the earlier studies by Lourdusamy et al (2002) who reported that, sodium chlorite is the most effective bleaching agent for gomphrena flowers at a concentration of 10% sodium chlorite is an oxidant particularly adapted for synthetic fiber bleaching (polyamidic, acrylic, polyester) and cellulosic (man-made and natural, particularly for linen) (Figure 2). The principal advantage of sodium chlorite is high degree of brightness (especially for acrylic fibers), negligible degradation of fibres (1 to 2% weight loss for cellulosic fibers and no attack to the polymeric chains in the synthetic fibers), lower environmental impact of wastewaters (Arulmurugan et al., 2007). In the field of synthetic organic chemistry, the most known use of sodium chlorite is in the efficient chemoselective oxidation of aldehydes to the corresponding carboxylic acid (Krapcho, 2006). While treatments involving sodium hydroxide combinations caused blackening in *S. indicum* and *A. auriculiformis* whereas, treatment combinations of sodium hypochlorite and calcium hypochlorite were not effective for any of the botanicals used. This was confirmed by Dubois and Joyce (2002) where materials bleached with sodium hypochlorite showed yellowing while calcium hypo chlorite showed cellulose damage.

## Conclusion

Based on the present study, treatment involving 20% sodium chlorite + 5% hydrochloric acid (cold water) proved superior bleaching properties with minimum time of 6 h with highest values for whiteness index and maximum scores for shape retention with minimum percentage of damage. Sodium chlorite effectively bleached the botanicals to required half whiteness in the present experiment. The advantage of bleaching of botanicals suitable for dry flower making with sodium chlorite in the presence of hydrochloric acid is that, the chemical is selective in mode of action to lignin and removes all colours quickly with minimal damage to cellulose.

## ACKNOWLEDGEMENT

This study was funded by the National Agricultural Innovation Project - ICAR- NAIP – TNAU – "A value chain on flowers for domestic and export markets". The authors, therefore, acknowledge with thanks NAIP for the technical and financial support.

### REFERENCE

Arulmurugan P, Thiyagarajan G, Ramesh K (2007). Dry flower technology. Science Entrepreneur: P.12.

Dubois P, Joyce DC (2002). Bleaching ornamental plant materials: a brief review, Austr. J. Exper. Agric. 32:785-90.

Dubois P, Joyce DC (2005). Drying cut flowers and foliage. Farm note: 10/89.

Fabian I (2001). The reactions of transition metal ions with chlorine (III). Co ord. Chem. 216-217:449-472.

Hassan EBM (2003). Sodium chlorite bleaching of textiles. J. Sci. Ind. Res. 62(3):669.

Joyce DC (1998). Dried and preserved ornamental plant material: not new, but often overlooked and under rated, Proc. 3rd international symposium on new floricultural crops. Acta Hort. 454:133-145.

Khristova P, Tomkinson J, Lioyd Jones G (2003). Bleaching cotton with peroxides. Ind. J. Fibre. Text. Res. 18(2):101.

Krapcho AP (2006). Uses of sodium chlorite and sodium bromate in organic synthesis: A Review. Org. Prep. Proc. Int. 38:177-216.

Lourdusamy DK, Vadivel E, Manavalan RSA (2002). Studies on bleaching of annual flowers for dry floral crafts. South Indian Hort. 50(4-6):658-660.

Masschelein WJ (1979). Chlorine di oxide - Chemistry and environmental impact of oxychlorine compounds. Ann. Arbor. Science. Michigan, USA.

Muthukumaran C (2009). Indian floriculture industry: opportunities and challenges, Cab calling: pp. 49-53.

Naresh MS, Deepak VA (2006). Yellowing of white fabrics and garments. http://www.sarex.com/sarexnew/textile_chemical/articles/pdf/ART124.pdf.

Palisoc JG, Camalate QJ (2002). Improvement of Bleaching and Dyeing Techiques for Labtang (*Anamirta cocculus*), Lukmoy (*Raphidophora merrilli*), Banana (*Musa sapientum*) leaf sheath, Buri (*Corypha elata*) leaves, Karagumoi (*Padanus simplex*) and Seagrass (*Actinoscirpus grossus*). http: // www. Fprdi.dost.gov.ph.

Samanta AK, Deepali SG (2007). Hydrogen peroxide and potassium per- oxo-sulphate combined room temperature bleaching of jute, cotton and jute – cotton union. Ind. J. Fibre. Text. Res. 32:221-231.

Sharon B (2006). Preserving Flowers and Foliage. http://www.ca.uky.edu/ agc/ pubs/ ho/ho70/ho70.pdf.

Shatalov AA, Pereira H (2005). Process of bleaching. Bioresour. Technol. 96(8):865.

Yogita C (2000). Studies on standardization of bleaching and dyeing technology of flowers. M.Sc. (Hort.) Thesis submitted to Tamil Nadu Agricultural University, Tamil Nadu.

Ziaie SY, Mohammadi J, Rezayati CP, Khajeheian MB (2008). Bleaching of rice pulp. Bioresour. Technol. 99(9):3568.

# Influence of spacing and fertilizer levels on the leaf nutrient contents of Bhendi (*Abelmoschus esculentus* L. Moench) under drip fertigation system

G. Rajaraman[1] and L. Pugalendhi[2]

[1]Department of Vegetable Crops, Horticultural College and Research Institute, Coimbatore- 641 003, Tamil Nadu, India.
[2]Tapioca and Castor Research Station, Tamil Nadu Agricultural University, Yethapur - 636 119, Salem District, Tamil Nadu, India.

**Field experiments were conducted during 2010 to 2012 to study the effect of spacing and fertilizer levels on the leaf nutrient contents of Bhendi (*Abelmoschus esculentus* L. Moench) under drip fertigation system at Horticultural College and Research Institute, Tamil Nadu Agricultural University, Coimbatore. The treatments consisted two spacings ($M_1$-60 x 45 cm and $M_2$ - 60 x 30 cm) and eight drip fertigation levels. The study revealed that drip fertigation at 125% recommended dose of fertilizers (250:125:125 kg NPK/ha) as water soluble fertilizer combined with Azophosmet (0.5% at750 ml/ha) and humic acid (0.4% at2.5 litre/ha) under wider spacing registered the highest leaf nutrient status like nitrogen (2.31%), phosphorus (0.53%), potassium (1.33%), calcium (0.59%), magnesium (1.46%), zinc (45.55 ppm), manganese (102.81 ppm) and boron content (16.70 ppm), whereas, the leaf copper content (22.31 ppm) was the highest under 100% recommended dose of fertilizers as water soluble fertilizer combination with Azophosmet and humic acid under wider spacing.**

**Key words:** Spacing, fertilizer levels, Bhendi, drip fertigation, leaf nutrients.

## INTRODUCTION

Bhendi (*Abelmoschus esculentus* L. Moench) is an important vegetable crop belonging to the family Malvaceae. It is also referred as ladies finger or okra. It is an annual vegetable crop and generally propagated through seeds. It is a cheap and nutritious vegetable. Hundred grams of consumable unripe bhendi contains 10.4 g dry matter, 3100 calories of energy, 1.8 g protein, 90 mg calcium, 110 mg iron, 0.1 mg carotene, 0.01 mg thiamine, 0.08 mg riboflavin, 0.08 mg niacin and 18 mg vitamin C (Thamburaj and Narendra, 2001). The dry seeds contain 13 to 22% edible oil and 20 to 24% protein.

It is essential to provide optimum plant population density per unit area by adjusting the spacing levels in bhendi crop unlike in normal spacing the plants grown in closer spacing exhibited more vertical growth but give less yield and poor quality for need of sufficient space, light, nutrient and moisture due to heavier plant population pressure (Dhanraj et al., 2001). Whereas the plants grown in the wider spacing exhibit more horizontal and continuous vegetative growth due to less population pressure per unit area but they also give less yield per unit area (Anilkumar, 2004). However the plants grown under normal spacing will have optimum population density per unit area which provides optimum conditions for luxuriant crop growth and better plant canopy area due to maximum light interception, photosynthetic activity, assimilation and accumulation of more photosynthates into plant system and hence they produce more yields with best quality traits (Mazumdar et al., 2007). Similar to spacing, judicious application of balanced and adequate nutrients play vital role in deciding the ultimate success of yield and quality of bhendi crop by realizing higher yield and the best quality.

The growth, yield and quality of crop are largely influenced by the fertility status of the soil (or) soil fertility status apart from genetic potential of the variety. Altering the soil nutrients and fertility status by providing balanced and adequate major nutrients like nitrogen, phosphorus and potassium as per the crop requirement is one of the easiest way to boost up crop productivity of bhendi. Since the interception in the supply of major nutrients even for a brief period is determined by pattern of crop growth and development which may produce less yield and poor quality and it cannot be corrected or altered at later stages of the crop growth even by supplying with heavier doses of major nutrients (Dwivedi et al., 2002). It is estimated that losses of water and applied nutrients in the conventional method of usage of water and fertilizers are more than 30 to 40%.

So, there is a need for proper fertigation management in order to ensure maximum crop productivity. It is well known that organic manures, inorganic fertilizers and biofertilizers are essential to increase the yield of vegetable crops. Therefore, it is imperative that the judicious combinations of chemical fertilizers, organics as well as biofertilizers are to be utilized properly and effectively not only as source of the nutrients but also for increasing nutrient use efficiency without adversely disturbing the soil health.

With this context, this research will closely monitor the leaf nutrient content of hybrid bhendi under different doses of fertilizers and spacing by utilizing drip fertigation. This research will also offer information on spatial variations in relation to drip fertigation system.

### MATERIALS AND METHODS

A field trial to study the influence of fertilizer levels and spacing on the leaf nutrient contents of Bhendi (*A. esculentus* L. Moench) under drip fertigation system was conducted at Thoppur village in Dharmapuri district of Tamil Nadu, India during Rabi 2010 to 2012 and Summer 2011. The experiment was laid out in split plot design (SPD) with three replications. Recommended dose of fertilizers – (NPK at 200:100:100 kg/ha), liquid bio inoculant (Azophosmet at 0.5%) at 750 ml/ha, liquid bio stimulant (Humic acid at 0.4%) at 2.5 litre/ha at 15 and 30 DAS through drip irrigation. Bio inoculant and bio stimulant were purchased from the Department of Agricultural Microbiology, Tamil Nadu Agricultural University. The treatments consisted of two levels of spacing in main plots and eight levels of drip fertigation in sub plots. The treatments consisted spacing of $M_1$ (60 x 45 cm) and $M_2$ (60 x 30 cm), drip fertigation levels of $S_1$ - Drip fertigation with WSF at 125% RDF + Azophosmet + Humic acid, $S_2$ - Drip fertigation with WSF at 100% RDF + Azophosmet + Humic acid, $S_3$ - Drip fertigation with WSF at 75% RDF + Azophosmet + Humic acid, $S_4$ - Drip fertigation with WSF at 100% RDF, $S_5$ - Drip fertigation with SF at 125% RDF + Azophosmet + Humic acid, $S_6$ - Drip fertigation with SF at 100% RDF + Azophosmet + Humic acid, $S_7$ - Drip fertigation with SF at 75% RDF + Azophosmet + Humic acid, $S_8$ - Drip fertigation with SF at 100% RDF.

In WSF 100% NPK application through drip fertigation system. In SF 100% phosphorus applied as basal and 100% nitrogen and potassium applied through drip fertigation. The water soluble fertilizer sources for supplying NPK through drip irrigation were urea, poly feed, MAP and Multi-k. The straight fertilizer sources for supplying nitrogen and potassium through drip irrigation were urea and MOP and the 100% phosphorus applied as SSP as basal. Leaf nutrient contents viz., Nitrogen, Phosphorus. Potassium, Calcium, Magnesium, Copper, Manganese, Zinc, Iron and Boron was estimated by the following methods.

## RESULTS AND DISCUSSION

Efficient use of fertilizers in any crop plant was necessary for optimum growth and yield. So knowledge about the availability of nutrients in the soil is very essential. A clear understanding of specific nutrient requirement of the crop during various stages of growth will substantially reduce the possible wastage of applied nutrients and improve both potentiality of the plant and nutrient use efficiency. During vegetative stage, the plant vigorously absorbs nutrients to build up the plant framework (Pflunger and Mengel, 1972). Plant analysis serves as an elegant tool for understanding the growth and physiology of the plant at various growth stages (Hartz and Hochmuth, 1996).

Among the spacing, leaf nitrogen content was higher at wider spacing. At vegetative stage, the content was higher (Tables 1 and 2). The mean leaf nitrogen content was higher in wider spacing ($M_1$) and it was 1.83%, whereas in closer spacing ($M_2$) it was 1.45%. Among the different levels of drip fertigation, the 125% recommended dose of fertilizers as water soluble fertilizer along with Azophosmet and humic acid ($S_1$) registered the maximum leaf nitrogen content (2.17%), whereas the minimum leaf nitrogen content was registered in 75% recommended dose of fertilizers as straight fertilizer along with Azophosmet and humic acid, $S_7$ (1.10%) at vegetative stage. The interaction effect was maximum in the 125% recommended dose of fertilizers as water soluble fertilizer with Azophosmet and humic acid ($M_1S_1$) which showed the maximum value (2.31%) under wider spacing, whereas the minimum interaction effect was observed in $M_2S_7$ (1.02%) in closer spacing at vegetative stage. The leaf nitrogen content was higher at vegetative stage and declined thereafter. Increased nitrogen content in different plant parts are due to the higher availability in the root zone, uptake and accumulation of nitrogen, which may take place gradually with the advancement of crop growth phase.

Similar findings were also reported by Colla et al. (2001) and Umamaheswarappa et al. (2005). The principle physiological function of humic acid is that it reduces the oxygen deficiency in plants which result in

**Table 1.** Effect of spacing and fertilizer levels on the nutrient content (N, P, K, Ca and Mg) of Bhendi under drip fertigation system.

| Treatment | Leaf nutrient content | | | | | | | | | | | | | | |
|---|---|---|---|---|---|---|---|---|---|---|---|---|---|---|---|
| | Nitrogen content (percent) | | | Phosphorus content (percent) | | | Potassium content (percent) | | | Calcium content (percent) | | | Magnesium content (percent) | | |
| | $M_1$ | $M_2$ | Mean | $M_1$ | $M_1$ | $M_1$ | $M_1$ | $M_2$ | Mean | $M_1$ | $M_2$ | Mean | $M_1$ | $M_2$ | Mean |
| $S_1$ | 2.31 | 2.02 | 2.17 | 0.53 | 0.46 | 0.50 | 1.33 | 1.25 | 1.29 | 0.59 | 0.55 | 0.57 | 1.46 | 1.35 | 1.41 |
| $S_2$ | 2.16 | 1.78 | 1.97 | 0.45 | 0.43 | 0.44 | 1.20 | 1.15 | 1.18 | 0.54 | 0.50 | 0.52 | 1.38 | 1.31 | 1.35 |
| $S_3$ | 1.81 | 1.29 | 1.55 | 0.36 | 0.33 | 0.35 | 1.03 | 0.98 | 1.01 | 0.46 | 0.42 | 0.44 | 1.25 | 1.19 | 1.23 |
| $S_4$ | 1.89 | 1.53 | 1.71 | 0.38 | 0.36 | 0.37 | 1.08 | 1.03 | 1.06 | 0.49 | 0.45 | 0.47 | 1.30 | 1.24 | 1.27 |
| $S_5$ | 2.02 | 1.68 | 1.85 | 0.41 | 0.39 | 0.40 | 1.14 | 1.07 | 1.11 | 0.52 | 0.47 | 0.50 | 1.34 | 1.28 | 1.31 |
| $S_6$ | 1.72 | 1.22 | 1.48 | 0.33 | 0.31 | 0.32 | 0.97 | 0.92 | 0.95 | 0.44 | 0.38 | 0.41 | 1.20 | 1.14 | 1.17 |
| $S_7$ | 1.17 | 1.02 | 1.10 | 0.24 | 0.23 | 0.24 | 0.87 | 0.84 | 0.86 | 0.37 | 0.33 | 0.35 | 1.11 | 1.03 | 1.07 |
| $S_8$ | 1.57 | 1.10 | 1.34 | 0.29 | 0.27 | 0.28 | 0.93 | 0.89 | 0.91 | 0.41 | 0.35 | 0.38 | 1.16 | 1.06 | 1.11 |
| Mean | 1.83 | 1.45 | 1.64 | 0.37 | 0.35 | 0.36 | 1.07 | 1.02 | 1.04 | 0.48 | 0.43 | 0.45 | 1.28 | 1.20 | 1.24 |
| | **SEd** | **CD (0.05%)** | | **SEd** | **CD (0.05%)** | | **SEd** | **CD (0.05%)** | | **SEd** | **CD (0.05%)** | | **SEd** | **CD (0.05%)** | |
| M | 0.01348 | 0.05801 | | 0.00281 | 0.01211 | | 0.01176 | 0.05060 | | 0.00243 | 0.01044 | | 0.00772 | 0.03323 | |
| S | 0.02997 | 0.06140 | | 0.00520 | 0.01065 | | 0.01506 | 0.03085 | | 0.00819 | 0.01678 | | 0.01796 | 0.03679 | |
| M X S | 0.04188 | 0.09557 | | 0.00743 | 0.01763 | | 0.02313 | 0.06086 | | 0.01110 | 0.02394 | | 0.02498 | 0.05415 | |
| S X M | 0.04239 | 0.08683 | | 0.00736 | 0.01507 | | 0.02130 | 0.04362 | | 0.01158 | 0.02373 | | 0.02540 | 0.05506 | |

uptake of larger amount of nitrogen reported by Bhuma (2001) and Kandasamy (1985). The very high ion exchange capacity of humates may bring about better utilization of applied nitrogen by making the nitrogen available in root zone (Seitz, 1960; Virghine, 2003). Another reason for enhanced nitrogen content may be the acid functionality of humic acid, which stimulated the nitrate uptake by the plants. Humic substances would have induced the activities of enzymes such as invertase and nitrate reductase and thereby helped in increased assimilation of nitrogen by plants (Ferretti et al., 1991).

Phosphorus plays key role in the plants energy transfer system. The data on leaf phosphorus as influenced by spacing and different levels of drip fertigation. Under wider spacing ($M_1$), the maximum leaf phosphorus content (0.37%) was observed and in the closer spacing ($M_2$) the minimum leaf phosphorus content was 0.35% at flowering stage. Different levels of drip fertigation showed significant differences. Application of 125% recommended dose of fertilizers as water soluble fertilizer along with Azophosmet and humic acid ($S_1$) registered the maximum leaf phosphorus content of 0.50%, whereas the minimum leaf phosphorus content was registered in 75%recommended dose of fertilizers as straight fertilizer along with Azophosmet and humic acid, $S_7$ (0.24%) at flowering stage. The maximum interaction effect was found in 125%recommended dose of fertilizers as water soluble fertilizer along with Azophosmet and humic acid ($M_1S_1$) under wider spacing (0.53%). The lower interaction effect was found (0.23%) in $M_2S_7$ in the closer spacing at flowering stage. This finding was in accordance with Balasubramanian (2008).

The application with humic acid recorded the highest level of phosphorus content at the flowering stage of the crop growth. The nutrients present in humic acid might have increased the synthesis of cytokinin and auxin in the root tissues, which would have resulted in the better mobilization of assimilates from the source. Similar findings were recorded by Ken (2000) in onion.

Potassium being a protoplasmic factor is an essential plant nutrient. Many enzymes are activated by potassium and potassium is also involved in photo and oxidative phosphorylation thus augmenting the energy required for pod

**Table 2.** Effect of spacing and fertilizer levels on the nutrient content (micro nutrients) of Bhendi under drip fertigation system.

| Treatment | Leaf nutrient content | | | | | | | | | | | | | | |
|---|---|---|---|---|---|---|---|---|---|---|---|---|---|---|---|
| | Iron content (ppm) | | | Zinc content (ppm) | | | Manganese content (ppm) | | | Copper content (ppm) | | | Boron content (ppm) | | |
| | $M_1$ | $M_2$ | Mean | $M_1$ | $M_1$ | $M_1$ | $M_1$ | $M_2$ | Mean | $M_1$ | $M_2$ | Mean | $M_1$ | $M_2$ | Mean |
| $S_1$ | 261.55 | 256.68 | 259.11 | 45.55 | 40.55 | 43.06 | 102.81 | 97.84 | 100.33 | 19.63 | 18.28 | 18.96 | 16.70 | 15.49 | 16.10 |
| $S_2$ | 255.59 | 252.09 | 253.84 | 41.19 | 38.08 | 39.64 | 98.40 | 94.76 | 96.58 | 22.31 | 19.02 | 20.67 | 16.02 | 15.29 | 15.65 |
| $S_3$ | 250.84 | 241.78 | 246.31 | 35.87 | 33.24 | 34.56 | 94.43 | 91.91 | 93.18 | 17.97 | 17.15 | 17.56 | 14.69 | 13.93 | 14.31 |
| $S_4$ | 252.60 | 246.05 | 249.32 | 37.80 | 34.98 | 36.39 | 94.94 | 93.09 | 94.02 | 18.61 | 17.56 | 18.09 | 14.90 | 14.36 | 14.63 |
| $S_5$ | 254.31 | 250.47 | 252.39 | 38.09 | 37.72 | 37.90 | 96.22 | 93.76 | 94.99 | 19.07 | 18.05 | 18.56 | 15.51 | 14.73 | 15.12 |
| $S_6$ | 249.29 | 238.59 | 243.94 | 34.65 | 32.84 | 33.74 | 93.92 | 90.89 | 92.41 | 17.91 | 16.83 | 17.37 | 14.60 | 13.82 | 14.21 |
| $S_7$ | 244.54 | 232.34 | 238.44 | 31.26 | 29.99 | 30.63 | 91.98 | 89.84 | 90.91 | 17.01 | 16.56 | 16.79 | 13.58 | 13.31 | 13.45 |
| $S_8$ | 247.40 | 238.03 | 242.71 | 32.19 | 31.30 | 31.75 | 92.63 | 90.32 | 91.48 | 17.47 | 16.75 | 17.12 | 14.22 | 13.80 | 14.01 |
| Mean | 252.02 | 244.50 | 248.26 | 37.07 | 34.84 | 35.96 | 95.67 | 92.80 | 94.23 | 18.75 | 17.53 | 18.14 | 15.03 | 14.34 | 14.69 |
| | **SEd** | **CD (0.05%)** | | **SEd** | **CD (0.05%)** | | **SEd** | **CD (0.05%)** | | **SEd** | **CD (0.05%)** | | **SEd** | **CD (0.05%)** | |
| M | 2.53079 | 5.48650 | | 0.26029 | 1.11994 | | 0.51041 | 2.19617 | | 0.01028 | 0.04423 | | 0.03349 | 0.14408 | |
| S | 2.98848 | 6.12178 | | 0.46458 | 0.95168 | | 1.65865 | 3.39768 | | 0.23874 | 0.48904 | | 0.23070 | 0.47258 | |
| M X S | 4.69405 | NS | | 0.66743 | 1.59603 | | 2.25277 | 4.41263 | | 0.31599 | 0.64804 | | 0.30702 | 0.66558 | |
| S X M | 4.22634 | NS | | 0.65702 | 1.34589 | | 2.34568 | 4.75134 | | 0.33762 | 0.69161 | | 0.32626 | 0.70729 | |

growth (Ghanta and Mitra, 1993). The results showed that both spacings and fertilizer levels had significant variation on leaf potassium. Under the wider spacing ($M_1$) at flowering stage the leaf potassium content was found maximum (1.07%), whereas in closer spacing ($M_2$) the leaf potassium content was found minimum (1.02%). Application of 125% recommended dose of fertilizers as water soluble fertilizer along with Azophosmet and humic acid ($S_1$) registered the maximum leaf potassium content (1.29%), whereas the treatment $S_7$ registered the minimum leaf potassium content (0.86%) at flowering stage. The interaction effect was found maximum (1.33%) in $M_1S_1$ (125% recommended dose of fertilizers as water soluble fertilizer along with Azophosmet and humic acid) under the wider spacing, whereas the interaction effect was low in $M_2S_7$ in closers pacing (0.84%) at flowering stage. Fontes et al. (2000) and Dangler and Lacascio (1990) opined that application of N and K in combination with drip irrigation increased the potassium content and yield by the way of maximizing the mobility of the nutrients around the root zone. These results were in agreement with those obtained by Kavitha (2005) in tomato.

The different levels of drip fertigation showed significant differences. Under wider spacing ($M_1$) leaf Ca content was maximum with 0.48%, whereas in closer spacing ($M_2$) the leaf Ca content was found minimum (0.43%) and which was observed at vegetative stage (Table 3). At vegetative stage, leaf Ca content was the maximum at 125% recommended dose of fertilizers as water soluble fertilizer along with Azophosmet and humic acid ($S_1$) with 0.57%. The minimum leaf Ca content was noticed at 75% recommended dose of fertilizers as straight fertilizer with Azophosmet and humic acid ($S_7$) with 0.35%. The maximum interaction effect was found in 125% recommended dose of fertilizers as water soluble fertilizer along with Azophosmet and humic acid ($M_1S_1$) under wider spacing (0.59%). The lower interaction effect was found (0.33%) in $M_2S_7$ under closer spacing at vegetative stage. The improved content of calcium by the highest amount of nutrients along with humic acid may be attributed to (i) complexation of calcium by various functional groups of humic acid, (ii) altered membrane permeability of humic acid, which allows active nutrient uptake process and (iii) enhanced cuticle and cell membrane penetration. Similar results were obtained by David et al. (1994) and Virgine (2003) in tomato.

Among the spacings, the wider spacing ($M_1$)

**Table 3.** Methods of chemical analysis on the nutrient contents of Bhendi under drip fertigation system.

| S/N | Determination | Method | Reference |
|---|---|---|---|
| 1. | Triple acid extract $HNO_3$; $H_2SO_4$; $HClO_4$ | Wet digestion of known quantity of plant material with 12 ml of triple acid and made up to desired volume | Piper (1966) |
| 2. | Diacid extract $H_2SO_4$ : $HClO_4$ (4:1) | Wet digestion of 0.5 g of plant material with 10 ml of diacid mixture and made up to desired volume | Piper (1966) |
| 3. | Nitrogen | Microkjeldahl method | Piper (1966) |
| 4. | Phosphorus | Vandamolybdate method | Piper (1966) |
| 5. | Potassium | Flame photometry method | Piper (1966) |
| 6. | Calcium, Magnesium, Copper and Manganese | Versenate titration method | Jackson (1973) |
| 7. | Zinc and Iron | Triple acid extract fed into atomic absorption spectrophotometer | Jackson (1973) |
| 8. | Boron | Colorimetric micro determination | Naftel (1986) |

recorded maximum leaf Mg content of 1.28%, whereas in closer spacing ($M_2$) the leaf Mg content was found minimum (1.20%) and was observed at flowering stage. Considering the treatments and stage of growth, the treatment $S_1$ (125% recommended dose of fertilizers as water soluble fertilizer plus Azophosmet and humic acid) recorded the maximum leaf Mg content of 1.41%. The minimum leaf Mg content was noticed at 75% recommended dose of fertilizers as straight fertilizer with Azophosmet and humic acid ($S_7$) with 1.07% at flowering stage. The maximum interaction effect was found in 125% recommended dose of fertilizers as water soluble fertilizer along with Azophosmet and humic acid ($M_1S_1$) under wider spacing (1.46%). The lower interaction effect was found (1.03%) in $M_2S_7$ in the closer spacing at flowering stage. This may be due to altered membrane permeability by nutrients, which would have allowed higher nutrient absorption process and enhanced the cuticle and cell membrane penetration Singh et al. (1995) humic acid by the formation of complex compounds with magnesium probably would have played a decisive role in the transport of magnesium in tissues (Mylona and McCants, 1980). Similar findings were reported by Pal and Sengupta (1985) and David et al. (1994).

All the drip fertigation treatments had significant influence on leaf Fe content. Among the drip fertigation treatments, the leaf Fe content was higher under wider spacing ($M_1$) with 252.02 ppm than under closer spacing ($M_2$). The leaf Fe content was lowers with 244.50 ppm at flowering stage. At flowering stage, application of 125% recommended dose of fertilizers as water soluble fertilizer in combination with Azophosmet and humic acid ($S_1$) recorded the maximum leaf Fe content of 259.11 ppm. The minimum leaf Fe content was registered in the treatment applied with 75% recommended dose of fertilizers as straight fertilizer with Azophosmet and humic acid ($S_7$) with values of 238.44 ppm. The interaction effect showed non-significant variation among the spacing and drip fertigation treatments. Enhanced solubilisation and increased extractability of iron and reduction of non-available higher oxidase forms to available form may account for its increased iron content. The results of the present study are in accordance with the findings of Dursun et al. (1999) in tomato and Demir et al. (1999) in cucumber.

All the drip fertigation treatments had significant influence on leaf Zn content. Among the different drip fertigation treatments, the leaf Zn content was maximum under wider spacing ($M_1$) with 37.07 ppm than under closer spacing ($M_2$). The leaf Zn content was minimum with 34.84 ppm at flowering stage (Table 3). Application of 125% recommended dose of fertilizers as water soluble fertilizer in combination with Azophosmet and humic acid ($S_1$) recorded the maximum leaf Zn content of 43.06 ppm, whereas the minimum leaf Zn content was registered in the treatment applied with 75% recommended dose of fertilizers as straight fertilizer with Azophosmet and humic acid ($S_7$) with values of 30.63 ppm at flowering stage. The maximum interaction effect was found in 125% recommended dose of fertilizers as water soluble fertilizer along with Azophosmet and humic acid ($M_1S_1$) under wider spacing (45.55 ppm). The minimum interaction effect was found (29.99 ppm) in $M_2S_7$ in the closer spacing at flowering stage. The better photosynthetic efficiency coupled vegetative growth would have encouraged nutrient absorption of nutrients. It is in confirmation with the findings of More and Shinde (1991). The results may also be attributed to the 3 x $10^4$ times increase in zinc solubility due to humic acid which contains fulvic acid and act as chelation of native zinc by chelating agent. Singaravel et al. (1998) opined that the beneficial influence of humic acid in soil would have prevented the formation of insoluble complexes of zinc and facilitated its uptake by plants. Similar findings were recorded by Chen et al. (2001) in cucumber and Virgine (2003) in tomato.

All the drip fertigation treatments had significant influence on leaf Mn content. Among the different drip fertigation treatments the leaf Mn content was higher under wider spacing ($M_1$) with 95.67 ppm than under closer spacing ($M_2$). The leaf Mn content was lower with 92.80 ppm at flowering stage. Application of 125% recommended dose of fertilizers as water soluble fertilizer in combination with Azophosmet and humic acid ($S_1$) recorded the maximum leaf Mn content of 100.33 ppm, whereas the minimum leaf Mn content was registered in the treatment applied with 75% recommended dose of fertilizers as straight fertilizer along with Azophosmet and humic acid ($S_7$) with values of 90.91 ppm at flowering stage. The interaction effect was found in 125% recommended dose of fertilizers as water soluble fertilizer along with Azophosmet and humic acid ($M_1S_1$) which produced the maximum leaf Mn under wider spacing (102.81 ppm). The lower interaction effect, 89.84 ppm in ($M_2S_7$) was found under closer spacing at flowering stage.

All the drip fertigation treatments had significant influence on leaf Cu content. Among the spacings, the leaf Cu content was higher under wider spacing ($M_1$) with 18.75 ppm than under closer spacing ($M_2$). The leaf Cu content was lower with 17.53 ppm at vegetative stage. Application of 100% recommended dose of fertilizers as water soluble fertilizer in combination with Azophosmet and humic acid ($S_2$) recorded the maximum leaf Cu content of 20.67 ppm at vegetative stage. The minimum leaf Cu content was registered in the treatment, 75% recommended dose of fertilizers as straight fertilizer with Azophosmet and humic acid ($S_7$) with values of 16.79 ppm at vegetative stage. The maximum interaction effect was found in 100% recommended dose of fertilizers as water soluble fertilizer along with Azophosmet and humic acid ($M_1S_2$) under wider spacing (22.31 ppm). The minimum interaction effect was found (16.56 ppm) in $M_2S_7$ in the closer spacing at vegetative stage.

The results showed that wider spacing ($M_1$) exhibited higher leaf B content of 15.03 ppm than under closer spacing ($M_2$). The leaf B content was lower with 14.34 ppm at flowering stage. Significant differences were noticed in all the drip fertigation treatments and spacings. The treatment $S_1$ (125% recommended dose of fertilizers as water soluble fertilizer plus Azophosmet and humic acid) registered the maximum boron content of 16.10 ppm. The minimum leaf boron content was observed at $S_7$ (75% recommended dose of fertilizers as straight fertilizer with Azophosmet and humic acid) with 13.45 ppm at flowering stage. Considering the treatments and stage of growth, the treatment application of 125% recommended dose of fertilizers as water soluble fertilizer plus Azophosmet and humic acid under wider spacing ($M_1S_1$) recorded the maximum leaf boron content of 16.70 ppm at flowering stage. . The increased manganese and boron contents in the leaf might be due to the effect of higher level of nutrients availability which resulting in increased intake of micronutrients from soil (Samson and Visser, 1989). Similar findings were recorded by Adani et al. (1998) in tomato.

It can be concluded from the foregoing discussion that drip fertigation at higher percentage of recommended dose of fertilizer as water soluble fertilizer along with Azophosmet and humic acid was higher nutrient content in leaves at all the stages of crop growth under wider spacing (60 x 45 cm).

**Abbreviations: RDF,** Recommended dose of fertilizer; **DAS,** Days after sowing; **NPK,** Nitrogen, Phosphorus and Potassium; **SF,** Straight fertilizer; **WSF,** Water soluble fertilizer; **SSP,** Single super phosphate; **MAP,** Mono ammonium phosphate; **SPD,** Split plot design; **MOP,** Muriate of potash.

## REFERENCES

Adani F, Genevini P, Zaccheo P, Zocchi G (1998). The effect of commercial humic acid on tomato plant growth and mineral nutrition. J. Plant Nutr. 21(3):561-585.

Anilkumar K (2004). Standardization of seed production techniques in fenugreek. M.Sc. Thesis, University of Agricultural Sciences, Dharwad, India.

Balasubramanian P (2008). Comparative analysis of growth, physiology, nutritional and production changes of tomato (*Lycopersicon esculentum* Mill.) under drip, fertigation and conventional systems. Ph.D. Thesis, Tamil Nadu Agricultural University, Coimbatore, India.

Bhuma M (2001). Studies on the impact of humic acid on sustenance of soil fertility and productivity of green gram. M.Sc. Thesis, Tamil Nadu Agricultural University, Coimbatore, India.

Chen Y, Magen H, Clapp CE (2001). Plant growth stimulation by humic substances and their complexes with iron. Proc. International Fertilizer Society. Israel. P. 14.

Colla G, Temperini O, Battistelli A, Moscatello S, Projetti S, Saccardo F, Brancaleone MP (2001). Nitrogen and celery. Coltura Protette 30:68-70.

Dangler JM, Lacascio SJ (1990). Yield of trickle irrigated tomatoes as affected by time of N and K application. J. Am. Soc. Hort. Sci. 115(4):585-589.

David PP, Nelson PV, Sanders DC (1994). Humic acid improves growth of tomato seedlings in solution culture. J. Plant. Nutr. 17(1):173-184.

Demir K, Gunes A, Inal A, Alpaslan M (1999). Effects of humic acids on the yield and mineral nutrition of cucumber (*Cucumis sativus* L.) grown with different salinity levels. Acta Hort. 492:95-103.

Dhanraj R, Prakash Om, Ahlawat IPS (2001). Response of rench bean (Phesolus vulgaris) varieties to plant density and nitrogen application. Indian J. Agron. 46:277-281.

Dursun A, Guvenc I, Tuzel Y, Tuncay O (1999). Effects of different levels of humic acid on seedling growth of tomato and eggplant. Acta Hort. 491:235-240.

Dwivedi YC, Kushwah S, Sengupta SK (2002). Studies on nitrogen, phosphorus and potash requirement of dolichos bean. JNKVV Res. J. 36(1-2):47-50.

Ferretti M, Ghisi RS, Nardi, Passera C (1991). Effect of humic substances on photosynthetic sulphate assimilation in maize seedlings. J. Soil Sci. 71:249-252.

Fontes PCR, Sampaio RA, Finger FL (2000). Fruit size, mineral composition and quality of trickle irrigated tomatoes as affected by potassium rates. Pesqui. Agropecuaria Bras. 35(1):21-25.

Ghanta PK, Mitra SK (1993). Effect of micronutrients on growth, flowering, leaf nutrient content and yield of banana cv. Giant Governor. Crop Res. 6:284-287.

Hartz TK, Hochmuth GJ (1996). Fertility management of drip irrigated vegetables. Hort. Tech. 6(3):168.

Jackson ML (1973). Soil Chemical Analysis. Prentice Hall of India Private Limited, New Delhi, India. P. 498.

Kandasamy S (1985). Studies on the effect of lignite humic acid on soil fertility and yield of black gram. M.Sc. Thesis, Tamil Nadu Agricultural University, Coimbatore, India.

Kavitha M (2005). Studies on effect of shade and fertigation on growth and yield of tomato (*Lycopersicon esculentum* Mill.) Hybrid Ruchi. Ph.D. Thesis, Tamil Nadu Agricultural University, Coimbatore, India.

Ken D (2000). Onions and garlic stand a better chance with humics. On the Horizon. 1(4):1-2.

Mazumdar SN, Moninuzzaman M, Rahman SMM, Basak NC (2007). Influence of support systems and spacing on hyacinth bean production in the eastern hilly area of Bangladesh. Leg. Res. 30(1):1-9.

More NB, Shinde PM (1991). Effects of different sources and methods of iron application on the uptake of iron, phosphorus and calcium by tomato cultivars. J. Maharastra Agric. Univ. 16 (2):172-176.

Mylona VA, Mc Cants CB (1980). Effect of humic acid and fulvic acid on growth of tobacco, root initiation and elongation. Plant Soil. 54:485-490.

Naftel JA (1986). Colorimetric micro determination of boron. Ind. Eng. Chem. Anal. Ed. 11:407-409.

Pal S, Sengupta MB (1985). Nature and properties of humic acid prepared from different sources and its effects on nutrient availability. Plant Soil. 88(1):71-79.

Pflunger R, Mengel K (1972). The Photochemical activity of chloroplasts from various plants with different potassium nutrition. Plant Soil. 36:417-425.

Piper CS (1966). Soil and Plant Analysis. Hans Publishing, Bombay, India. P. 236.

Samson G, Visser SA (1989). Surface active effects of humic acids on Potato cell membrane properties. Soil Biol. Biochem. 21:343-347.

Seitz DS (1960). Ion exchange resins from lignite coal. Ind. Eng. Chem. 52:313-314.

Singaravel R, Govindasamy R, Balasubramanian TN (1998). Influence of humic acid, nitrogen and Azospirillum on the yield and nutrient uptake by sesame. J. Indian Soc. Soil Sci. 46(1):145-146.

Singh KG, Gulshan M, Mukesh S (2005). Economic evaluation of drip irrigation system for sequential cropping cauliflower and hybrid chilli. In: International conference on plasticulture and precision farming, November 17-21, New Delhi, India. P. 207.

Thamburaj S, Narendra S (2001). Vegetables, tuber crops and species, Directorate of information and publications of agriculture, ICAR, NewDelhi, pp. 222-237.

Umamaheswarappa P, Krishnappa KS, Venkatesha Murthy P, Adivapper N, Pitachaimuthu M (2005). Effect of NPK on dry matter accumulation and primary nutrient content in leaf of bottle gourd cv. Arka Bahar. Crop Res. 30(2):181-186.

Virgine JS (2003). Effect of lignite humic acid on soil fertility, growth, yield and quality of Tomato. M.Sc. Thesis, Tamil Nadu Agricultural University, Coimbatore, India.

# Antioxidants and chlorophyll in cassava leaves at three plant ages

**Anderson Assaid Simão , Mírian Aparecida Isidro Santos, Rodrigo Martins Fraguas, Mariana Aparecida Braga, Tamara Rezende Marques, Mariene Helena Duarte, Claudia Mendes dos Santos, Juliana Mesquita Freire and Angelita Duarte Corrêa**

Chemistry Department, Biochemistry Laboratory, Federal University of Lavras – UFLA, PO Box 3037, Zip Code 37200.000, Lavras, MG, Brazil.

**The aim of this study was to quantify antioxidant substances and chlorophyll content, as well as to measure the antioxidant activity in cassava leaf flour (CLF) of different cultivars at several plant ages, in order to lead to a higher utilization of these leaves, and consequently to an enhancement of this agricultural by-product. The contents of antioxidant substances (vitamin C, polyphenols and β-carotene) were regarded as high and increased as the plants matured. The chlorophyll content decreased with plant maturity and presented a negative correlation with antioxidant substances, which indicates that the highest antioxidant levels are found when the plant presents low chlorophyll levels. CLF showed a high antioxidant activity when the lipid oxidation inhibition method (β-carotene/linoleic acid) was used, and moderate when the free radical capture method (ABTS) was used. The main contribution to the CLF antioxidant activity seems to be provided by vitamin C, which presented the best correlation with the ABTS test. Out of the ages studied, that of 14 months presented the highest antioxidant levels; Mocotó and Pão da China cultivars stood out the most.**

**Key words:** Cassava leaf flour, antioxidant activity, antioxidant substances.

## INTRODUCTION

Cassava (*Manihot esculenta* Crantz) is a perennial, bushy plant, of the Euphorbiaceae family. Originating on the American continent, probably in Central Brazil, it is cultivated all over the world, mainly in poor areas, where its cultivation constitutes one of the main agricultural activities, having a high social importance as a main carbohydrate source for more than 500 million people, essentially in developing countries (Brazilian Company of Agricultural Research-EMBRAPA, 2010). In recent years, the aerial part of the plant, which had been treated as an agricultural by-product, but that nutritionally, presents great potential for human and animal consumption, has been gaining prominence. Those leaves are rich in proteins and vitamins A and C (Corrêa et al., 2004; Wobeto et al., 2006), and minerals, especially Mg, Fe, Zn and Mn (Wobeto et al., 2006), obtained at a low cost, when compared to conventional leafy vegetables. Furthermore, their use can provide an extra income to various producers that live on the cassava culture.

In the search for natural substances that can bring benefits, mainly to health, aggregating value to this agricultural by-product can contribute to a greater use of the cassava leaves in feeding and in many other applications.

Experimental and epidemiological studies have been demonstrating that people who consume foods rich in antioxidants could have a reduced risk of many diseases, such as cancer, cardiovascular diseases, chronic diseases and aging, among others.

Antioxidants are substances that combat free radicals, which are extremely reactive species that cause oxidation of various biomolecules present in the organism. Moreover, besides the problems found with the synthetic antioxidants used in food conservation and their high production costs, it has also been demonstrated, by toxicological studies, that they can provoke undesirable effects in human and animal organism. Therefore, the search for natural substances that is efficient antioxidant sources, and less expensive than synthetic antioxidants, has been the objective of various studies (Gan et al., 2010; Surveswaran et al., 2007; Wojdylo et al., 2007).

In plant leaves, chlorophyll, the photosynthetic pigment, is directly related to their nutritional state. The amount of that pigment has been used as an evaluation index of the nutritional state for several types of cultures (Argenta et al., 2001). Variables, such chlorophyll and antioxidant levels, can be directly correlated, demanding studies to confirm that correlation and, after being confirmed, can serve as a parameter for obtaining antioxidant substances in cassava leaves.

Therefore, this work was conducted with the objective of quantifying antioxidant substances, chlorophyll and to measure the antioxidant activity in cassava leaf flour of different cultivars, at several plant ages.

## MATERIALS AND METHODS

### Samples

Mature leaves of four cassava cultivars, 'Mocotó ', 'Ufla ', 'Pão da China ' and 'Ouro do Vale', obtained from the Agriculture Department of Federal University of Lavras, were picked in the morning, from plants at three different ages (TPA), in December (10 months), February (12 months) and April (14 months), in three repetitions. Intact leaves, free from diseases and pests, were selected, washed in running water and distilled water and, soon afterwards, dried in a forced-air oven, for 48 h, at temperatures ranging from 30 to 35°C. After drying, they were ground (without petioles) in a Willy type mill and the cassava leaf flours (CLF) were stored in hermetically sealed flasks under refrigeration, until the analyses were performed.

### Analyses

#### *Determination of antioxidant substances*

**Vitamin C:** The extraction of vitamin C for chromatographic analyses was conducted in oxalic acid, according to Strohecker and Henning (1967). A Shimadzu LC 200A liquid chromatography with UV-VIS detector, wavelength detection at 254 nm; quaternary pump, degasser, and automatic injection was used. A $C_{18}$ Nucleosil (250 × 4.6 mm × 5 µm) column and a $C_{18}$ (15 × 3.2 mm × 7.5 µm) pre-column were used. As movable phase a pH 6.7 buffer was used, containing sodium acetate 0.04 mol $L^{-1}$, EDTA 0.05 mol $L^{-1}$, tributyl ammonium phosphate isocratic mode, flow rate 0.6 ml $min^{-1}$, 15 min running time for each sample.

**Phenolic compounds:** The extraction of phenolic compounds was carried out with 1 g of sample in 50 ml of 50% methanol, under reflux for three consecutive times, at 80°C, and the extracts were collected, evaporated up to 25 ml and submitted to phenolic compound measurement, using Folin-Denis reagent, and tannic acid as a standard (Association of Official Analytical Chemists - AOAC, 2011).

**β-carotene:** The extraction of β-carotene was carried out with 0.5 g of sample in a 40 ml extraction solution of isopropyl alcohol:hexane 3:1. The content was transferred to a 125 ml separation funnel wrapped in aluminum, where the volume was completed with distilled water. It was left at rest for 30 min, followed by washing of the material and discard of the aqueous phase. This operation was repeated three more times. The content was filtered with cotton sprayed with anhydrous sodium sulphate 99% to a 25 ml volumetric flask wrapped with aluminum, where 5 ml of acetone were added and the volume completed with hexane. Then, absorbance readings were taken in a spectrophotometer at four wavelengths (453, 505, 645 and 663 nm) (Nagata and Yamashita, 1992) and the results expressed in mg 100 $g^{-1}$, calculated by the formula:

$$\beta\text{-carotene (mg 100 g}^{-1}) = 0.216\ A_{663} - 1.22\ A_{645} - 0.304\ A_{505} + 0.452\ A_{453}.$$

#### *Antioxidant activity*

**Extract preparation:** For the obtention of the extracts, the CLFs were maintained under maceration using 50% ethanol, 1:40 (w/v), for 30 min and soon afterwards centrifuged at 2,500 g, for 15 min. The supernatant was collected and the precipitate was again submitted to the previously described extraction process, substituting 50% ethanol for 70% acetone; the supernatants were collected and then subjected to the detection of antioxidant activity by the methods described subsequently.

**ABTS method:** The methodology used was developed by Rufino et al. (2007). Four different dilutions were made from the obtained extracts for the assays and subsequent construction of analytical curves. Six point analytical curves were made with trolox (6-hydroxy-2,5,7,8-tetramethylchroman-2-carboxylic acid) (100 to 2,000 µmol $L^{-1}$) and with ascorbic acid (10 to 200 mg $mL^{-1}$), in addition to tests for comparison with the patterns BHT (butylhydroxytoluene–synthetic antioxidant) and with rutin and quercetin, that are flavonoids with proven antioxidant activity; these standards were prepared at a concentration of 200 mg $L^{-1}$.

**β-carotene/linoleic acid method:** The methodology used was developed by Rufino et al. (2006). The test was applied to the extracts, at a concentration of 10,000 mg $L^{-1}$.

For the preparation of the β-carotene/linoleic acid solution system, 50 µl of β-carotene diluted in chloroform (20 g $L^{-1}$) were used, to which 40 µl of linoleic acid were added, as well as 530 µl of tween 20 (emulsifier) and, for solubilization, 1 ml of chloroform. In a flask covered with aluminum for protection against light, chloroform was evaporated in a rotary-evaporator and 100 ml of oxygen saturated water (distilled water treated with oxygen for 30 min) were added, and the combination was agitated until the solution system presented a yellow-orange coloration. In test tubes, 2.5 ml of that solution system were added to 0.2 ml of each sample dilution used for the test. Control tubes were made containing 2.5 ml of the solution system with 0.2 ml of 2,6-di-tert-butyl-p-cresol (BHT), quercetin and rutin, all at the concentration of 200 mg $L^{-1}$. In laboratory tests, it was found that the concentration of 200 mg $L^{-1}$ BHT is the one that provides the greatest protection for the system, when compared to others; therefore, its use is suggested. After

**Table 1.** Antioxidant levels in cassava leaf flour of four cultivars, at three plant ages.

| Antioxidant substances | Cultivar | 10 months | 12 months | 14 months |
|---|---|---|---|---|
| Vitamin C (mg 100 $g^{-1}$) | Mocotó | $281.53±14.84^{aB}$ | $294.77±5.47^{aB}$ | $521.19±31.60^{bA}$ |
| | Ufla | $149.29±3.84^{dC}$ | $249.68±1.32^{cB}$ | $474.05±30.07^{cA}$ |
| | Pão da China | $244.74±8.58^{bC}$ | $276.97±11.50^{bB}$ | $568.64±7.71^{aA}$ |
| | Ouro do Vale | $171.44±2.58^{cC}$ | $310.88±26.16^{aB}$ | $463.69±2.40^{cA}$ |
| Polyphenols (mg $g^{-1}$) | Mocotó | $29.48±0.56^{aC}$ | $52.12±1.13^{bB}$ | $55.06±0.30^{aA}$ |
| | Ufla | $29.14±0.19^{aC}$ | $54.53±1.58^{aA}$ | $50.14±0.66^{bB}$ |
| | Pão da China | $29.18±0.30^{aB}$ | $55.57±1.54^{aA}$ | $56.21±1.39^{aA}$ |
| | Ouro do Vale | $16.46±0.16^{bC}$ | $31.73±0.86^{cB}$ | $36.26±0.67^{cA}$ |
| β-carotene (mg $100g^{-1}$) | Mocotó | $50.20±1.60^{bB}$ | $70.31±0.24^{aA}$ | $65.92±2.08^{aA}$ |
| | Ufla | $53.08±3.21^{bB}$ | $70.07±3.59^{aA}$ | $65.60±1.39^{aA}$ |
| | Pão da China | $65.90±0.83^{aB}$ | $72.72±1.09^{aA}$ | $61.63±1.38^{aB}$ |
| | Ouro do Vale | $51.82±4.59^{bB}$ | $68.79±4.74^{aA}$ | $51.44±6.67^{bB}$ |

Data are the mean of three replicates ± standard deviation. Lowercase letters in columns compare among cultivars and uppercase on the lines compare among ages. Same letters do not differ among themselves by the Scott-Knott test at 5% probability.

homogenization, their readings were taken in a spectrophotometer at 470 nm, using water to calibrate the spectrophotometer; this was considered to be the reading at time zero (initial). The tubes were placed in a water bath, at 40°C and readings were taken after 2 h.

#### Chlorophyll determination

The chlorophyll determination was conducted with Minolta SPAD-502 equipment, directly on the leaf still attached to the plant, moments before the collection, according to the methodology described by Argenta et al. (2001). Each SPAD reading is equivalent to the result obtained by the measurement performed on six leaves, with five measurements taken on each leaf, which were performed at points situated in half to two thirds of the sampled leaf length. The results were expressed in SPAD units, which are equal to the average of thirty readings.

#### Statistical analysis

The experiment was conducted in a completely randomized design, in a 4 × 3 factorial outline, with four cassava cultivars and three plant ages, and three repetitions. The statistical analysis was conducted using R (version 2.15.2) statistical software (R Core Team, 2012). Averages were compared by the Scott-Knott test, at 5% probability. Pearson correlation analysis was conducted among the antioxidant substances, antioxidant tests and chlorophyll, using the Statistical Analysis System program (1999).

## RESULTS AND DISCUSSION

### Antioxidant substances

Table 1 shows vitamin C, polyphenol and β-carotene levels of CLF in the four cultivars, at TPA. It was observed that vitamin C levels increased with plant maturity, for all cultivars. Among the cultivars, 'Pão da China' (14 months) presented the highest vitamin C level. Carvalho et al. (1989) found an increase in vitamin C levels for cassava leaves dried at 45°C, up to 14 months of age. It was also certified by Wobeto et al. (2006) who, analyzing five cassava cultivars, in TIP, observed an increase in vitamin C with plant maturity. These results are consistent with the observations made in the present study.

Higher vitamin C levels were found in this study, when compared to Wobeto et al. (2006) who, analyzing cultivars Mocotó and Ouro do Vale at 12 months of age, found levels of 55.72 and 64.12 mg 100 $g^{-1}$ of dry matter (DM) for Mocotó and Ouro do Vale, respectively. In the present study, it was found, at the same age, levels of 294.77 and 310.88 mg 100 $g^{-1}$ DM for Mocotó and Ouro do Vale, respectively. These differences may have occurred as a result of the analysis method used, which, in this study, was the chromatographic method, while Wobeto et al. (2006) used the colorimetric method, besides other factors, such as manuring, climatic conditions, among others.

When compared with other unconventional foods, CLF at 14 months presented average levels of vitamin C superior to those found in 100 g, of carrot leaves (203.70 mg) (Pereira et al., 2003). However, in relation to the fruits, those levels are also superior to those found in 100 g of fresh matter (FM) for orange (66 mg) and papaya (149 mg) (Hernández et al, 2006). Nevertheless, they are inferior to those of fruits considered rich in vitamin C, such as acerola, that contains 1,500 mg in 100 $g^{-1}$ FM. With the discovery of the antioxidant action of this vitamin, ingestion of substances with high vitamin C content has been recommended. As such, those plants are shown as good vitamin C sources with potential for use as antioxidants.

It was possible to verify that, in general, polyphenol levels increased with plant maturity. Cultivars Pão da China and Mocotó, at 14 months of age, did not differ significantly amongst themselves and they presented the highest polyphenol levels. Ouro do Vale, at the three analyzed ages, was that which presented significantly lower polyphenol levels.

Wobeto et al. (2007), studying CLF from five cultivars, in TIP, also observed an increase in polyphenol levels with plant maturity. In that study, the authors also observed higher polyphenol levels for cultivar Ouro do Vale (61.49 mg 100 $g^{-1}$ DM) and lower for Mocotó (44.13 mg 100 $g^{-1}$ DM) at 12 months of age, when compared to those recorded in this study at the same age, for Ouro do Vale (31.73 mg 100 $g^{-1}$ DM) and Mocotó (52.12 mg 100 $g^{-1}$ DM). These different results are probably due to factors related to manuring, climatic conditions, among others.

Polyphenol levels found in this study, independent of age and cultivar, are higher than those found in mg 100 $g^{-1}$ FM for some vegetables, such as broccoli (0.68), onion (1.13) and cabbage (0.67) and for some fruits, such as pineapple (0.85), banana (2.16), orange (1.14), papaya (0.15) and mango (1.10) according to Faller and Fialho (2009). These are within the range related by Asolini et al. (2006) who, analyzing phenolic compounds in plants used as tea, found levels between 15 (mate) to 56 (lemongrass) mg $g^{-1}$ DM.

Phenolic compounds act as antioxidants, due to their redox properties that allow them to act as reducing agents, hydrogen donors and metal chelators. Besides their role as antioxidants, these compounds present a wide spectrum of medicinal properties, such as antiallergic, anti-inflammatory, anti-bacterial and anti-thrombotic, plus they present cardioprotective and vasodilator effects (Balasundram et al., 2006), showing a broad field of application for the phenolics in these plants. It can be noticed that there is an increase in β-carotene levels until the age of 12 months and that those levels did not differ significantly from those at 14 months, except for Ouro do Vale and Pão da China cultivars, that presented a reduction in β-carotene levels between 12 and 14 months of age.

Wobeto et al. (2006), analyzing five cassava cultivars, in TIP, also found, at 12 months of age, higher β-carotene levels. However, these authors observed a decline in these levels with plant maturity.

Lower β-carotene levels were found in this study, when compared to Wobeto et al. (2006) who, analyzing Mocotó and Ouro do Vale cultivars at 12 months of age, found levels of 126.57 and 124.24 mg 100 $g^{-1}$ DM for Mocotó and Ouro do Vale, respectively. In the present study, we found, at the same age, levels of 70.31 and 68.79 mg 100 $g^{-1}$ DM for Mocotó and Ouro do Vale, respectively. Corrêa (2004), analyzing different forms of cassava leaf drying (sun-dried, shade-dried and oven-dried at 30°C and 40°C), observed that these forms of drying cause significant differences in β-carotene levels for CLF from Baiana cultivar. The highest levels are found for the oven-drying process at 30°C (84.83 mg 100 $g^{-1}$ DM) and the lowest ones for the shade-dried leaves (64.88 mg 100 $g^{-1}$ DM). Consequently, the different levels of this substance can be probably attributed to plant age, cultivars and also to the forms of drying.

The average β-carotene levels in the CLF studied, independent of cultivar and plant age, ranged from 50.20 to 72.72 mg 100 $g^{-1}$ DM and, when compared to those found in 100 g DM of other green leaves, unconventionally used, are comparable to those of sweet potato (40 to 120 mg) (Almazan and Begun, 1996), superior to those of carrot (8.70 mg) (Pereira et al., 2003) and lower than those of peanut leaves (100 to 140 mg) (Almazan and Begun, 1996). However, in relation to the levels in 100 g FM of green vegetables, these are higher than those of lettuce (1.37 mg), watercress (5.26 mg), green onion (2.31 mg) and parsley (3.82 mg) (Campos et al., 2003).

The consumption of natural antioxidants, such as vitamin C, polyphenols and β-carotene present in most plants, has been associated to a lower incidence of diseases caused by free radicals. Consequently, the levels of these substances found in these CLFs can contribute to their antioxidant capacity.

## Antioxidant activity

The CLF antioxidant activity (AA) determined by the ABTS method for four cultivars, at TPA, is shown in Table 2.

It was possible to observe that AA by the ABTS method increased with plant maturity for all cultivars, in equivalent of trolox, as well as ascorbic acid, and it can be explained by an increase in vitamin C and polyphenol levels with cultivar maturity, which is substances that have a strong action in capturing ABTS. The highest AA was observed at 14 months for Pão da China cultivar, in equivalent of trolox (680.62 μmol $L^{-1}$ $g^{-1}$), as well as ascorbic acid (102.42 mg $g^{-1}$), probably due to higher levels of vitamin C and phenolic compounds in their leaves at this age. However, Ouro do Vale presented the lowest AA, in equivalent of trolox, as well as ascorbic acid, at TPA, which can be explained by lower levels of antioxidants found in the flour of that cultivar.

It is observed that the CLF antioxidant potential by the ABTS method, when compared to BHT and rutin standards, in equivalent of trolox, as well as ascorbic acid, is considered of moderate potential, because at 14 months it reached on average, almost 50% of the potential of those standards, except for Ouro do Vale cultivar. In relation to quercetin, cultivar potential at any age was very inferior to the antioxidant potential of this standard.

The good antioxidant potential shown by CLF is

**Table 2.** Antioxidant activity of cassava leaf flour of four cultivars, by the ABTS method, at three plant ages.

| Cultivar | μmol trolox $L^{-1}$ $g^{-1}$ | | | mg of vitamin C $g^{-1}$ | | |
|---|---|---|---|---|---|---|
| | 10 months | 12 months | 14 months | 10 months | 12 months | 14 months |
| Mocotó | 238.62±1.05$^{bC}$ | 320.25±1.02$^{cB}$ | 636.12±1.41$^{cA}$ | 38.54±0.04$^{bC}$ | 49.11±0.10$^{cB}$ | 100.83±0.25$^{cA}$ |
| Ufla | 359.85±0.86$^{aC}$ | 444.65±0.59$^{aB}$ | 658.71±1.84$^{bA}$ | 50.11±0.12$^{aC}$ | 52.07±0.20$^{aB}$ | 101.20±0.03$^{bA}$ |
| Pão da china | 240.25±0.68$^{bC}$ | 338.15±0.67$^{bB}$ | 680.62±0.30$^{aA}$ | 38.79±0.03$^{bC}$ | 50.45±0.11$^{bB}$ | 102.42±0.58$^{aA}$ |
| Ouro do Vale | 155.90±2.39$^{cC}$ | 266.83±0.64$^{dB}$ | 327.27±0.08$^{dA}$ | 24.51±0.09$^{cC}$ | 44.60±0.13$^{dB}$ | 50.85±0.18$^{dA}$ |
| BHT | | 1,418.81±97.98 | | | 259,55±3.27 | |
| Quercetin | | 7,491.59±119.04 | | | 942,96±1.89 | |
| Rutin | | 1,055.50±13.39 | | | 161,29±0.30 | |

Data are the mean of three replicates ± standard deviation. Lowercase letters in columns compare among cultivars and uppercase on the lines compare among ages. Same letters do not differ among themselves by the Scott-Knott test at 5% probability.

**Table 3.** Antioxidant activity of the cassava leaf flour of four cultivars, at three plant ages, in % inhibition by the β-carotene/linoleic acid method.

| Cultivar | % inhibition* | | |
|---|---|---|---|
| | 10 months | 12 months | 14 months |
| Mocotó | 93.04±3.14$^{aA}$ | 94.05±0.20$^{aA}$ | 89.81±0.19$^{aB}$ |
| Ufla | 85.17±0.96$^{bA}$ | 86.47±0.33$^{bA}$ | 84.61±0.66$^{bA}$ |
| Pão da china | 81.47±1.27$^{cB}$ | 85.51±0.63$^{bA}$ | 79.79±1.28$^{cB}$ |
| Ouro do Vale | 82.79±0.35$^{cB}$ | 85.69±0.80$^{bA}$ | 78.87±2.25$^{cC}$ |
| BHT (200 mg $L^{-1}$) | 93.60±0.56 | | |
| Quercetin (200 mg $L^{-1}$) | 76.20±1.15 | | |
| Rutin (200 mg $L^{-1}$) | 19.21±0.88 | | |

Data are the mean of three replicates ± standard deviation. Lowercase letters in columns compare among cultivars and uppercase on the lines compare among ages. Same letters do not differ among themselves by the Scott-Knott test at 5% probability. *The test was applied to the extracts at a concentration of 10,000 mg $L^{-1}$.

evidenced when compared to other studies, in which, independently of the extract, surpassed that AA found by Kuskoski et al. (2005), in frozen fruit pulps, in μmol trolox $L^{-1}$ $g^{-1}$ FM: acerola (66.5), mango (11.8), grape (8.5), guava (7.2) and passion fruit (1.02); that of two grape peels in DM: 'Isabel' (89.22) and 'Niágara' (157.31) according to Soares et al. (2008) and that of wine agroindustrial residues: (98.9), according to Cataneo et al. (2008). It also surpassed that detected by Bouayed et al. (2007) in several parts of medicinal plants, in mg $g^{-1}$ vitamin C: 2.8 (*Alcea kurdica* flowers), 7.36 (*Valerian officinalis* root), 15.4 (*Stachys lavandulifolium* flowers), 19.2 (*Lavandula officinalis* leaves) and 19.3 (*Melissa officinalis* leaves), and the ones detected by Wojdylo et al. (2007) who, in 32 Polish herbs, verified potentials between 0.0045 (*Archangelica officinalis*) and 3.46 (*Syzygium aromaticum*) μmol trolox $g^{-1}$.

The lipidic oxidation inhibition results for CLF from four cultivars, by the β-carotene/linoleic acid method, at TPA after 2 h of reaction, are shown in Table 3. Plant maturity had little influence on the CLF AA, by the β-carotene/linoleic acid method. The greatest inhibition potential of lipid oxidation occurred at 12 months of age for all cultivars, which can be explained by higher levels of β-carotene (nonpolar antioxidant) in CLF at this age, since the β-carotene/linoleic acid method shows a better response to antioxidants with apolar character. Among them, 'Mocotó ' was that which presented the highest oxidation inhibition potential, at the TPA.

The inhibition potential of lipid oxidation presented by CLF was considered high, because, compared to the BHT standard, it was verified that Mocotó cultivar, at TPA, showed AA similar to that standard. However, other cultivars presented lower AA in relation to BHT. In relation to rutin and quercetin standards, the cultivars, at TPA, presented higher AA than those standards. Many authors relate antioxidant potential above 70% as optimum for lipid oxidation inhibition. All cultivars, at TPA, reached AA above 70% and, as such, present high lipid oxidation inhibition potential.

Melo et al. (2006) analyzing lipid oxidation inhibition potential of some plants, found an inhibition potential of 70% for broccoli in relation to BHT, but in other plants the potential was below 60%, emphasizing the high antioxidant

**Table 4.** Average levels of chlorophyll (SPAD units) for cassava leaf flour of four cultivars, at three plant ages.

| Cultivar | Months | | |
|---|---|---|---|
| | 10 | 12 | 14 |
| Mocotó | 565.60±3.05$^{dA}$ | 479.80±9.07$^{dB}$ | 429.60±8.68$^{dC}$ |
| Ufla | 719.80±7.63$^{bA}$ | 631.20±5.72$^{bC}$ | 677.00±11.51$^{bB}$ |
| Pão da China | 589.40±10.43$^{cA}$ | 491.60±1.52$^{cB}$ | 468.40±13.67$^{cC}$ |
| Ouro do Vale | 771.00±5.20$^{aA}$ | 743.20±4.38$^{aC}$ | 758.40±0.63$^{aB}$ |

Data are the mean of three replicates ± standard deviation. Lowercase letters in columns compare among cultivars and uppercase on the lines compare among ages. Same letters do not differ among themselves by the Scott-Knott test at 5% probability.

**Table 5.** Correlation among antioxidant substances with antioxidant activity and chlorophyll in cassava leaf flour at three plant ages.

| Antioxidant substances | Vitamin C | Polyphenols | β-carotene |
|---|---|---|---|
| ABTS (trolox) | 0.78* | 0.75* | 0.28$^{ns}$ |
| ABTS (vit C) | 0.84* | 0.70* | 0.25$^{ns}$ |
| β-carotene/linoleic acid | -0.20$^{ns}$ | 0.18$^{ns}$ | 0.20$^{ns}$ |
| Chlorophyll | -0.40* | -0.71* | -0.45* |

*Significant at 5% probability; ns = not significant.

potential of CLF in relation to other plants.

## Chlorophyll

The SPAD-502 equipment has been investigated as an instrument for fast diagnosis of the nutritional state of various cultures, adding advantages such as simplicity of use, besides enabling a non-destructive evaluation of the foliar tissue substituting, with good precision, the traditional chlorophyll level determination in plants (Argenta et al., 2001).

Chlorophyll reading results for CLF at TPA, using the SPAD-502 portable meter, are in Table 4. It was verified that the highest chlorophyll levels were found at 10 months of age and that there was a tendency of reduction of those levels with plant maturity, Ouro do Vale, at TPA, was the one which presented the highest chlorophyll levels. Mocotó cultivar, at TPA, was that which showed significantly lower chlorophyll levels.

## Correlation among antioxidant substances with AA and chlorophyll

Results of the correlations among antioxidant substances (Table 1) with AA (Tables 2 and 3) and with chlorophyll (Table 4) in CLF, at TPA, are shown in Table 5. The ABTS antioxidant tests (equivalent to trolox and vitamin C) presented a high positive correlation with vitamin C and with polyphenols, but there was no correlation with β-carotene. For the β-carotene/linoleic acid test, there was no correlation with antioxidant substances. The antioxidant substances presented a negative correlation with chlorophyll, that is, when antioxidant levels increased, chlorophyll contents decreased in the plant. This is easily explained for β-carotene, because, with the increase in maturity, chlorophyll begins to disappear and carotenoids protrude.

Results of the correlations indicate that vitamin C and polyphenols are the antioxidant substances that most contributed to the increase in AA for CLF; vitamin C stood out, therefore it presented the highest correlation coefficients, 0.78 and 0.84, for equivalent of trolox and vitamin C, respectively. β-carotene levels did not contribute to the CLF AA increase. The negative correlation among the antioxidant substances levels and chlorophyll indicate that antioxidant substances are found in a higher amount when cassava leaves present low chlorophyll levels.

Cataneo et al. (2008) and Soares et al. (2008) also observed a positive correlation between total phenols and AA by the ABTS test, equivalent to trolox, in studies performed with wine agroindustrial residues and grape skins of two varieties, respectively, indicating that polyphenols are substances with a high antioxidant potential and are one of the main antioxidants present in medicinal plants and food.

Kubola and Siriamornpun (2008) and Melo et al. (2008) found no correlation between the β-carotene/linoleic acid test and polyphenol contents in fruits. Duarte-Almeida et al. (2006) found that acerola, rich in vitamin C, showed a low AA by the the β-carotene/linoleic acid test, while the DPPH test (method of scavenging free radicals, which has the same principle of the ABTS test), showed a high AA, which demonstrates that vitamin C little contributes to AA in the β-carotene/linoleic acid test, confirming the results of the present study, in which a negative correlation between antioxidants (polyphenols and vitamin C) was also observed, as the β-carotene/linoleic acid test, which can be explained by the fact that the β-carotene/linoleic acid test shows a better response to nonpolar antioxidants, thus polar antioxidants (vitamin C and polyphenols), and has little contribution in the inhibition of lipid oxidation.

## Conclusions

CLFs are rich in antioxidant substances and show high antioxidant capacity in the protection of lipid peroxidation; however, they present moderate capacity in the capture of free radicals. Cassava leaves at 14 months of age present the highest antioxidant levels, and Mocotó and Pão da China cultivars stood out. The highest antioxidant levels (Vitamin C, polyphenols and β-carotene) are found when the cassava leaf presents low chlorophyll levels.

## ACKNOWLEDGMENTS

The expresses their thanks to CAPES, for the doctoral scholarship, and FAPEMIG, for financial support and scientific initiation scholarship.

### REFERENCES

Almazan AM, Begum F (1996). Nutrients and antinutrient in peanut greens. J. Food Compos. Anal. 9(43):375-383.

Argenta G, Silva PRF, Forsthofer EL, Strieder ML (2001). Relationship of reading of portable chlorophyll meter with contents of extractable chlorophyll and leaf nitrogen in maize. Rev. Bras. Fisiol. Veg. 13(2):158-167.

Asolini FC, Tedesco AM, Carpes ST (2006). Antioxidant and Antibacterial Activities of Phenolic Compounds from Extracts of Plants Used as Tea. Bra. J. Food Technol. 9(3):209-215.

Association of Oficcial Anlytical Chemists (AOAC) (2011). Official methods of analysis of the association of the analytical chemists (18 ed). Washington, DC. USA.

Balasundram N, Sundram K, Sammar S (2006). Phenolic compounds in plants and agri-industrial by-products: Antioxidant activity, occurrence, and potential uses. Food Chem. 99(1):191-203.

Bouayed J, Piri K, Rammal H, Dicko A, Desor F, Younos C, Soulimani S (2007). Comparative evaluation of the antioxidant potential of some Iranian medicinal plants. Food Chem. 104(1):364-368.

Brazilian Company of Agricultural Research (EMBRAPA). Disponível em: <http://sistemasdeprodução.cnptia.embrapa.br/FONTESHTML/Mandioca/Mandioca centrosul/importância.htm>. Acesso em: 15 mai, 2010.

Campos FM, Santana HM, Stringheta PC, Chaves JBP (2003). Levels of Beta Carotene in Leafy Vegetables Prepared in Commercial Restaurants in Viçosa, Brazil. Bra. J. Food Technol. 6(2):163-169.

Carvalho VD, Chagas SJR, Morais AR, Paula MB (1989). Efeito da época de colheita na produtividade e teores de vitamina C e β-caroteno da parte aérea de cultivares de mandioca (*Manihot esculenta* Crantz). Revista Brasileira de Mandioca 8(1):25-35.

Cataneo CB, Caliari V, Gonzaga L, Kuskoski LM, Fett R (2008). Antioxidant activity and phenolic content of agricultural by-products from wine production. Semina Ciênc. Agrár. 29(1):93-102.

Core Team R (2012). R: A language and environment for statistical computing. Viena: R Foundation for Statistical Computing; 2012. ISBN 3-900051-07-0. Available: http://www.R-project.org/.

Corrêa AD, Santos SR, Abreu CMP, Jokl L, Santos CD (2004). Removal of polyphenols of the flour cassava leaves. Ciênc. Tecnol. Aliment 24(2):159-164.

Duarte-almeida JM, Santos RJ, Genovese MI, Lajolo FM (2006). Evaluation of antioxidant activity using system b-carotene/linoleic acid and sequestration method DPPH radical. Ciênc. Tecnol. Aliment 26(2):446-452.

Faller ALK, Fialho E (2009). Polyphenol availability in fruits and vegetables consumed in Brazil. Rev. Saúde Públ. 43(2):211-218.

Gan R, Xu XR, Song FL, Kuang L, Li HB (2010). Antioxidant activity and total phenolic content of medicinal plants associated with prevention and treatment of cardiovascular and cerebrovascular diseases. J. Med. Plant. Res. 4(22):2438-2444.

Hernández Y, Lobo MG, González M (2006). Determination of vitamin C in tropical fruits: a comparative evaluation of methods. Food Chem. 96(4):654-664.

Kubola J, Siriamornpum S (2008). Phenolic contents and antioxidant activities of bitter gourd (*Momordica Charantia* L.) leafs stem and fruit fraction extracts in vitro. Food Chem. 1110(4):881-890.

Kuskoski EM, Asuero AG, Troncoso NA, Filho JM, Fett R (2005). Aplicación de diversos métodos químicos para determinar a actividad antioxidant en pulpa de frutos. Ciênc. Tecnol. Aliment 25(4):726-732

Melo EA, Maciel MIS, Lima VLAG, Araujo CR (2008). Total phenolic content and antioxidant capacity of frozen fruit pulps. Alimentos e Nutrição 19(1):67-72.

Melo EA, Maciel MIS, Lima VLAG, leal FLL, Caetano ACS, Nascimento RJ (2006). Antioxidant capacity of vegetables commonly consumed. Ciênc. Technol. Aliment 26(3):639-644.

Nagata M, Yamashita I (1992). Simple method for simultaneous determination of chlorophyll and carotenoids in tomatoes fruit. J. Japan. Soc. Food Sci. Technol. 39(10):925-928.

Pereira GIS, Pereira RGFA, Barcelos MFP, Morais AR (2003). Carrot leaf chemical evaluation aiming its use in human feeding. Ciênc. Agrotec. 27(4):852-857.

Rufino MSM, Alves RS, Brito ES, Filho JM, Moreira AVB (2006). Determination of total antioxidant activity in fruit by the method β-caroteno/ácido linoleic. Fortaleza: Embrapa Agroindústria tropical.

Rufino MSM, Alves RS, Brito ES, Morais SM (2007). Determination of total antioxidant activity in fruits by capturing the free radical ABTS $^{+}$. Fortaleza: EMBRAPA Agroindústria tropical.

Soares M, Welter L, Kuskoski EM, Gonzaga L, Fett R (2008). Phenolic compounds and antioxidant activity in skin of Niagara and Isabel grapes. Rev. Bras. Frutic. 30(1):59-64.

Statistical Analysis System (SAS) (1999). Users quide: statistcs : versão 6. 4. ed. Cary. 2:1686.

Strohecker R, Henning HM (1967). Analisis de vitaminas: metodos comprobados. Madrid: Paz Montalvo. P. 428.

Surveswaran S, Cai Y, Corke H, Sun M (2007). Systematic evaluation of natural phenolic antioxidants from 133 Indian medicinal plants. Food Chem. 102(3):938-953.

Wobeto C, Corrêa AD, Abreu CMP, Santos CD, Abreu JR (2006). Nutrients in the cassava (*Manihot esculenta* Crantz) leaf powder at three ages of the plant. Ciência e Tecnologia de Alimentos. 26(4):865-869.

Wobeto C, Corrêa AD, Abreu CMP, Santos CD, Abreu JR (2007). Antinutrients in the cassava (*Manihot esculenta* Crantz) leaf powder at three ages of the plant. Ciênc. Tecnol. Aliment 27(1):108-112.

Wojdylo A, Osmianski J, Czermerys R (2007). Antioxidant activity and phenolic compounds in 32 selected herbs. Food Chem. 105(3):940-949.

# 28

# Water relations, gas exchange characteristics and water use efficiency in maize and sorghum after exposure to and recovery from pre and post-flowering dehydration

**Abuhay Takele[1] and Jill Farrant[2]**

[1]Melkassa Agricultural Rresearch Center, P. O. Box 436, Nazret, Ethiopia.
[2]University of Cape Town, Department of Molecular and Cell Biology, Private bag 7701, Rondebosch, Cape Town, South Africa.

**The effect of dehydration and rehydration at pre and post-flowering stages on the water relations, gas exchange characteristics (stomata conductance, photosynthesis and Fv/Fm (quantum efficiency) ratio), pigment compositions (chlorophyll and carotenoid contents) and water use efficiency on maize (cv Melkassa-2) and sorghum (cv Macia) were investigated with the objectives of understanding the physiological basis of drought resistance mechanisms and investigating whether there were differential responses in some of the physiological traits of drought resistance and recovery upon rehydration of maize and sorghum. The study was carried out in a controlled environment growth chamber under constant environmental conditions (12/12 h day/night, 28-32/17°C day/night temperature, 60-80% RH and PPFD of 1200-1400 μ mol $m^{-2}s^{-1}$). Both species showed reduced $g_s$ (stomatal conductance) in response to dehydration to reduce water loss over a range of relative water contents during both developmental stages. In maize, stomata appeared to be closed earlier and completely, while partial stomatal closure at relatively higher relative water contents appeared to have occurred in sorghum. $g_s$ recovery occurred following pre-flowering rehydration to the control level in both species only. The response of all other gas exchange characteristics ($P_n$ and Fv/Fm) and water use efficiency followed similar trends to that of $g_s$ both in maize and sorghum at pre and post-flowering dehydration and rehydration. Dehydration also led to a decrease in Fv/Fm ratios as compared to the control plants in both species. Both species, however, exhibited similar rates of Fv/Fm ratios during pre and post-flowering dehydration. Fv/Fm ratios appeared to be affected more during post than pre-flowering dehydration in both species. Fv/Fm ratios of both species were recovered following pre-flowering rehydration but only maize recovered from post-flowering rehydration.**

**Key words:** Chlorophyll, carotenoid, dehydration, electrolyte leakage, rehydration, stomatal conductance.

## INTRODUCTION

Water deficit is the most common adverse environmental factor limiting crop production in the dry land areas of Africa. In these areas, maize and sorghum are very important staple crops. They are grown across a range of agro-ecological zones where shortages of water resulting from low and erratic rainfall is a major constraint for crop production (Rosenow et al., 1997).

Crop plants when exposed to water deficits undergo physiological, morphological and biochemical changes in order to survive. The changes that occur at various levels of plant organization (cellular, molecular, etc.) in response to drought stress are considered to be adaptation mechanisms (Turner, 1997). Several workers have examined the response of different crops to water deficits and have identified various traits that confer drought resistance in cereals (Blum, 1989; Ludlow and Muchow, 1990; Turner, 1997). These include maintenance of high water potential, control of stomatal behaviour and osmotic adjustment under drought conditions (Blum, 1988; Ludlow and Muchow, 1990). Genotypes differ in their ability to recover upon rehydration and the ability of a genotype to recover from stress is closely related to its hydration status prior to recovery (Malabuyot et al., 1985).

Genetic variation in leaf water potential, stomatal conductance and photosynthetic rate have been reported in several crop species (Peng and Kreig, 1992). These traits might be used to select superior cultivars or crop species with the ability to maintain high plant water status, high stomatal conductance and maintenance of photosynthetic rate under water deficit conditions. The selection of such physiological traits in the improvement programs of crop plants, however, requires the establishment of significant association between various traits and drought resistance. More detailed understanding of the physiological adaptations enabling superior performance of crop species and/or genotypes under drought stress and/or required for the maintenance of physiological activities for growth and productivity during periods of recovery from stress upon rehydration will ultimately help in the selection and promotion of drought tolerant crop species. This is particularly relevant for crops such as sorghum and maize, which are predominantly grown in marginal rainfall regions of the world.

This study was, therefore, carried out with the objectives of understanding the physiological basis of drought resistance mechanisms in maize and sorghum and investigating whether there were differential responses in some of the physiological traits of drought resistance and recovery upon rehydration of maize and sorghum. Measurements were taken on plant water relations, gas exchange characteristics, Fv/Fm and water use efficiency of maize and sorghum after exposure to and recovery from pre and post-flowering dehydration.

## MATERIALS AND METHODS

### Growth conditions and treatment

Maize seeds (*Zea mays* L.) (cv *Melkassa*-2) and sorghum (*Sorghum bicolour* Moench L.) (*cv* Macia) were obtained from the Maize Improvement Program for Moisture Stress Areas, Melkassa Agricultural Research Centre, Ethiopia and ICRISAT Centre, Bulawayo, Zimbabwe, respectively. The experiment was conducted in a controlled chamber under constant environmental conditions (12/12 h day/night, 28-32/17°C day/night temperature, 60-80% RH and PPFD of 1200-1400 µ mol $m^{-2}s^{-1}$) at the Department of Botany, University of Cape Town. The following was done for both species. To ensure emergence, 5 seeds were sown in plastic pots, each was 31 cm deep with an internal diameter of 18 cm. About 10 kg of sandy loam soil was used for each pot. Emergence occurred 5-7 days after planting (DAP). 20 Days after emergence, the pots were thinned to 2 seedlings of uniform size per pot. Plants were watered frequently to avoid the development of any moisture deficit. At 60 (pre-flowering) and 90 (post-flowering, grain filling stage) days after emergence, 2 watering treatments were applied: either maintained fully hydrated (control) or dehydrated treatments. Control plants were regularly watered to field capacity (F.C) to avoid any development of water stress and the dehydration was induced by withholding water for 20 days at each growth stages. At the end of each dehydration treatment, plants were rehydrated by soil watering (as for the control plants) for another 20 days and their recovery was studied. 3 pots of each species in each treatment represented 3 replications. 5 different samples were taken during the dehydration period at each different growth stage and during recovery, respectively. Each pot was given P and N at the rate of 0.80 g/pot (150 kg/ha) and 1.1 g/pot (200 kg/ha), respectively. Single super phosphate and lime ammonium nitrate were used as source of P and N, respectively.

At regular interval during the entire cycle (pre and post-flowering dehydration), water relations, gas exchange characteristics and water use efficiency were measured. The same measurements were performed on control plants which remained hydrated throughout.

### Plant water relations

Water status of both species was determined by measuring the relative water content. Relative water content was calculated using the method of Henson et al. (1981) as:

RWC (%) = [(FW-DW) / (TW-DW)] x 100

Where, FW represent fresh weight, DW is the dry weight and TW the turgid weight. Turgid weight was determined after floating leaf segments in distilled water in sealed vials for 24 h at room temperature, and oven dried at 70°C for 48 h.

### Gas exchange parameters

At each growth stages during dehydration and rehydration, the gas exchange physiology [stomatal conductance ($g_s$), photosynthesis ($P_n$) and transpiration (E)] of fully expanded intact leaves of upper canopy were recorded using a portable infrared gas analysis (IRGA) system (LCA-3, the Analytical Development Corporation, Hoddeston, England). Photosynthetic water use efficiency (WUE) was estimated as the ratio of photosynthesis rate to transpiration rate.

Chlorophyll fluorescence was measured using the leaves for the measurement of gas exchange, by a modulated portable fluorometer (os-500:opti-sciences, USA) and by calculating the quantum efficiency of leaves at various stages of dehydration and recovery during rehydration. The initial $F_0$, and maximum fluorescence, $F_M$, using a saturating light intensity of approximately 4 µ mol photons $m^{-2}s^{-1}$ and duration of 1 s, was measured. $F_v$ was

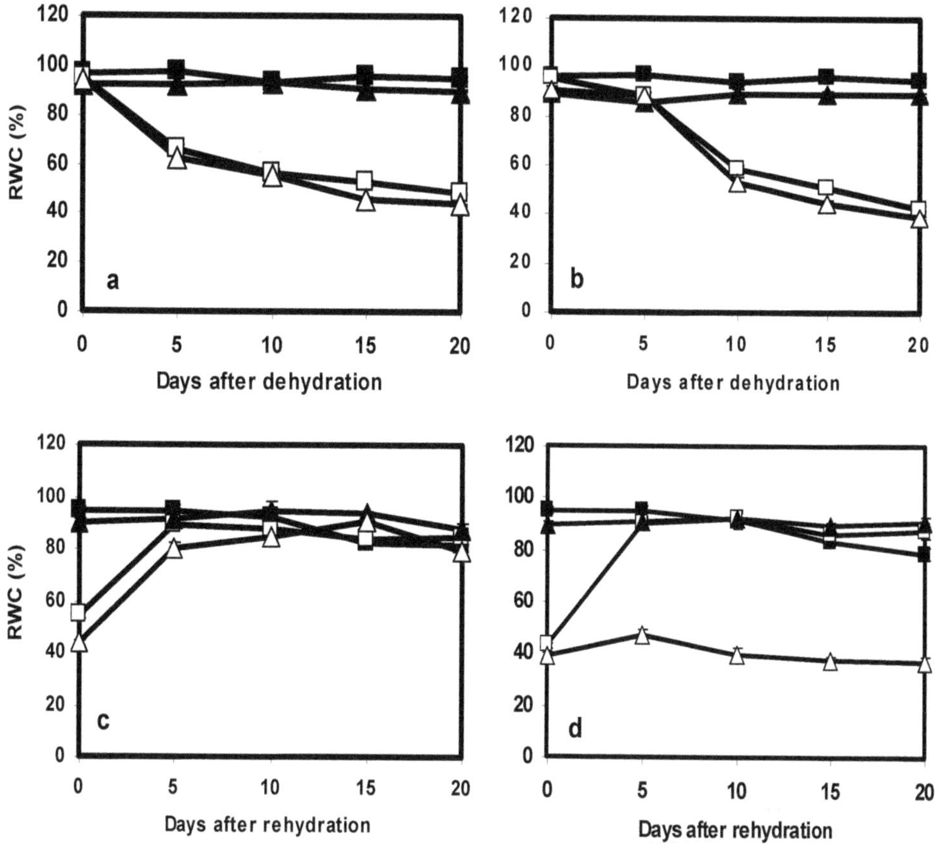


**Figure 1.** Changes in mean relative water content (%) of maize (a, c) and sorghum (b, d) during pre (■, □) and post-flowering (▲, Δ) dehydration (a, b) and rehydration (c, d), respectively. ■ and ▲ represent control and □ and Δ represent dehydration/rehydration treatments. Vertical bars denote standard errors of means (n=3).

obtained by subtracting $F_0$ from $F_M$ and $F_v/F_M$ was calculated.

### Statistical analysis

Statistical analyses were carried out using STATISTICA for windows version 6.0, Statsoft, Inc, USA. The results presented were the mean of 3 replicates. In all figures, the scores of the mean were calculated and significances between treatments as well as between the 2 species were tested by factorial analysis of variance and Duncan's multiple range tests at the 5% level of significance. Standard errors were represented as vertical bars.

## RESULTS

### Leaf relative water content (RWC %)

Dehydration caused a significant ($P<0.05$) decrease in the relative water content of both species during both pre and post-flowering stages (Figure 1a and b). The difference between the 2 species was significant for the first 5 days after withholding water. Relative water content of dehydrated maize plants decreased rapidly for the first 10 days and then slowly but steadily. In contrast to maize, there was almost no change in the RWC of sorghum plants for the first 5 days under the prevailing dehydration conditions but thereafter, there was a sharp decrease to the level equivalent to the values of maize plants. Mean relative water content during dehydration was reduced from the initial full turgor value of 94 and 93% to 46 and 41% in maize and sorghum at the end of dehydration, respectively.

Visual observations indicated that in maize, leaf rolling was displayed within 5 days (corresponding to relative water content of 82%) of pre-flowering dehydration and then as the dehydration process progressed, leaf rolling was more tightened. In contrast, visible leaf turgidity was maintained in sorghum for at least 12 days after initiation of dehydration and then leaf rolling was initiated at approximately relative water content of 76%. During post-flowering dehydration, leaf rolling as a dehydration avoidance mechanism was not displayed in either species.

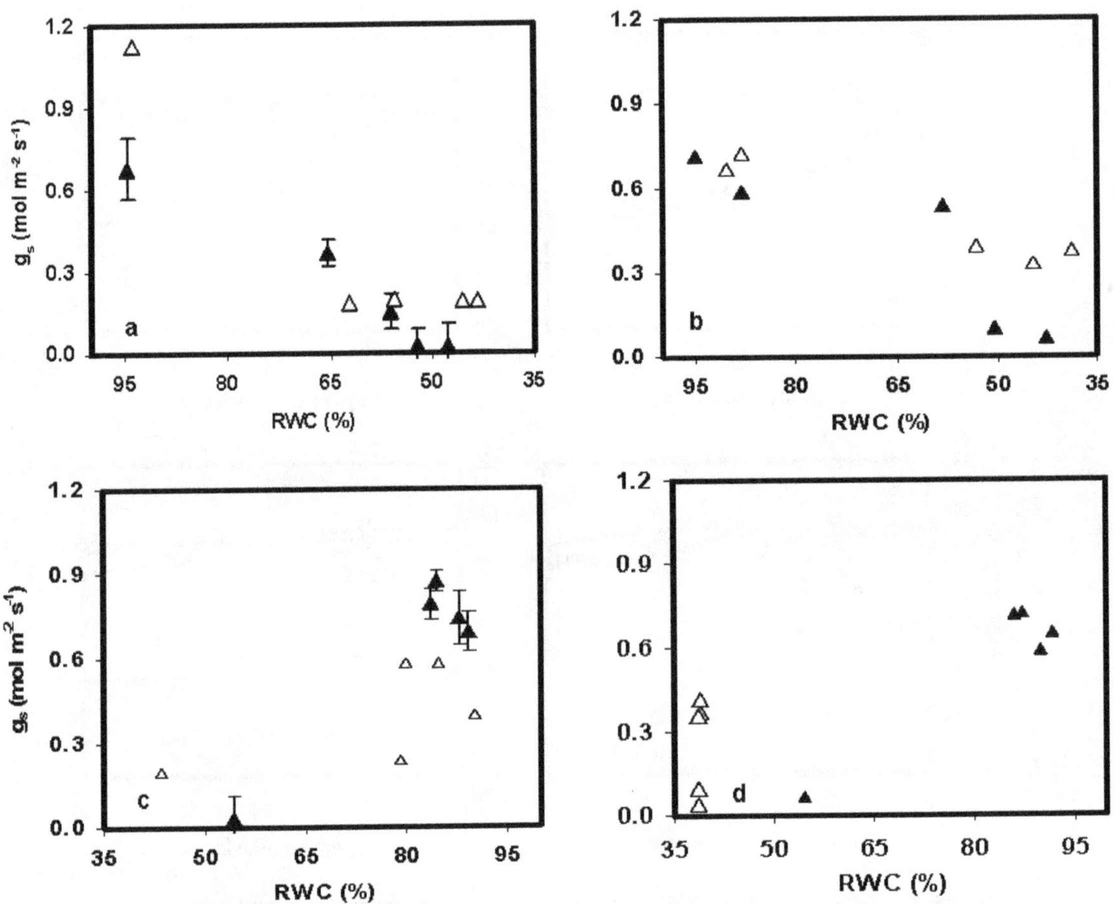


**Figure 2.** Changes in mean stomatal conductance (mol $m^{-2}$ $s^{-1}$) of maize (a and c) and sorghum (b and d) during pre (▲) and post-flowering (Δ) dehydration (a, b) and rehydration (c, d) as related to RWCs, respectively. Vertical bars denote standard errors of means (n=3).

Relative water content of both crops returned to control values within 5 days of rehydration during pre-flowering rehydration, but only maize rose to full recovery during post-flowering rehydration (Figure 1c and d).

## Gas exchange characteristics

### Stomatal conductance

Dehydration during both pre and post-flowering stages significantly decreased stomatal conductance of both species (Figures 2a and b). Differences in stomatal conductance of both species were observed between pre and post-flowering dehydrated plants. In maize, differences occurred at the beginning of the measurement period approximately at RWC of 95% and between RWC of 53 and 48% when conductance of post-flowering dehydrated plants was significantly ($p<0.05$) higher than pre-flowering dehydrated plants. Differences in stomatal conductance of sorghum between pre and post-flowering dehydrated plants became evident towards the end of the treatment period between RWC of 58 and 39% at which point the post-flowering dehydrated plants showed significantly ($p<0.05$) higher stomatal conductance values than pre-flowering dehydrated plants (Figure 2b).

Corresponding to the decrease in RWC, the patterns of changes of stomatal conductance differed between species during pre and post-flowering dehydration. Stomatal conductance of maize that underwent post-flowering dehydration, showed a dramatic decrease between RWC of 95 and 64% within 5 days of withholding water after which it remained without any significant change between RWC of 64% and 48% at the final stage of dehydration. In contrast to maize, after an initial decrease, approximately between RWC of 95 and 88%, stomatal conductance of sorghum plants during pre-flowering dehydration showed no remarkable change between RWC of 88 and 58% and then conductance

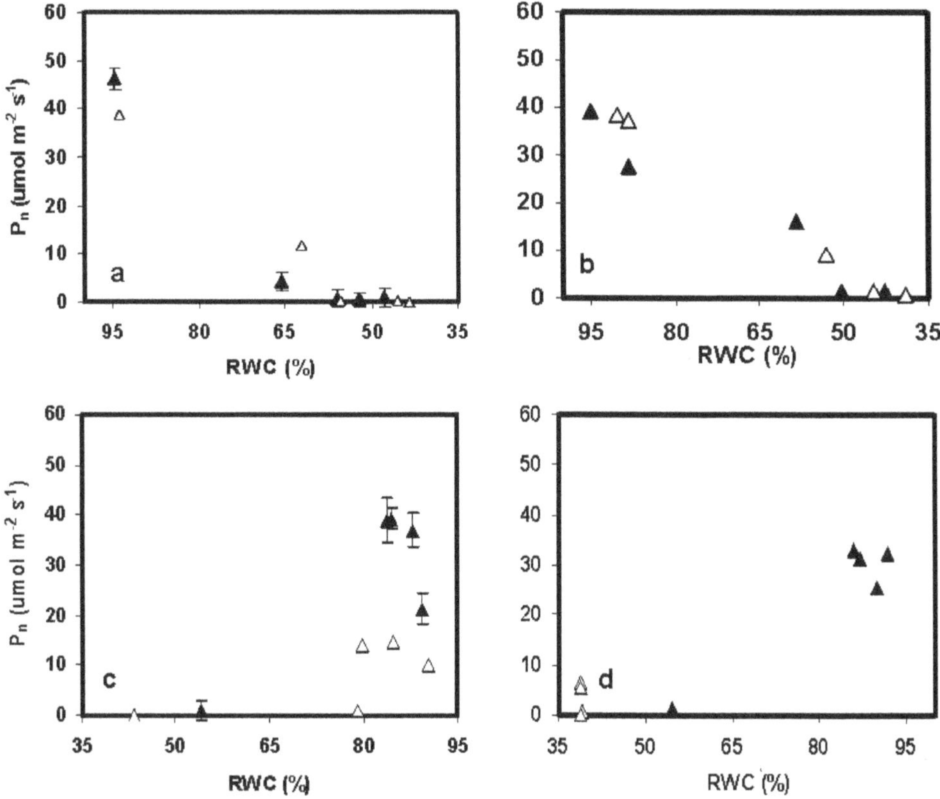


**Figure 3.** Changes in mean photosynthesis rate (μmol $m^{-2}$ $s^{-1}$) of maize (a, c) and sorghum (b, d) during pre (▲) and post-flowering (Δ) dehydration (a, b) and rehydration (c, d) as related to RWCs, respectively. Vertical bars denote standard errors of means (n=3).

decreased most markedly until the end of the experiment. Between RWC of approximately 88 and 53%, the decrease in stomatal conductance of post-flowering dehydrated sorghum was rather gradual and thereafter conductance showed little change until RWC reached 39% at the final phase of dehydration.

After 5 days of rehydration, stomatal conductance of maize fully recovered at about 88% RWC following pre-flowering rehydration (Figure 2c). Recovery in the stomatal conductance of pre-flowering rehydrated sorghum was rather slow and full recovery occurred at approximately 90% of RWC (%) at the final phase (20 days after rehydration began) of rehydration (Figure 2d). Approximately 55% of stomatal conductance in maize that underwent post-flowering rehydration resumed at 80% of RWC 5 days after rehydration began but stomatal conductance showed a steady decrease 10 days after post-flowering rehydration. There was no change in stomatal conductance of sorghum from that measured at the end of post-flowering dehydration for the first 10 days after rehydration commenced but thereafter, conductance decreased to almost zero at the end of the rehydration period.

### Photosynthesis rate ($P_n$)

Dehydration during pre and post-flowering stages caused a considerable decrease in photosynthesis rate of both species (Figures 3a and b). Differences in the patterns of changes in photosynthesis rate were noted between pre and post-flowering dehydrated plants with changes in RWC in both species. In maize, differences between pre and post-flowering dehydrated plants were evident approximately between RWC of 95 and 65% when photosynthesis rate of pre-flowering dehydrated plants decreased more markedly than post-flowering dehydrated plants (Figure 3a). In sorghum, photosynthesis rate of pre-flowering dehydrated plants decreased continually until the end of the dehydration period whereas photosynthesis rate of plants that underwent post-flowering dehydration was maintained without changes for the first 5 days of dehydration approximately between RWC of 93 and 88%. With further decrease in RWC, photosynthesis rate was negligible after 10 days of dehydration (Figure 3b). Significant ($p<0.05$) difference between species in the sequence of changes in photosynthesis rate was also observed in

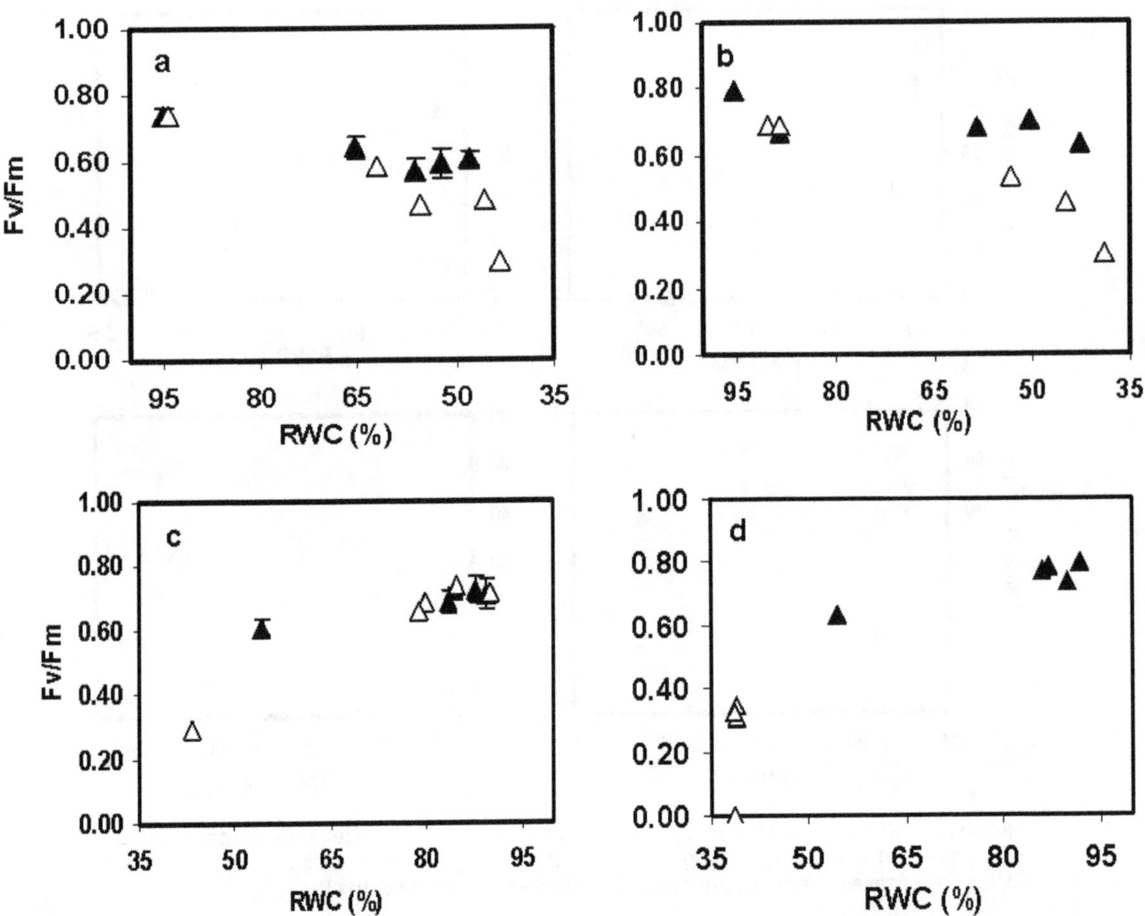


**Figure 4.** Changes in mean quantum efficiency of maize (a, c) and sorghum (b, d) during pre (▲) and post-flowering (Δ) dehydration (a, b) and rehydration (c, d) as related to RWCs, respectively. Vertical bars denote standard errors of means (n=3).

response to pre and post-flowering dehydration.

In maize, as RWC dropped from 95 to 55% within 5 days after the initiation of dehydration, the decrease in photosynthesis rate ranged from 46.2 μmol $m^{-2}$ $s^{-1}$ to 0.7 μmol $m^{-2}$ $s^{-1}$ during pre-flowering dehydration and from 39.0 μmol $m^{-2}$ $s^{-1}$ to 0.4 μmol $m^{-2}$ $s^{-1}$ during post-flowering dehydration. In contrast, at similar RWC, the decrease in the photosynthesis rate of sorghum was from 39.2 μmol $m^{-2}$ $s^{-1}$ to 16.3 μmol $m^{-2}$ $s^{-1}$ and from 38.3 μmol $m^{-2}$ $s^{-1}$ to 9.1μmol $m^{-2}$ $s^{-1}$ during pre and post-flowering dehydration, respectively.

There was a noticeable difference in the patterns of response in photosynthesis rate between pre and post-flowering rehydrated plant in both species (Figures 3c and d). Photosynthesis rate of pre-flowering dehydrated maize fully resumed 10 days after rehydration began and that of post-flowering dehydrated plants showed an initial increasing trend following rehydration, but with further increase in RWC (%), photosynthesis rate rather decreased until the end of the rehydration period. Sorghum on the other hand exhibited only 75% recovery until the end of rehydration period but post-flowering rehydrated plants did not show any noticeable change until the end of the rehydration period (Figures 3c and d).

## Quantum efficiency of photosystem II (Fv/Fm)

Dehydration during pre and post-flowering stages caused a significant ($p<0.05$) decrease in Fv/Fm ratio of both species as compared to the well watered plants, indicating that dehydration caused a direct effect on the PSII photochemistry (Figures 4a and b). There were slight differences in the patterns of the changes in Fv/Fm ratio between pre and post-flowering dehydration in both maize and sorghum.

During pre-flowering dehydration, there was a gradual but consistent decrease in the Fv/Fm ratio of both species. In contrast, during post-flowering dehydration, when RWC decreased below 65%, there was a large and

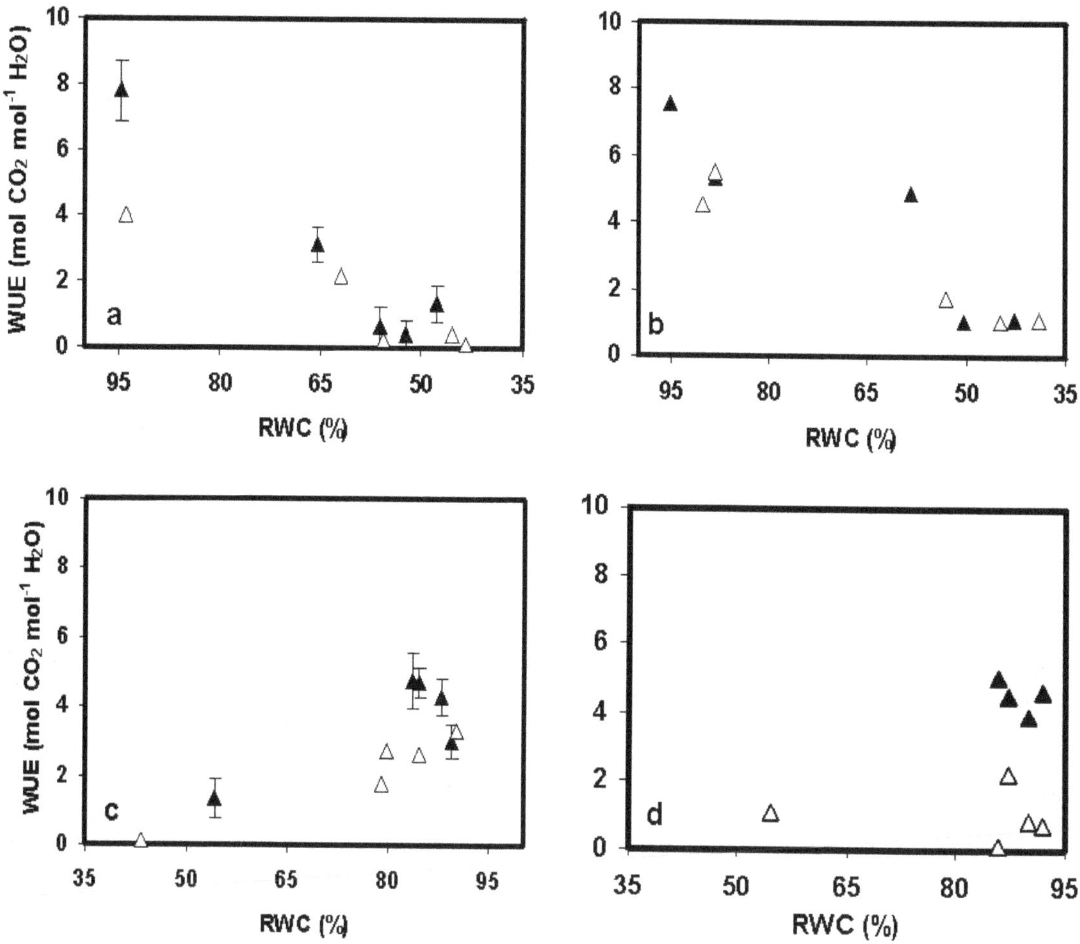


**Figure 5.** Changes in mean water use efficiency ($CO_2$ $mol^{-1}$ $H_2O$) of maize (a, c) and sorghum (b, d) during pre (▲) and post-flowering (Δ) dehydration (a, b) and rehydration (c, d) as related to RWCs, respectively. Vertical bars denote standard errors of means (n=3).

much faster decrease. The magnitude of the effect of pre and post-flowering dehydration on Fv/Fm ratio was similar for the 2 species. When Fv/Fm ratio was expressed in relation to the initial value, there were 18 and 20% reduction during pre-flowering dehydrated maize and sorghum at the end of the dehydration cycle, respectively. In contrast, during post-flowering dehydration, the reduction in Fv/Fm ratios was 60 and 56% in maize and sorghum at the end of the dehydration period, respectively.

Maize recovered to the control level as soon as rehydration began during both pre and post-flowering rehydration (Figure 4c). Similar to maize, Fv/Fm of sorghum was restored to the pre water stress level within 5 days of pre-flowering rehydration, but there was no change in the Fv/Fm ratio from that measured during post-flowering dehydration until day 15 and thereafter, Fv/Fm rather decreased with further rehydration (Figure 4d).

### Water use efficiency

Dehydration significantly (P=0.05) decreased photosynthetic water use efficiency (the ratio of $P_n$ to transpiration rate) in both species (Figure 5a and b). The decrease in water use efficiency was mainly due to greater dehydration induced reduction in $P_n$ than transpiration. There was a significant difference between the tested species for water use efficiency in response to dehydration. Dehydration induced decrease in water use efficiency in maize by 63% and that of sorghum by only 37%. Time course of the changes over the period of dehydration in both species followed similar patterns to those of $P_n$ and transpiration rate. When a pattern of response over the duration of dehydration was considered, there was a dramatic decrease in water use efficiency in maize until day 15 of the onset of dehydration. In contrast, water use efficiency in dehydrated sorghum plants was up to the level of the control plants in the first phase but there was a rapid decrease 5 days after imposition of dehydration. Mean of water use efficiency was not significantly different between pre and post-flowering dehydration. The magnitude of the changes in water use efficiency over the duration of dehydration period was similar for both pre and post-flowering stages.

Recovery of water use efficiency upon rehydration was slow for both species and in maize it attained fully hydrated level after 15 days of rehydration after which there was a decrease while in sorghum, approximately 50% of water use efficiency was recovered after 20 days of rehydration (Figures 5c and d).

## DISCUSSION

### Differences in physiological responses to dehydration between maize and sorghum

The 2 crops differed in the rate of change in RWC immediately after withholding water (Figure 1). As evapotranspiration from the limited volume of soils in the pots of maize became high, there was a sharp decrease in the soil water contents and the leaves experienced a shortage of water supply. RWC decreased quickly both during pre and post-flowering dehydration (Figure 1). The fast rate of development of plant water deficits in maize than sorghum induced early leaf rolling (data not shown) and stomatal closure (Figure 2) in the former than the latter.

Variation in stomatal conductance is a more sensitive indicator for selecting desirable cultivars and crop species. This study indicated that both maize and sorghum showed reduced stomatal conductance to reduce water loss continuously over a range of RWC and duration of pre and post-flowering dehydration. However, there was a difference in the extent of response of stomatal conductance between maize and sorghum with changes in RWC. During both pre and post-flowering dehydration, drought resistance in maize was achieved through complete stomatal regulation, while partial stomatal closure at relatively higher values of RWC appeared to have occurred in sorghum (Figures 2a, b and inset). Dehydration-induced stomatal regulation in maize was also earlier reported (Premachandra et al., 1992; Stikic and Davies, 2000). However, despite complete stomatal regulation and early onset of leaf rolling, maize exhibited fast rate of decrease in RWC. Under dehydration conditions, stomatal closure could be triggered by both changes in chemical signalling and/or hydraulic status of the plant (Tardieu and Davies, 1992). Chemical messages, mainly ABA originating from roots and transferred to the leaf through the transpiration stream may cause stomatal closure regardless of the $\psi_L$ (leaf water potential) of the plant (Davies and Zhang, 1991).

Decrease in stomatal conductance could improve the stability of yield, because it reduces water loss and lowers the probability of exhausting soil water before maturity. Alternatively, since stomata controls gas exchange, dehydration avoidance achieved through stomatal closure in plants reduces productivity (Ludlow and Muchow, 1990). The parallel decrease in gas exchange characteristics (photosynthesis rate and Fv/Fm ratios) in both species at approximately between RWC of 93 and 65% during both pre and post-flowering dehydration strongly support stomatal closure as the major factor in reducing photosynthesis (Chaves, 1991; Cornic, 1994). The greater decrease in photosynthetic rate (Figures 3a and b) of maize was probably attributed to the much faster decrease in RWC which induced early leaf rolling and effective stomatal closure. Sorghum on the other hand was able to use water more efficiently by maintaining relatively higher RWC, delayed leaf rolling for extended period and stomata remained partially open altogether allowing a relatively higher photosynthesis rate. Stomatal closure is considered to be responsible for the decrease in photosynthetic rate in several crop species exposed to moderate water deficits (Cornic, 1994) and there was no indication of damage to chloroplast reactions (Sharkey and Seemann, 1989).

Under more severe dehydration however, reduced photosynthesis rate is generally considered to be due to non-stomatal factors (Lawlor, 1995; Lawlor and Cornic, 2002). The decrease in photosynthesis rate of both maize and sorghum to almost zero below RWC of approximately 65% during the late phase of dehydration period, and subsequent lack of recovery upon rehydration may suggest that factors other than stomatal limitation might have been involved. Several studies suggest that the ratio of Fv/Fm gives a direct estimate of the yield of PSII photochemistry (Kicheva et al., 1994; Liang et al., 1997). A sustained decrease in Fv/Fm is believed to indicate the occurrence of photo inhibitory damage, in response to many environmental stresses including water deficit stress (Maxwell and Johnson, 2000). In this study, the observed decrease in the ratio of Fv/Fm in pre and post-flowering dehydrated maize and sorghum (Figures 4a, b and inset) supported the idea that dehydration during both developmental stages in maize and sorghum had a direct effect on the PSII photochemistry. The results of this study are in agreement with the findings of Massacci et al. (1996) who reported that the inhibition of photosynthesis rate by dehydration was due to non-stomatal factor in field grown sweet sorghum.

Stomatal regulation controls the exchange of water and carbon between the leaf and the atmosphere and thus affects water use efficiency (Blum, 1988). In this study, the faster rate of decrease observed in the water use efficiency (Figure 5a) of maize may have been attributed to the early and complete stomatal closure which in turn resulted to a greater decrease in $P_n$ than transpiration.

Sorghum by retaining green leaf area for an extended period during pre and post-flowering dehydration was better able to maintain relatively higher water use efficiency than maize. In this case, higher water use efficiency was due to increased photosynthetic rate during both pre and post-flowering dehydration (Figure 3). Therefore in the present study water use efficiency, which indicates the tissue water relation of a species, suggests differences in adaptation strategies to dehydration between the 2 species.

**Differences between pre and post-flowering growth stages in response to dehydration in maize and sorghum**

The expression of drought tolerance in crop plants is dependent on the stage of development at which the stress occurs (Blum, 1988; Tuinstra et al., 1997). For example, in sorghum developmentally specific patterns of drought tolerance have been identified and symptoms of susceptibility during each stage have been characterized (Rosenow et al., 1997). It has been proposed that growth stage had a major effect on stomatal sensitivity to dehydration and this has been demonstrated in maize and sorghum (Ackerson and Krieg, 1977; Garrity et al., 1984) at which stomatal response was totally insensitive during the reproductive stage. This change in stomatal sensitivity with crop age was suggested to be due to osmotic adjustment which would allow the plant to maintain cell turgor and open stomata under low leaf water potentials (Ackerson and Krieg, 1977). However, the results of the present study indicated that stomatal response of both maize and sorghum at both pre and post-flowering dehydration were sensitive to the decrease in the relative water contents over the duration of dehydration cycle (Figure 2). This is in agreement with Massacci et al. (1996) who reported that stomata did not show decreasing sensitivity to drought stress during plant development in field grown sweet sorghum.

So far, the influence of dehydration on Fv/Fm ratio has usually been examined in the early developmental stages of plants and experimental data are scarce for comparison with the results from dehydration at pre and post-flowering stages, although the available reportsindicate that the way Fv/Fm changes with dehydration strongly depend with plant age (David et al., 1998; Massacci et al., 1996). In the present study, dehydration during pre and post-flowering stage exerted differing effects on the Fv/Fm ratio of both species with the effect being much more pronounced below RWC of 65% during the late phase of post-flowering dehydration (Figures 4a and b). Hence the contribution of non-stomatal factors to explain drought induced depression in photosynthesis may be expected to increase with plant age, and our results are in accordance with this hypothesis. The result reported in this study is consistent with the findings of Massacci et al. (1996).

**Differences in the physiological responses to rehydration between maize and sorghum**

Despite the fact that rehydration following pre and post-flowering dehydration is a determinant to stabilize grain yield in cultivated crop plants, there is a lack of literature concerning the effect of rehydration on the physiological response of crop plants at pre and post-flowering stages. The understanding of the recovery of gas exchange characteristics (stomatal conductance and photosynthesis rate) and the processes which controls it, is therefore poorly understood at different developmental stages in general and at pre and post-flowering stages in particular at which water deficits exerts the greatest loss of grain yield. The rehydration experiment indicates that the ability of stomatal conductance to recover after stress relief decreases with plant age. As shown in Figures 2c and d, although maize attained full recovery, the rate of recovery of stomatal conductance during pre-flowering rehydration was slow and sorghum did not attain full recovery. These findings are in accordance with the findings of Ludlow et al. (1980) who reported that recovery of stomatal conductance upon rewatering was slow and incomplete. During post-flowering stage, although there was an initial recovery, rehydration accentuated the negative effect of dehydration on stomatal conductance of both maize and sorghum (Figure 2). These results indicate that under the condition of dehydration that prevailed in this work, rehydration of the species under investigation at the later reproductive stage was apparently deleterious than pre-flowering stage. It appeared that except during post-flowering rehydration in sorghum where RWC did not improve upon rehydration, the rate of recovery of stomatal conductance was not determined by the rate at which RWC recovered during pre and post-flowering rehydration in maize and pre-flowering rehydration in sorghum, since full recovery of RWC was attained during the periods. According to Ludlow et al. (1980) the slow rate of recovery of stomatal conductance to the level of control plants results from accumulation of abscisic acid. This could be the reason why there was a slow rate of recovery in both species during pre-flowering rehydration. The absence of recovery during post-flowering rehydration in both maize and sorghum may be associated with the harmful effect of rehydration at the maturity stage of both species. The lack of recovery of stomatal conductance also exerted an influence on the recovery of photosynthesis rate (Figure 3) and water use efficiency (Figure 5).

During pre and post-flowering stage in maize and during pre-flowering stage in sorghum, when plants were fully rehydrated, the Fv/Fm ratio r eturned to the control level and the photosynthetic apparatus and cell membranes were repaired completely (Figures 4c and d). However, after an apparent initiate repair, the rapid decrease in Fv/Fm ratio (Figure 4d) during post-flowering rehydration in sorghum suggest that rapid rehydration may be as harmful to the photosynthetic apparatus as the dehydration itself during post-flowering stage.

## Conclusion

The results of this study pointed out the need for using integrated traits when evaluating drought resistance of plants. The results showed that the maize cv Melkassa-2 and sorghum cv MACIA had a remarkable array of contrasting behaviour in response to pre and post-flowering dehydration and rehydration. Differences in maintenance of RWC may be related to performance under dehydration, particularly when crop species of different adaptation were compared under stress conditions of varied intensity. This was observed particularly during moderate water deficits when sorghum exhibited relatively higher RWC during both pre and post-flowering dehydration. In conclusion, sorghum appeared to be more resistant to moderate pre and post-flowering dehydration than maize; this can be attributed to its greater capability to maintain relatively higher RWC and consequently delay leaf rolling, maintain stomata partially open, maintain $P_n$ at a reduced rate and relatively higher water use efficiency. Both species, however, were found to be susceptible to severe pre and post-flowering dehydration.

This study will help in the understanding of some adaptive mechanisms developed by maize and sorghum and contribute to the identification of useful traits for breeding programs.

However, further studies are necessary under field conditions to clarify the adaptive responses in both maize and sorghum during pre and post-flowering dehydration and the capacity to return to normal physiology during post-stress rehydration.

**Abbreviations: Fv/Fm,** Quantum efficiency ratio; $\mathbf{g_s}$, stomatal conductance; $\mathbf{P_n}$, photosynthesis.

## ACKNOWLEDGEMENTS

The Ethiopian Agricultural Research Organization is greatly acknowledged for providing the financial support for this study. ICRISAT centre, Bulawayu, Zimbabwe and Melkassa Agricultural Research Centre, Nazreth, Ethiopia are also acknowledged for supplying sorghum and maize seeds, respectively.

## REFERENCES

Ackerson RC, Kreig DR (1977). Stomatal and non stomatal regulation of water use in cotton, corn and sorghum. Plant Physiol, 60:850-853.

Blum A (1988). Plant breeding for stress environments, CRC, Florida, P. 223.

Blum A (1989). Osmotic adjustment and growth of barley genotypes under drought stress. Crop Sci. 29:230-233.

Cornic G (1994). Drought stress and high light effects on leaf photosynthesis. In Baker, N. and J. R. Bowyer (eds) Photoinhibition of photosynthesis pp. 297-313. BIOS Scientific Publishers LTD, Oxford.

Chaves MM (1991). Effect of water deficits on carbon assimilation. J. Exp. Bot. 42:1-16.

David MM, Coelho D, Barrote I, Corria MJ (1998). Leaf age effects on photosynthetic activity and sugar accumulation in droughted and rewatered *Lupinus albus* plants. Aust. J. Plant Physiol. 25:299-306.

Davies WJ, Zhang J (1991). Root signals and the regulation of growth and development of plants in drying soil. Ann. Rev. Plant Physiol. Plant Mol. Biol. 42:55-76.

Garrity DP, Sullivan CY, Watts DG (1984). Changes in grain sorghum stomatal and photosynthetic response to moisture stress across growth stages. Crop Sci. 24:441-446.

Henson IE, Mahalakshmi V, Bidinger FR, Alagarswamy G (1981). Genotypic variation in Pearl millet (*Pennisetum americanum* L. Leeke) in the ability to accumulate ABA in response to water stress. J. Exp. Bot. 32:899- 904.

Kicheva MI, Tsonev TD, Popova LP (1994). Stomatal and nonstomatal limitations to photosynthesis in two wheat cultivars subjected to water stress. Photosynthetica 30:107-116.

Lawlor DW (1995). The effects of water deficit on photosynthesis. In: N. Smirnoff (eds.) Environment and plant metabolism: flexibility and acclimation. BIOS Scientific Publishers Limited. Oxford.

Lawlor DW, Cornic G (2002). Photosynthetic carbon assimilation and associated metabolism in relation to water deficits in higher plants. Plant Cell Environ. 25:275-294.

Liang L, Zhang J, Wong MH (1997). Can stomatal closure caused by xylem ABA explain the inhibition of leaf photosynthesis under soil drying? Photosynthesis Res. 51:149-159.

Ludlow MM, Muchow RC (1990). A critical evaluation of traits for improving crop yields in water limited environments. Adv. Agron. 43:107-153.

Ludlow MM, Ng TT, Ford CW (1980). Recovery after water stress of leaf gas exchange in *Panicum maximum var. trichoglume*. Aust. J. Plant Physiol. 7:299-313.

Malabuyot JA, Aragon EL, De Dataa SK (1985). Recovery from drought induced desiccation at the vegetation growth stages in direct seeded rainfed rice. Field Crops Res. 10:105-112.

Massacci A, Battistelli A, Loretto F (1996). Effects of drought stress on photosynthetic characteristics, growth and sugar accumulation on field grown sweet sorghum. Aust. J. Plant Physiol. 23:331-340.

Maxwell K, Johnson GN (2000). Chlorophyll fluorescence a practical guide. J. Exp. Bot. 51:659-668.

Peng S, Kreig DR (1992). Gas exchange traits and their relationship to water use efficiency of grain sorghum. Crop Sci. 32:386-391.

Premachandra GS, Saneoka H, Fujita K, Ogata S (1992). Osmotic adjustment and stomatal response to water deficits in maize. J. Exp. Bot. 43:1451-1456.

Rosenow DT, Ejeta G, Clark LE, Gilbert ML, Henzell RG, Borrell AK, Muchow RC (1997). Breeding for pre and post-flowering drought stress resistance in sorghum. In: Proceedings of the International Conference on genetic improvement of sorghum and pearl millet. September 22-27, 1996. Holyday Inn plaza, Lubbock, Texas, pp. 97-95.

Sharkey TD, Seemann JR (1989). Mild water stress effects on carbon reduction cycle intermediate, ribulose bisphosphate carboxylase activity, and spatial homogeneity of photosynthesis in intact leaves. Plant Physiol. 89:1060-1065.

Stikic R, Davies WJ (2000). Stomatal relations of two different maize lines to osmotically induced drought stress. Biologia Plant 43:399-405.

Tardieu F, Davies WJ (1992). Stomatal responses to abscisic acid is a function of current plant water status. Plant Physiol. 98:540-545.

Tuinstra MR, Grote EM, Goldsbrough PB, Gebisa E (1997). Genotypic analysis of post-flowering tolerance and components of grain development in sorghum bicolour (L.) Moench. Mol. Breed. 3:439-448.

Turner NC (1997). Further Progress in Crop Water Relations. Adv. Agron. 58:293-338.

# Allelopathic influence of the aqueous extract of *Jatropha curcas L.* leaves on wild *Cichorium intybus*

**Paulo André Cremonez, Armin Feiden, Reginaldo Ferreira Santos, Doglas Bassegio, Eduardo Rossi, Willian Cézar Nadaleti, Jhonatas Antonelli and Fabiola Tomassoni**

UNIOESTE – Western Paraná State University – Post graduation Program, Master Course of Energy in Agriculture. Street Universitária, 2069, CEP: 85.819-130 Faculdade, Cascavel, PR, Brazil.

**The communication presents an evaluation of the allelopathic effect of the aqueous extract of physic nut (*Jatropha curcas L.*) leaves on the early development of wild chicory (*Cichorium intybus*) seedlings. The experiment was performed in Palotina City, State of Paraná, Brazil. The physic nut leaves were harvested and then crushed in a blender using the proportion of 1 L of distilled water for each 200 g of leaves resulting in a crude extract of 100%. The experiments were conducted in 6 trials with 8 repetitions. The dilutions of 80, 40, 20, 10, 5 and 0% (only distilled water) were made using the crude extract. Data was subjected to statistical analysis (regression at 1% of probability). Aqueous extract of physic nut leaves showed an inhibitory allelopathic effect on the development of wild chicory (sugarloaf variety) which increased with enhanced concentration, presenting negative linear tendency, except for the variable fresh weight of shoots.**

**Key words:** *Jatropha curcas* L., *Cichorium intybus,* allelopathy.

## INTRODUCTION

The allelopathic effects can be observed on plants/microorganism however, it becomes more evident in vegetables. The allelopathic effect is a natural interference in which the plant produces substances and metabolites that may benefit or harm other plants/organisms when released (Corsato et al., 2010; Gliessman, 2000).

During the intense process of development of agriculture, synthetic herbicides have played a very important role in the control of weeds, however, due to the emergence of weeds that are resistant to herbicides, and due to its impact on the environment, it increases the pressure and study to reduce or eliminate these materials/products of the food production process (Duke et al., 2002).

In agriculture, the allelopathic effects may interfere in the productivity of cultivars, and they can also be used to control undesirable plants, reducing the production costs, resulting into less use of agrochemicals and safeguarding against the adverse impact it may cause environmentally (Silva et al., 2012).

The tolerance to metabolites is a characteristic of specific species, where some are more sensitive like *Lactuca sativa* L. (lettuce) and *Cucumis sativus* L. (cucumber). These plants are considered indicators of allelopathic activity due to their characteristics such as uniform and quick germination, and sensitivity to submit significant results by applying small concentrations of allelopathic substances (Ferreira and Áquila, 2000).

Physic nut (*Jatropha curcas* L.) belongs to the Euphorbiaceae, the family of cassava and castor beans, being a multiuse tree that is used as a living fence and in the production of soap, it also contains certain medicinal properties in the treatment of arthritis, gout and toothache. In addition, its oil is destined to biodiesel production (Openshaw, 2000; Van Valkenburg and Bunyapraphatsara, 2002). As pointed out by Beltrão et al. (2005), physic nut is a perennial and monoecious species, having high oil content in its seeds and low production cost, besides being resistant to water stress, so it is indicated to be used in the semiarid northeast. The oil content in seeds varies between 30 and 37% (He et al., 2009).

The cultivation of *Jatropha curcas* L. does not interfere in food security and it is considered as viable option in the field intercropped with other crops. This plant is used in Latin America and West Africa as medicinal herb, besides its extract is widely used as insecticide, because of its proven toxicity to some insects, with other control methods as consequence of its easy retrieval and application. Its cultivation appears as an additional option among the areas of renewable energy not only for their oil content, but also for their by-products: like glycerin and pie, be used as biofertilizers or in biogas production (Sousa et al., 2009; Servin et al., 2006; Fernandes, 2012; Achten et al., 2008).

Using the root extract of *Jatropha*, positive and negative allelopathic interferences can be obtained in several cultures, depending on the vegetal species being treated therewith. Usually, when these substances interfere negatively on other plants, this inhibition occurs at the stage of seedlings, being corn and turnip good examples of affected plants (Abugre and Sam, 2010; Silva et al., 2012).

The utilization of extract obtained from the leaf of *J. curcas* can cause different interference from what was observed with the extract of roots. According to the same author, in order to observe the allelopathic effect from the roots, it is required a significantly concentration of the extract, while in the case of the leaves, small concentrations are required. Lemos et al. (2009) confirm that a small concentration of leaf extract can cause negative allelopathic effects on vegetables such as lettuce, from the stage of germination (rate and speed), until reducing the roots of seedlings, directly interfering on their development. Applications of high concentrations of leaf extract of *Jatropha* in soil cause reduction in root length of *Tagetes erecta* L., however, low concentrations of extract are shown as stimulating for the plant (He et al., 2009).

*C. intybus* is a species of plant with diploid characteristics of the Asteraceae family, which has hundreds of species that are used as horticultural products for salads. Different varieties of this species have been cultivated especially in northwestern Europe, India, South Africa and Chile. Due to its high content of inulin and fructans, it is widely employed in the production of functional foods, providing health benefits (Bremer, 1994; Kaur and Gupta, 2002; Pool-Zobel, 2005). The presence of weeds in crops and vegetable gardens is one of the main problems faced by organic producers, greatly increasing the production costs (Ferreira et al., 2007).

Due to the importance of cultivation and easy perception of the allelopathic effect in vegetables, the aim of this study was to evaluate the positive or negative allelopathy of the application of different concentrations of aqueous extract of leaves of *Jatropha curcas* L. on the early development of seedlings of the chicory species (*Cichorium intybus*) of the sugarloaf variety.

## MATERIALS AND METHODS

The experiment was performed in Palotina city, State of Paraná, Brazil. The region is geographically located at 24°29'4" south latitude, 53°84'2" west longitude (Greenwich), with humid subtropical climate. The experiments were conducted from February of 2013 to April of 2013. The extract of physic nut leaves was prepared with green and healthy leaves just harvested and then milled with a blender. Then, cleavage of the material was performed through a sieve of 1 mm, after that, the parts were soaked in water yielding a final concentration of 200 g L-1.

This extract was diluted in distilled water to lower concentrations of 80, 40, 20, 10, 5 and 0% (and only distilled water served as control group). The design was completely randomized where each treatment was applied with 3 repetitions using 16 plants of wide chicory (*C. intybus*) of sugarloaf variety. The seeds were sown at a depth of 0.2-0.4 mm in specific trays for the vegetable seedling production. The sprinkling was performed 4 times per day, with a manual sprinkler and a volume of 5 mL for each cell during the period of 30 days till the harvest. It was conducted in a manner that it was exposed to sunlight mainly in the morning, and not exposed to rain, nor to any other environmental conditions.

Data pertaining to different parameters of the plant like its length, number of leaves, fresh and dry weight of aerial portion was recorded. The statistical analysis (ANOVA) was performed by a statistical software known as ASSISTAT 7.5 and the comparison among the averages of the treatment was performed with the application of the Tukey test at 5% of probability and the regression at 1% of probability.

## RESULTS AND DISCUSSION

Table 1 shows that the aqueous extract of *J. curcas* has not just influenced linearly the fresh matter of air portion ($p<0.05$) but also the air portion, the number of leaves the root system and the dry mass have decreased linearly with the increase in concentrations (Lemos et al., 2009; Sanderson et al., 2013).

Increasing concentration of the aqueous extract has led to a reduction of the air portion in plant, according to the adjusted linear regression equation ($Y = -0.0329x + 4.436$ $r^2 = 0.65$). Sanderson et al. (2013) observed that when the concentrations of the aqueous extract of *J. curcas* has gone from 0 to 20% there was no allelopathic effect on the air portion of the plant, however, it can be seen a declining propensity. Reichel et al. (2013) observed that the application of aqueous extract of unsterilized

**Table 1.** Effect of the aqueous extract of physic nut (*Jatropha curcas* L.) on evaluated variables of wild chicory (*Chicorium intybus* L.).

| Treat. (%) | Air portion (cm) | Root lenght (cm) | Leaves | Fresh weight/ aerial portion (g) | Dry weight/ aerial portion (g) |
|---|---|---|---|---|---|
| 0 | $3.9125^{bc}$ | $5.8000^{ab}$ | $5.2500^{a}$ | $0.1325^{bc}$ | $0.0183^{a}$ |
| 5 | $5.1045^{a}$ | $5.9125^{a}$ | $4.6250^{abc}$ | $0.0815^{cd}$ | $0.0128^{abc}$ |
| 10 | $4.8456^{ab}$ | $4.7825^{b}$ | $5.0000^{ab}$ | $0.1841^{ab}$ | $0.0139^{ab}$ |
| 20 | $2.6334^{d}$ | $3.0125^{cd}$ | $4.0000^{cd}$ | $0.0540^{d}$ | $0.0103^{bc}$ |
| 40 | $3.3918^{c}$ | $3.4500^{c}$ | $4.2500^{bcd}$ | $0.1991^{a}$ | $0.0073^{bc}$ |
| 80 | $1.8424^{e}$ | $2.1443^{d}$ | $3.3750^{d}$ | $0.0502^{d}$ | $0.0054^{c}$ |
| L.R. | 146.395**(1) | 155.963**(2) | 37.898**(3) | $3.3370^{n.s.}$ | 32.2495**(4) |
| Q.R. | 30.6733** | $0.1173^{n.s.}$ | $0.7313^{n.s.}$ | $5.9480^{n.s.}$ | $0.0099^{n.s.}$ |
| CV (%) | 13.66 | 16.5 | 14.3 | 31.45 | 44.55 |
| O. Average | 3.58 | 4.14 | 4.42 | 0.12 | 0.0113 |

Averages followed by the same letter do not differ significantly between each other by the Tukey test, at 5% of probability. (**) = significant at 1% probability; (*) = significant at 5% probability; ($^{n.s.}$) = not significant. VC (%) = Variation coefficient. L.R. Linear regression. Q.R. Quadratic regression. (1) $y = -0.0329x + 4.436$ $r^2 = 0.65$; (2) $y = -0.0446x + 5.3326$ $r^2 = 0.74$; (3) $y = -0.0202x + 4.94$ $r^2 = 0.78$; (4) $y = -0.0001 + 0.017$ $r^2 = 0.93$.

*J. curcas* has given a reduction of 73.3% in the length of the air portion of the wheat seedlings.

The aqueous extract of physic nut has shown a strong inhibitory effect on the root development of wild chicory seedlings (Table 1). A linear declining effect with the increase of the extract concentration is in agreement with Lemos et al. (2009), who worked with concentration going from 0 to 100% of aqueous extract of physic nut, and observed a negative effect, besides morphological changes on root with thickening and lack of absorption zone. Sanderson et al. (2013) did not find in their studies an inhibitory effect of the aqueous extract of physic nut on the root system of lettuce, but they used lower concentrations (0, 1, 5, 10 and 15%). Reichel et al. (2013) found the aqueous extract of *J. curcas* of 20, 25, 30 and 35% to stimulate the root growth in the wheat plantation CD104. Abugre and Sam (2010) found an inhibition of the growth of the *Zea mays* seedlings at high concentrations of *J. curcas* root extract. Bonamigo et al. (2009) have reported an allelopathic effect of the aqueous extract of the root in the early development of soy and canola.

Abugre and Sam (2010) have noticed a huge amount of phenolic compounds in the leaf extract of physic nut which are mainly responsible for the the allelopathic effect and their high concentrations can inhibit the growth of seedlings in beans (*Phaseolus vulgaris L.*), corn (*Z. mays L.*), tomato (*Solanum lycopersicum L.*) and okra crops.

The allopathic effect is also evident in the number of leaves per plant suffering negative interference due to the increased concentrations of aqueous extract. The fresh mass of the plantation was not adjusted by the tested model, showing no inhibitory effect for this variable. Several studies indicate that *J. curcas* root and stalk extract also showed allelopathic effects on some of the cultivated species (Abugre et al., 2011; Rejila and Vijayakumar, 2011).

Reichel et al. (2013) observed in a study with the extract of physic nut leaves (*J. curcas*) on early wheat development (*Triticum aestivum L.*), using concentrations of (5, 10, 15, 20, 25, 30 and 35%), a possible allelopathic action, however, the authors emphasize that alcoholic extracts of the physic nut leaves affect more the wheat seedling growth than the crude aqueous extract.

## Conclusion

It was possible to conclude that the aqueous extract of physic nut showed an inhibitory allelopathic effect on the development of wild chicory sugarloaf variety, in all the analyzed variables, presenting negative linear tendency, except for the variable fresh weight of shoots.

### REFERENCES

Abugre S, Sam SJQ (2010). Evaluating the allelopathic effect of *Jatropha curcas* aqueous extract on germination, radicle and plumule length of crops. Int. J. Agric. Biol. Faisalabad 12(5):769-772.

Abugre S, Apetorgbor AK, Antwiwaa A, Apetorgbor MM (2011). Allelopathic effects of ten tree species on germination and growth of four traditional food crops in Ghana. J. Agric. Technol. 7(3):825-834.

Achten WMJ, Verchot L, Franken YJ, Mathijs E, Singh VP, Aerts R, Muys B (2008). Jatropha biodiesel production and use. Biomass Bioenergy 32(12):1063-1084.

Beltrão NEM, Severino LS, Suinaga FA, Veloso JF, Junqueira N, Fidelis M, Gonçalves NP, Saturnino HM, Roscoe R, Gazzoni D, Duarte JO, Drumond MA, Anjos JB (2005). Recomendação técnica sobre o plantio de pinhão-manso no Brasil. Disponível em: <http://www.cpao.embrapa.br/portal/noticias/Position%20Paper.pdf> Acesso: maio/2013.

Bonamigo T, Siberte PSS, Da Silva J, Poliszuk MCC, Fortes AMT (2009). Efeito alelopático faça extrato de Raiz de pinhão-manso na germinação e Desenvolvimento inicial de soja e canola. In: XII Congresso Brasileiro de Fisiologia Vegetal, Fortaleza, CE. Sociedade Brasileira de Fisiologia Vegetal, SBFV, P. 1.

Bremer K (1994). Asteraceae: Cladistics and classification. Portland: Timber Press.

Corsato JM, Fortes AMT, Santorum M, Leszczynski R (2010). Efeito alelopático do extrato aquoso de folhas de girassol sobre a germinação de soja e picão-preto. Ciências Agrárias, Londrina, 31(2):353-360.

Duke SO, Dayan FE, Rimando AM, Schrader KK, Aliotta G, Oliva A, Romagni JG (2002). Chemicals from nature for weed management. Weed Sci. 50(2):138-151.

Fernandes TS (2012). Bioatividade de extratos aquosos de pinhão roxo *Jatropha gossypiifolia* L. sobre *Spodoptera frugiperda* (J. E. SMITH). Programa de Pós-graduação em Agronomia, Universidade Federal do Piauí. Teresina-PI, 58 P.

Ferreira AG, Aquila MEA (2000). Alelopatia: uma área emergente da ecofisiologia. Revista Brasileira de Fisiologia Vegetal, Campinas, 12:175-204.

Ferreira MC, Souza JRP, Faria TJ (2007). Potenciação alelopática de extratos vegetais na germinação e no crescimento inicial de picão-preto e alface. Ciência e Agrotecnol. 31(4):1054-1060.

Gliessman SR (2000). Agroecologia: processos ecológicos em agricultura sustentável. Porto Alegre: UFRGS, P. 653.

He CZ, Zhong L, He HF, Li D, Xu H (2009). Allelopathic Effect of Water Extracts from Leaves of *Jatropha curcas* on It's Seed Germination. Seed 6:1.

He Y, Guo X, Lu R, Niu B, Pasapula V, Hou P, Cai F, Xu Y, Chen F (2009). Changes in morphology and biochemical indices in browning callus derived from *Jatropha curcas* hypocotyls. Plant Cell Tiss. Org. 98(1):11-17.

Kaur N, Gupta AK (2002). Applications of inulin and oligofructose in health and nutrition. J. Biosci. 27(7):703-714.

Lemos JM, Meinerz CC, Bertuol P, Corteza O, Guimarães VF (2009). Efeito Alelopático do Extrato Aquoso de Folha de Pinhão Manso (*Jatropha curcas* L.) sobre a Germinação e Desenvolvimento Inicial de Alface (*Lactuca sativa* cv. *Grand Rapids*). Rev. Bras. Agroecol, 4(2):2529-2532.

Openshaw K (2000). A review of *Jatropha curcas*: an oil plant of unfulfilled promise. Biomass and Bioenergy, Silver Spring. 19(1):1-15.

Pool-Zobel BL (2005). Inulin-type fructans and reduction in colon cancer risk: Review of experimental and human data. Br. J. Nutr. 93(S):73-90.

Reichel T, Barazetti JF, Stefanello S, Paulert R, Zonetti PD (2013). Alelopatia de extratos de folhas de pinhão-manso (*Jatropha curcas* L.) no desenvolvimento inicial do trigo (*Triticum aestivum* L.). Idesia 31(1):45-52.

Rejila S, Vijayakumar N (2011). Allelopathic effect of *Jatropha curcas* on selected intercropping plants (Green Chilli and Sesame). J. Phytol. 3(5):01-03.

Servin SCN, Torres OJM, Matias JEF, Agulham MA, Carvalho FA, Lemos R, Soares EWS, Soltoski PR, Freitas ACT (2006). Ação do extrato de *Jatropha gossypiifolia* L. (Pinhão Roxo) na cicatrização de anastomose colônica: Estudo experimental em ratos. Acta Cir. Bras. 21(3):89-96.

Silva PSS, Fortes AMT, Pilatti DM, Boiago NP (2012). Atividade alelopática do exsudato radicular de *Jatropha curcas* L. sobre plântulas de *Brassica napus* L., *Glycinemax* L., *Zea mays* L. e *Helianthus annuus* L. Insula Revista de Botânica. Florianópolis 41:32-41.

Sousa AH, Faroni LRDA, Pereira MDP, Almeida JPM, Silva FN (2009). Atividade inseticida de genótipos de pinhão manso para insetos praga de produtos armazenados. I Congresso Brasileiro de Pesquisas de Pinhão Manso. Brasília-DF, pp. 1-4.

Van Valkenburg JLCH, Bunyaprahatsara N (2002). Plant Resources of South-East Asia: Medicinal and Poisonous Plants. Prosea, Bogor. 12(2):324.

# The effects of flurprimidol concentrations and application methods on *Ornithogalum saundersiae* Bak. grown as a pot plant

**Piotr Salachna and Agnieszka Zawadzińska**

Laboratory of Ornamental Plants at the Department of Horticulture, West Pomeranian University of Technology in Szczecin, Papieża Pawła VI 3A, 71-459 Szczecin, Poland.

***Ornithogalum saundersiae* is an ornamental bulb plant originating from South Africa that is mostly grown for cut flowers. The aim of the study was to assess the possibility of producing *O. saundersiae* as a pot plant using flurprimidol. A retardant in the form of the Topflor 015 preparation was used at concentrations of 15 and 30 mg·dm$^{-3}$ as a pre-plant bulb soak, a soil drench or a foliar spray. Retardant applied as a foliar spray at a concentration of 30 mg·dm$^{-3}$ inhibited plant growth most effectively (by 52.5%), caused flowering delay and reduced the inflorescence and flower diameter. The plants obtained from the bulbs soaked in the retardant solution at a concentration of 30 mg·dm$^{-3}$ had the smallest plant diameter and the shortest leaves. The use of flurprimidol resulted plants with an increased relative chlorophyll content. Retardant used as a foliar spray increased the stomatal conductance, regardless of the solution concentration.**

**Key words:** Geophytes, plant growth retardants, relative chlorophyll content (SPAD), stomatal conductance ($g_s$), Topflor.

## INTRODUCTION

*Ornithogalum saundersiae* Bak., also called the giant chincherinchee, is an attractive bulb plant which naturally occurs in South Africa. It has ornamental white flowers arranged in several dozens on large inflorescences, which remain for a long time. The bulbs of this plant contain a lot of biologically active compounds, including OSW-1 saponins with anticancer activity (Morzycki and Wojtkielewicz, 2005). This species is mostly grown in Kenya and Israel for cut flowers and the value of their sales at the Dutch flower exchange has been growing steadily (Kariuki and Kako, 1999; Anonymous, 2009). So far, no scientific studies have been published on the cultivation of *O. saundersiae* as a pot plant.

The use of plant growth retardants (PGRs) to obtain low and compact plants allows for continuous broadening of the pot plant assortment to include new species and cultivars (Criley, 2005; Mello et al., 2012). Flurprimidol from the pyrimidine group [alpha-(1-methylethyl)-alpha-(4-(trifluoromethoxyphenyl)-5-pyrimidinemethanol] is used to reduce unwanted plant growth, while enhancing the quality of ornamentals (Currey et al., 2012; Fair et al., 2012; Sprzaczka and Laskowska, 2013). According to the literature (Whipker et al., 2011; Miller, 2013), the influence of flurprimidol on the morphological features and yield of plants varies and it largely depends on the species, concentration and the application method.

So far, no information has been available on the use of plant growth retardants in the cultivation of *O. saundersiae*, and this is the reason why research was undertaken to determine the influence of flurprimidol on the morphological features, the relative chlorophyll content, the stomatal conductance and the flowering of *O. saundersiae* grown in pots.

## MATERIALS AND METHODS

The studies were carried out in the Laboratory of Ornamental Plants at the Department of Horticulture West Pomeranian University of Technology in Szczecin (53° 25' N, 14° 32' E). *O. saundersiae* bulbs with a circumference of 12 to 14 cm were used in the experiment. Before planting, the bulbs were stored at 25°C. The bulbs were planted singly on 20 May 2011 in pots with a circumference of 15 cm filled with a medium consisting of a peat substrate with a pH of 6.5 with the addition of the multi-ingredient Hydrocomplex fertiliser (N18%-P11%-K18% plus microelements) at a dose of 5 $g·dm^{-3}$. The plants were grown in a greenhouse under natural photoperiod conditions with temperatures at 22 to 26°C during the day and 16 to 18°C at night with a relative humidity of 65 to 80%.

The Topflor 015 SL preparation (SePRO Corporation, USA) contained 1.5% of flurprimidol and was used in the experiment. Retardant was applied as (1) a pre-plant bulb soak (for 60 min), (2) a soil drench (in the 5th week of the cultivation using 100 ml of the solution per pot) or (3) a foliar spray (in the 5th week of cultivation using 100 ml of the solution per plant). In each variant, two concentrations of the flurprimidol solution were used: 15 and 30 $mg·dm^{-3}$. The control plants were not treated with the retardant. Total plant height (measured from the soil line to the uppermost part of the inflorescences), plant diameter, number of leaves, leaf length and width, number of inflorescences per plant, number of flowers in inflorescence, inflorescence diameter, flower diameter were recorded at flowering stage (one flower per plant fully opened).

The relative chlorophyll content in the leaves was measured in dimensionless units, the so-called SPAD readings using a Chlorophyll Meter SPAD-502 (Minolta, Japan). This measurement consists of the determination of the quotient of the light absorption connected with the chlorophyll presence (the wavelength of 650 nm) and absorption through the leaf tissue (the wavelength of 940 nm). The measurements were performed at the flowering stage using three leaves from each plant. Three readings were performed for each leaf and the average value was calculated.

The stomatal conductance was determined using an SC1 porometer (Dekagon Devices, USA). The measurements were performed only on sunny days between 10 to 12 a.m. at the flowering flowering stage on three developed leaves of each plant, three readings per leaf. The experiment was conducted as a one-factor experiment with a full randomisation system with four replications of 20 plants each. The results of the measurements were verified using a variance analysis model and the ANALWAR 4.6 software based on Microsoft Excel. The confidence semi-intervals were calculated on the basis of Tukey's test at a significance level of $\alpha=0.05$.

## RESULTS

In the conducted experiment, a significant influence of flurprimidol on the height and diameter of the plants and the length and width of *O. saundersiae* leaves was found. The retardant reduced the plant height as a result of foliar spray with the solution at a concentration of 30 $mg·dm^{-3}$. These plants were 52.5% shorter than the untreated control. The plants drenched and sprayed with a retardant solution had the greatest diameter, while plants obtained from the bulbs soaked in the retardant solution at a concentration of 30 $mg·dm^{-3}$ had the smallest plant diameter and the shortest leaves. It was shown that the plants whose bulbs were soaked in flurprimidol, regardless of the solution concentration, formed the widest leaves (Table 1).

Flurprimidol had a significant influence on the days to flowering, the diameter of inflorescences and flowers (Table 2). The plants sprayed with the solution at a concentration of 30 $mg·dm^{-3}$ were delayed of flowering by 25 days. An analysis of the inflorescence diameter revealed that the greater inflorescences were obtained after using the retardant as a soil drench at a concentration of 30 $mg·dm^{-3}$, while the inflorescences of the plants sprayed with the retardant were smaller. As a result of foliar spray with flurprimidol, the plants had smaller flower diameter, regardless of the retardant concentration (Table 2).

The relative chlorophyll content in the leaves of the plants treated with the retardant was significantly higher, on average by 10.1 SPAD units, as compared to control plants. The form of retardant application and the solution concentration did not influence the relative chlorophyll content in the leaves (Table 3). The flurprimidol treatment influenced the stomatal conductance of the leaves. The plants sprayed with the retardant, regardless of the concentration, were characterised by a higher stomatal conductance of the leaves, on average by 7.60 mmol $H_2O\ m^{-2}s^{-1}$ than the untreated control (Table 3).

## DISCUSSION

A lot of ornamental plants are too high, so methods are searched to limit their height by the use of retardants. On the basis of the results obtained in the experiment, it was shown that flurprimidol effectively inhibited the growth of *O. saundersiae*. The results are consistent with the data presented by other authors, who obtained lower plants by using flurprimidol in the cultivation of tulips (Fair et al., 2012; Sprzaczka and Laskowska, 2013) and lilies (Currey et al., 2012). In this research, it was shown that flurprimidol applied as a foliar spray at a concentration of 30 $mg·dm^{-3}$ inhibited plant growth most effectively. Similar results were obtained by Pobudkiewicz and Treder (2006), who reported that *Lilium* 'Mona Lisa' plants sprayed with flurprimidol were significantly lower than plants drenched with the retardant.

According to Whipker et al. (2004), plants treated with growth retardants are more compact, owing to which a larger number of plants can be placed per square metre, thus increasing the profitability of production. In the conducted experiment, stocky plants with a significantly

**Table 1.** Effect of flurprimidol on plant height, plant diameter, no. of leaves, leaf length and leaf width of *Ornithogalum saundersiae*.

| Flurprimidol treatment | | Plant height | Plant diameter | No. of leaves | Leaf length | Leaf width |
|---|---|---|---|---|---|---|
| Application method | Concentration ($mg \cdot dm^{-3}$) | (cm) | (cm) | | (cm) | (cm) |
| Control | 0 | $133^{a}$ | $47.5^{b}$ | $6.25^{a}$ | $55.2^{a}$ | $5.68^{c}$ |
| Bulb soak | 15 | $89.7^{b}$ | $43.2^{b}$ | $7.25^{a}$ | $36.0^{c}$ | $7.38^{a}$ |
| | 30 | $85.7^{b}$ | $39.5^{c}$ | $7.50^{a}$ | $29.3^{c}$ | $7.50^{a}$ |
| Soil drench | 15 | $88.5^{b}$ | $55.2^{a}$ | $6.75^{a}$ | $46.0^{b}$ | $6.13^{b}$ |
| | 30 | $79.2^{b}$ | $61.0^{a}$ | $7.00^{a}$ | $45.5^{b}$ | $6.38^{b}$ |
| Foliar spray | 15 | $86.5^{b}$ | $57.0^{a}$ | $6.25^{a}$ | $47.2^{b}$ | $6.38^{b}$ |
| | 30 | $63.2^{c}$ | $59.2^{a}$ | $7.00^{a}$ | $49.7^{ab}$ | $6.50^{b}$ |

Values are means of 4 replicates. Means followed by the same letter are not statistically different ($P < 0.05$) by Tukey's test.

**Table 2.** Effect of flurprimidol on number of days to flowering, number of inflorescence per plant, number of flowers in inflorescence, inflorescence diameter and flower diameter of *Ornithogalum saundersiae*.

| Flurprimidol treatment | | No. of days to flowering | No. of inflorescence per plant | No. of flowers in inflorescence | Inflorescence diameter (cm) | Flower diameter (cm) |
|---|---|---|---|---|---|---|
| Application method | Concentration ($mg \cdot dm^{-3}$) | | | | | |
| Control | 0 | $115^{c}$ | $1.00^{a}$ | $73.0^{a}$ | $12.7^{ab}$ | $3.52^{a}$ |
| Bulb soak | 15 | $123^{bc}$ | $1.00^{a}$ | $70.3^{a}$ | $13.0^{ab}$ | $3.50^{a}$ |
| | 30 | $126^{b}$ | $1.00^{a}$ | $68.7^{a}$ | $15.5^{a}$ | $3.53^{a}$ |
| Soil drench | 15 | $117^{c}$ | $1.00^{a}$ | $75.8^{a}$ | $13.5^{ab}$ | $3.70^{a}$ |
| | 30 | $112^{c}$ | $1.00^{a}$ | $75.3^{a}$ | $12.5^{abc}$ | $3.55^{a}$ |
| Foliar spray | 15 | $123^{bc}$ | $1.00^{a}$ | $76.7^{a}$ | $10.2^{bc}$ | $2.97^{b}$ |
| | 30 | $140^{a}$ | $1.00^{a}$ | $72.0^{a}$ | $9.38^{c}$ | $2.76^{b}$ |

Values are means of 4 replicates. Means followed by the same letter are not statistically different ($P < 0.05$) by Tukey's test.

smaller diameter and length of leaves were obtained as a result of soaking the bulbs in the retardant.

In this study, the foliar spray of the plants with flurprimidol delayed the flowering. A larger number of days to anthesis as a result of flurprimidol treatment was also observed in the cultivation of oriental lily 'Mona Lisa' (Pobudkiewicz and Treder, 2006) and *Dahlia variabilis* 'Purple Gem' (Whipker et al., 2011). In the experiment conducted, flurprimidol did not influence the number of inflorescences obtained per plant or the number of flowers in an inflorescence. The research by Currey et al. (2012) on the use of flurprimidol in pot cultivation of *Lilium longiflorum* 'Nellie White' also showed that the number of flowers did not depend on the treatment of plants with the retardant. In this research, the *O. saundesrae* plants had smaller flowers as a result of foliar spray the leaves with flurprimidol. Similarly, Pobudkiewicz and Treder

**Table 3.** Effect of flurprimidol on relative chlorophyll content and stomatal conductance of *Ornithogalum saundersiae* at flowering.

| Flurprimidol treatment | | Relative chlorophyll content (SPAD) | Stomatal conductance (mmol $H_2O$ $m^{-2}s^{-1}$) |
|---|---|---|---|
| Application method | Concentration ($mg \cdot dm^{-3}$) | | |
| Control | 0 | $45.1^b$ | $7.30^b$ |
| Bulb soak | 15 | $57.3^a$ | $9.42^b$ |
| | 30 | $55.6^a$ | $8.90^b$ |
| Soil drench | 15 | $52.1^a$ | $8.55^b$ |
| | 30 | $56.2^a$ | $8.01^b$ |
| Foliar spray | 15 | $53.5^a$ | $15.2^a$ |
| | 30 | $56.4^a$ | $14.6^a$ |

Values are means of 4 replicates. Means followed by the same letter are not statistically different ($P < 0.05$) by Tukey's test.

(2006) observed that a retardant applied as a foliar spray reduces the tepal size of oriental lily 'Mona Lisa'.

In the present study, the plants treated with flurprimidol contained more chlorophyll in leaves than the control plants. These results are confirmed by the research conducted by Thohirah et al. (2005), who reported that flurprimidol applied as a soil drench at concentration of 40 $mg \cdot dm^{-3}$ increased the chlorophyll content in leaves of *Curcuma alismatifolia*. The results presented in the study show that the foliar spray of plants with a flurprimidol solution at concentrations of 15 and 30 $mg \cdot dm^{-3}$ increased the stomatal conductance of the leaves, which implies a higher intensity of photosynthesis. According to the results presented in the few available publications on this subject, the stomatal conductance of leaves increases under the influence of retardants (Thetford et al., 1995), decreases (Premachandra et al., 1997) or remains unchanged (Nazarudin et al., 2012).

In summary, it can be concluded that *O. saundersiae* can be cultivated as an attractive pot plant. The use of flurprimidol, depending on the application form and concentration of the solution, does not only reduce the growth and modifies the appearance of *O. saundersiae*, but it also influences the flowering time, the relative chlorophyll content and the stomatal conductance of the leaves.

## REFERENCES


Anonymous (2009). Cut flowers and foliage: The EU market for summer flowers. CBI Market Information Database, pp. 7-8.

Criley RA (2005). Creating a potted, flowering *Hedychium* with growth retardants. Acta Hortic. 683:201-206.

Currey CJ, Lopez RG, Krug BA, McCall I, Whipker BE (2012). Substrate drenches containing flurprimidol suppress height of 'Nellie White' easter lilies. Hort. Technol. 22(2):164-168.

Fair BA, Whipker B, McCall I, Buhler W (2012). Height control of 'Hot Lips' hybrid sage to flurprimidol substrate drench. Hort. Technol. 22:539-541.

Kariuki W, Kako S (1999). Growth and flowering of *Ornithogalum saundersiae* Bak. Sci. Hortic. 81:57-70.

Mello SC, Matsuzaki RT, Campagnol R, Mattiuz CF (2012). Effects of plant growth regulators in ornamental kale (*Brassica oleracea* var. *acephala*). Acta Hortic. 937:245-251.

Miller WB (2013). Dark-stored flurprimidol solutions maintain efficacy over many weeks. HortScience 48(1):77-81.

Morzycki J, Wojtkielewicz A (2005). Synthesis of a highly potent antitumor saponin OSW-1 and its analogues. Phytochem. Rev. 4:259-277.

Nazarudin MR Ahmad, Tsan FY, Fauzi R Mohd (2012). Morphological and physiological response of *Syzygium myrtifolium* (Roxb.) Walp. to paclobutrazol. Sains Malays. 41(10):1187-1192.

Pobudkiewicz A, Treder J (2006). Effects of flurprimidol and daminozide on growth and flowering of oriental lily 'Mona Lisa'. Sci. Hortic. 110(4):328-333.

Premachandra GS, Chaney WR, Holt HA (1997). Gas exchange and water relations of *Fraxinus americana* affected by flurprimidol. Tree Physiol. 17(2):97-103.

Sprzaczka I, Laskowska H (2013). Evaluation of flurprimidol efficiency in pot cultivation of forced tulips. Acta Sci. Pol. Hortorum Cultus 12(2):25-33.

Thetford M, Warren SL, Blazich FA, Thomas JF (1995). Response of *Forsythia ×intermedia* 'Spectabilis' to Uniconazole. II. Leaf and stem anatomy, chlorophyll, and photosynthesis. J. Am. Soc. Hort. Sci. 120(6):983-988.

Thohirah LA, Ramlan MF, Kamalakshi N (2005). The effects of paclobutrazol and flurprimidol on the growth and flowering of *Curcuma roscoeana* and *Curcuma alismatifolia*. Malays. Appl. Biol. 34(2)1-5.

Whipker BE, McCall I, Buhler W, Krug B (2011). Flurprimidol preplant tuber soaks for dahlia growth control. Acta Hortic. 886:393-396.

Whipker BE, McCall I, Gibson JL, Cavins TJ (2004). Flurprimidol foliar sprays and substrate drenches control growth of 'Pacino' pot sunflowers. Hort. Technol. 14(3):411-414.

# Permissions

All chapters in this book were first published in AJAR, by Academic Journals; hereby published with permission under the Creative Commons Attribution License or equivalent. Every chapter published in this book has been scrutinized by our experts. Their significance has been extensively debated. The topics covered herein carry significant findings which will fuel the growth of the discipline. They may even be implemented as practical applications or may be referred to as a beginning point for another development.

The contributors of this book come from diverse backgrounds, making this book a truly international effort. This book will bring forth new frontiers with its revolutionizing research information and detailed analysis of the nascent developments around the world.

We would like to thank all the contributing authors for lending their expertise to make the book truly unique. They have played a crucial role in the development of this book. Without their invaluable contributions this book wouldn't have been possible. They have made vital efforts to compile up to date information on the varied aspects of this subject to make this book a valuable addition to the collection of many professionals and students.

This book was conceptualized with the vision of imparting up-to-date information and advanced data in this field. To ensure the same, a matchless editorial board was set up. Every individual on the board went through rigorous rounds of assessment to prove their worth. After which they invested a large part of their time researching and compiling the most relevant data for our readers.

The editorial board has been involved in producing this book since its inception. They have spent rigorous hours researching and exploring the diverse topics which have resulted in the successful publishing of this book. They have passed on their knowledge of decades through this book. To expedite this challenging task, the publisher supported the team at every step. A small team of assistant editors was also appointed to further simplify the editing procedure and attain best results for the readers.

Apart from the editorial board, the designing team has also invested a significant amount of their time in understanding the subject and creating the most relevant covers. They scrutinized every image to scout for the most suitable representation of the subject and create an appropriate cover for the book.

The publishing team has been an ardent support to the editorial, designing and production team. Their endless efforts to recruit the best for this project, has resulted in the accomplishment of this book. They are a veteran in the field of academics and their pool of knowledge is as vast as their experience in printing. Their expertise and guidance has proved useful at every step. Their uncompromising quality standards have made this book an exceptional effort. Their encouragement from time to time has been an inspiration for everyone.

The publisher and the editorial board hope that this book will prove to be a valuable piece of knowledge for researchers, students, practitioners and scholars across the globe.

# List of Contributors

**Arporn Krudnak**
School of Crop Production Technology, Institute of Agricultural Technology, Suranaree University of Technology, Nakhon Ratchasima, 30000, Thailand

**Sodchol Wonprasaid**
School of Crop Production Technology, Institute of Agricultural Technology, Suranaree University of Technology, Nakhon Ratchasima, 30000, Thailand

**Thitiporn Machikowa**
School of Crop Production Technology, Institute of Agricultural Technology, Suranaree University of Technology, Nakhon Ratchasima, 30000, Thailand

**Mingliang Yu**
Institute of Horticulture, Jiangsu Academy of Agricultural Sciences, Nanjing 210014, P. R. China

**Jianqing Chu**
College of Horticulture, Nanjing Agricultural University, Tongwei Road 6, Nanjing 210095, P. R. China
Jiangsu Fruit Crop Genetics Improvement and Seedling Propagation Engineering Center, Nanjing 210095, P. R. China

**Ruijuan Ma**
Institute of Horticulture, Jiangsu Academy of Agricultural Sciences, Nanjing 210014, P. R. China

**Zhijun Shen**
Institute of Horticulture, Jiangsu Academy of Agricultural Sciences, Nanjing 210014, P. R. China

**Jinggui Fang**
College of Horticulture, Nanjing Agricultural University, Tongwei Road 6, Nanjing 210095, P. R. China
Jiangsu Fruit Crop Genetics Improvement and Seedling Propagation Engineering Center, Nanjing 210095, P. R. China

**B. M. Baloyi**
Department of Soil Science, Plant Production and Agricultural Engineering, University of Limpopo, Private Bag X1106, Sovenga, 0727, South Africa

**V. I. Ayodele**
Department of Soil Science, Plant Production and Agricultural Engineering, University of Limpopo, Private Bag X1106, Sovenga, 0727, South Africa

**A. Addo-Bediako**
Department of Biodiversity, University of Limpopo, Private Bag X1106, Sovenga, 0727, South Africa

**S. V. Pawar**
Department of Plant Pathology, Marathwada Agricultural University, Parbhani, Maharashtra, India

**Utpal Dey**
Department of Plant Pathology, Marathwada Agricultural University, Parbhani, Maharashtra, India

**V. G. Munde**
Department of Plant Pathology, Marathwada Agricultural University, Parbhani, Maharashtra, India

**Hulagappa**
Department of Plant Pathology, University of Agricultural Sciences, Dharwad, Karnataka, India

**Anamika Nath**
Department of Plant Breeding and Genetics, Mahatma Phule Krishi Vidyapeeth, Rahuri, Maharashtra, India

**M. Seyyedi**
Ferdowsi University of Mashhad, Faculty of Agriculture, P. O. Box 91775-1163, Mashhad, Iran

**P. Rezvani Moghaddam**
Ferdowsi University of Mashhad, Faculty of Agriculture, P. O. Box 91775-1163, Mashhad, Iran

**R. Shahriari**
Ferdowsi University of Mashhad, Faculty of Agriculture, P. O. Box 91775-1163, Mashhad, Iran

**M. Azad**
Ferdowsi University of Mashhad, Faculty of Agriculture, P. O. Box 91775-1163, Mashhad, Iran

**E. Eyshi Rezaei**
Ferdowsi University of Mashhad, Faculty of Agriculture, P. O. Box 91775-1163, Mashhad, Iran

**M. H. Ligavha-Mbelengwa**
Department of Botany, University of Venda, Thohoyandou, Limpopo 0950, South Africa

**R. B. Bhat**
Department of Botany, University of Venda, Thohoyandou, Limpopo 0950, South Africa

**Zia Ullah**
Biotechnology Center, Agricultural Research Institute, Tarnab, Peshawar, Pakistan

**Sayed Jaffar Abbas**
Biotechnology Center, Agricultural Research Institute, Tarnab, Peshawar, Pakistan
Center of Biotechnology and Microbiology, University of Peshawar, Pakistan

**Nisar Naeem**
Biotechnology Center, Agricultural Research Institute, Tarnab, Peshawar, Pakistan

**Ghosia Lutfullah**
Center of Biotechnology and Microbiology, University of Peshawar, Pakistan

**Taimur Malik**
Biotechnology Center, Agricultural Research Institute, Tarnab, Peshawar, Pakistan

**Malik Atiq Ullah Khan**
Biotechnology Center, Agricultural Research Institute, Tarnab, Peshawar, Pakistan

**Imran Khan**
Biotechnology Center, Agricultural Research Institute, Tarnab, Peshawar, Pakistan

**Gowhar Ali**
Saffron Research Station, Pampore Sher-e-Kashmir University of Agricultural Sciences and Technology of Kashmir, Shalimar, 191121, Jammu & Kashmir, India

**Asif M. Iqbal**
Saffron Research Station, Pampore Sher-e-Kashmir University of Agricultural Sciences and Technology of Kashmir, Shalimar, 191121, Jammu & Kashmir, India

**F. A. Nehvi**
Saffron Research Station, Pampore Sher-e-Kashmir University of Agricultural Sciences and Technology of Kashmir, Shalimar, 191121, Jammu & Kashmir, India

**Sheikh Sameer Samad**
Saffron Research Station, Pampore Sher-e-Kashmir University of Agricultural Sciences and Technology of Kashmir, Shalimar, 191121, Jammu & Kashmir, India

**Shaheena Nagoo**
Saffron Research Station, Pampore Sher-e-Kashmir University of Agricultural Sciences and Technology of Kashmir, Shalimar, 191121, Jammu & Kashmir, India

**Sabeena Naseer**
Saffron Research Station, Pampore Sher-e-Kashmir University of Agricultural Sciences and Technology of Kashmir, Shalimar, 191121, Jammu & Kashmir, India

**Niyaz A. Dar**
Saffron Research Station, Pampore Sher-e-Kashmir University of Agricultural Sciences and Technology of Kashmir, Shalimar, 191121, Jammu & Kashmir, India

**Vinod Kumar Sharma**
Indian Agricultural Research Institute, Pusa Campus Regional Research Station Katrain Dist-Kullu Valley-175129, India

**Shailaja Punetha**
Department of Vegetable Science, G. B. Pant University of Agriculture and Technology Pantnagar-263145, India

**Brij Bihari Sharma**
Division of Vegetable Science, Indian Agricultural Research Institute, Pusa Campus New Delhi-110012, India

**Hui Song**
State Key Laboratory of Crop Stress Biology on Drought Regions, Northwest A&F University, Yangling, Shaanxi 712100, P. R. China

**Xiaoli Gao**
State Key Laboratory of Crop Stress Biology on Drought Regions, Northwest A&F University, Yangling, Shaanxi 712100, P. R. China

**Baili Feng**
State Key Laboratory of Crop Stress Biology on Drought Regions, Northwest A&F University, Yangling, Shaanxi 712100, P. R. China

**Huiping Dai**
College of Life Science, Northwest A&F University, Yangling, Shaanxi 712100, China

**Panpan Zhang**
State Key Laboratory of Crop Stress Biology on Drought Regions, Northwest A&F University, Yangling, Shaanxi 712100, P. R. China

**Jinfeng Gao**
State Key Laboratory of Crop Stress Biology on Drought Regions, Northwest A&F University, Yangling, Shaanxi 712100, P. R. China

**Pengke Wang**
State Key Laboratory of Crop Stress Biology on Drought Regions, Northwest A&F University, Yangling, Shaanxi 712100, P. R. China

**Yan Chai**
State Key Laboratory of Crop Stress Biology on Drought Regions, Northwest A&F University, Yangling, Shaanxi 712100, P. R. China

**R. Guzmán-Cruz**
División de Investigación y Posgrado, Facultad de Ingeniería, Universidad Autónoma de Querétaro, Cerro de las Campanas s/n, Col. Las Campanas, C.P. 76010 Querétaro, Qro., México

**E. Olvera-González**
Universidad Autónoma de Zacatecas, Jardín Juárez 147, Centro Histórico, Zacatecas, Zacatecas, México

**I. L. López-Cruz**
Posgrado en Ingeniería Agrícola y Uso Integral del Agua, Universidad Autónoma de Chapingo, C.P. 056230 Chapingo, Méx., México

**R. Montoya-Zamora**
División de Investigación y Posgrado, Facultad de Ingeniería, Universidad Autónoma de Querétaro, Cerro de las Campanas s/n, Col. Las Campanas, C.P. 76010 Querétaro, Qro., México

**I. O. Lawal**
Forestry Research Institute of Nigeria (FRIN), Ibadan, Oyo State, Nigeria

**P. O. Ige**
Forestry Research Institute of Nigeria (FRIN), Ibadan, Oyo State, Nigeria

**E. A. Awosan**
Forestry Research Institute of Nigeria (FRIN), Ibadan, Oyo State, Nigeria

**T. I. Borokini**
National Centre for Genetic Resources and Biotechnology (NACGRAB), Nigeria

**O. A. Amao**
Forestry Research Institute of Nigeria (FRIN), Ibadan, Oyo State, Nigeria

**Ajay Kumar Singh**
Research Station and KVK, Lohaghat (GBPUA&T, Pantnagar) Uttarakhand, 262 524, India

**D. K. Singh**
Research Station and KVK, Lohaghat (GBPUA&T, Pantnagar) Uttarakhand, 262 524, India

**Balraj Singh**
Research Station and KVK, Lohaghat (GBPUA&T, Pantnagar) Uttarakhand, 262 524, India

**Shailja Punetha**
Research Station and KVK, Lohaghat (GBPUA&T, Pantnagar) Uttarakhand, 262 524, India

**Deepak Rai**
Research Station and KVK, Lohaghat (GBPUA&T, Pantnagar) Uttarakhand, 262 524, India

**S. Ramesh Kumar**
Department of Horticulture, Vanavarayar Institute of Agriculture, Tamil Nadu Agricultural University, Manakkadavu, Pollachi-642 103, Tamil Nadu, India

**T. Arumugam**
Department of Horticulture, Agricultural College and Research Institute, Tamil Nadu Agricultural University, Madurai-625 104, Tamil Nadu, India

**V. Premalakshmi**
Department of Horticulture, Agricultural College and Research Institute, Tamil Nadu Agricultural University, Madurai-625 104, Tamil Nadu, India

**C. R. Anandakumar**
Centre for Plant Breeding and Genetics, Tamil Nadu Agricultural University, Coimbatore-641003, Tamil Nadu, India

**D. S. Rajavel**
Department of Crop Protection, Agricultural College and Research Institute, Killikulam, Tamil Nadu Agricultural University, Tuticorin- 628 252, Tamil Nadu, India

**Behzad Kaviani**
Department of Horticultural Science, Rasht Branch, Islamic Azad University, Rasht, Iran

**Afshin Ahmadi Hesar**
Department of Horticultural Science, Rasht Branch, Islamic Azad University, Rasht, Iran

**Alireza Tarang**
North Biotechnology Institute, Rasht, Guilan, Iran

**Sahar Bohlooli Zanjani**
North Biotechnology Institute, Rasht, Guilan, Iran

**Davood Hashemabadi**
Department of Horticultural Science, Rasht Branch, Islamic Azad University, Rasht, Iran

**Mohammad Hossein Ansari**
Department of Agronomy, Rasht Branch, Islamic Azad University, Rasht, Iran

**B. Sagrera**
IRTA, Environmental Torre Marimon E-08140 Caldes de Montbui, Spain

**C. Biel**
IRTA, Environmental Torre Marimon E-08140 Caldes de Montbui, Spain

**R. Savé**
IRTA, Environmental Torre Marimon E-08140 Caldes de Montbui, Spain

**I. M. Khan**
Division of Floriculture, Medicinal and Aromatic Plants, Sher-e- Kashmir University of Agricultural Science and Technology of Kashmir, Shalimar, Srinagar, India

**F. U. Khan**
Division of Floriculture, Medicinal and Aromatic Plants, Sher-e- Kashmir University of Agricultural Science and Technology of Kashmir, Shalimar, Srinagar, India

**M. Salmani**
Division of Floriculture, Medicinal and Aromatic Plants, Sher-e- Kashmir University of Agricultural Science and Technology of Kashmir, Shalimar, Srinagar, India

**M. H. Khan**
Division of Floriculture, Medicinal and Aromatic Plants, Sher-e- Kashmir University of Agricultural Science and Technology of Kashmir, Shalimar, Srinagar, India

**M. A. Mir**
Division of Floriculture, Medicinal and Aromatic Plants, Sher-e- Kashmir University of Agricultural Science and Technology of Kashmir, Shalimar, Srinagar, India

**Amir Hassan**
Division of Floriculture, Medicinal and Aromatic Plants, Sher-e- Kashmir University of Agricultural Science and Technology of Kashmir, Shalimar, Srinagar, India

**Anju Ahlawat**
Department of Botany and Plant Physiology, CCS Haryana Agricultural University, Hisar-125004, India

**S. K. Pahuja**
Forage Section, Department of Genetics and Plant Breeding, CCS Haryana Agricultural University, Hisar-125004, India

**H. R. Dhingra**
Department of Botany and Plant Physiology, CCS Haryana Agricultural University, Hisar-125004, India

**M. K. Tripathi**
Sunnhemp Research Station (CRIJAF, ICAR), Pratapgarh, U.P., India-230001, India

**Babita Chaudhary**
Sunnhemp Research Station (CRIJAF, ICAR), Pratapgarh, U.P., India-230001, India

**S. R. Singh**
Central Research Institute for Jute and Allied Fibres (ICAR), Barrackpore, W. B., India

**H. R. Bhandari**
Sunnhemp Research Station (CRIJAF, ICAR), Pratapgarh, U.P., India-230001, India

**Nidhi Lohani**
Regional Research Institute of Himalayan Flora, CCRAS, Tarikhet, Ranikhet-263 663, Uttarahand, India

**Lalit M. Tewari**
Department of Botany, D. S. B. Campus, Kumaun University, Nainital-263 201, Uttarakhand, India

**G. C. Joshi**
Regional Research Institute of Himalayan Flora, CCRAS, Tarikhet, Ranikhet-263 663, Uttarahand, India

**Ravi Kumar**
Regional Research Institute of Himalayan Flora, CCRAS, Tarikhet, Ranikhet-263 663, Uttarahand, India

**Kamal Kishor**
Department of Botany, D. S. B. Campus, Kumaun University, Nainital-263 201, Uttarakhand, India

**Sarkaut Salimi**
Islamic Azad University, Mariwan Branch, Mariwan, Iran

**R. Muthukrishnan**
Department of Soil Science and Agricultural Chemistry, Tamil Nadu Agricultural University, Coimbatore - 641 003, Tamil Nadu, India

**K. Arulmozhiselvan**
Department of Soil Science and Agricultural Chemistry, Tamil Nadu Agricultural University, Coimbatore - 641 003, Tamil Nadu, India

**M. Jawaharlal**
Department of Soil Science and Agricultural Chemistry, Tamil Nadu Agricultural University, Coimbatore - 641 003, Tamil Nadu, India

**Bo Teng**
National Engineering Laboratory for Clean Technology of Leather Manufacture, Institute of Life Sciences, Sichuan University, Chengdu, 610065, China

**Tao Zhang**
National Engineering Laboratory for Clean Technology of Leather Manufacture, Institute of Life Sciences, Sichuan University, Chengdu, 610065, China

**Ying Gong**
National Engineering Laboratory for Clean Technology of Leather Manufacture, Institute of Life Sciences, Sichuan University, Chengdu, 610065, China

**Wuyong Chen**
National Engineering Laboratory for Clean Technology of Leather Manufacture, Institute of Life Sciences, Sichuan University, Chengdu, 610065, China

**S. H. Sengar**
Department of Electrical and Other Energy Sources, College of Agricultural Engineering and Technology, DBSKKV, Dapoli, Dist: Ratnagiri-415712, India

**S. Kothari**
College of Technology and Engineering, Department of Renewable Energy Sources, Maharana Pratap University of Agriculture and Technology, Udaipur-313001, India

**Manickam Visalakshi**
Department of Floriculture and Landscaping, Horticultural College and Research institute, Tamil Nadu Agricultural University, Coimbatore-3, Tamil Nadu, India

**Murugiah Jawaharlal**
Department of Floriculture and Landscaping, Horticultural College and Research institute, Tamil Nadu Agricultural University, Coimbatore-3, Tamil Nadu, India

**Manickam Kannan**
Department of Floriculture and Landscaping, Horticultural College and Research institute, Tamil Nadu Agricultural University, Coimbatore-3, Tamil Nadu, India

**G. Rajaraman**
Department of Vegetable Crops, Horticultural College and Research Institute, Coimbatore- 641 003, Tamil Nadu, India

**L. Pugalendhi**
Tapioca and Castor Research Station, Tamil Nadu Agricultural University, Yethapur - 636 119, Salem District, Tamil Nadu, India

**Anderson Assaid Simão**
Chemistry Department, Biochemistry Laboratory, Federal University of Lavras - UFLA, PO Box 3037, Zip Code 37200.000, Lavras, MG, Brazil

**Mírian Aparecida Isidro Santos**
Chemistry Department, Biochemistry Laboratory, Federal University of Lavras - UFLA, PO Box 3037, Zip Code 37200.000, Lavras, MG, Brazil

**Rodrigo Martins Fraguas**
Chemistry Department, Biochemistry Laboratory, Federal University of Lavras - UFLA, PO Box 3037, Zip Code 37200.000, Lavras, MG, Brazil

**Mariana Aparecida Braga**
Chemistry Department, Biochemistry Laboratory, Federal University of Lavras - UFLA, PO Box 3037, Zip Code 37200.000, Lavras, MG, Brazil

**Tamara Rezende Marques**
Chemistry Department, Biochemistry Laboratory, Federal University of Lavras - UFLA, PO Box 3037, Zip Code 37200.000, Lavras, MG, Brazil

**Mariene Helena Duarte**
Chemistry Department, Biochemistry Laboratory, Federal University of Lavras - UFLA, PO Box 3037, Zip Code 37200.000, Lavras, MG, Brazil

**Claudia Mendes dos Santos**
Chemistry Department, Biochemistry Laboratory, Federal University of Lavras - UFLA, PO Box 3037, Zip Code 37200.000, Lavras, MG, Brazil

**Juliana Mesquita Freire**
Chemistry Department, Biochemistry Laboratory, Federal University of Lavras - UFLA, PO Box 3037, Zip Code 37200.000, Lavras, MG, Brazil

**Angelita Duarte Corrêa**
Chemistry Department, Biochemistry Laboratory, Federal University of Lavras - UFLA, PO Box 3037, Zip Code 37200.000, Lavras, MG, Brazil

**Abuhay Takele**
Melkassa Agricultural Rresearch Center, P. O. Box 436, Nazret, Ethiopia

**Jill Farrant**
University of Cape Town, Department of Molecular and Cell Biology, Private bag 7701, Rondebosch, Cape Town, South Africa.

**Paulo André Cremonez**
UNIOESTE - Western Paraná State University - Post graduation Program, Master Course of Energy in Agriculture. Street Universitária, 2069, CEP: 85.819-130 Faculdade, Cascavel, PR, Brazil

**Armin Feiden**
UNIOESTE - Western Paraná State University - Post graduation Program, Master Course of Energy in Agriculture. Street Universitária, 2069, CEP: 85.819-130 Faculdade, Cascavel, PR, Brazil

**Reginaldo Ferreira Santos**
UNIOESTE - Western Paraná State University - Post graduation Program, Master Course of Energy in Agriculture. Street Universitária, 2069, CEP: 85.819-130 Faculdade, Cascavel, PR, Brazil

**Doglas Bassegio**
UNIOESTE - Western Paraná State University - Post graduation Program, Master Course of Energy in Agriculture. Street Universitária, 2069, CEP: 85.819-130 Faculdade, Cascavel, PR, Brazil

**Eduardo Rossi**
UNIOESTE - Western Paraná State University - Post graduation Program, Master Course of Energy in Agriculture. Street Universitária, 2069, CEP: 85.819-130 Faculdade, Cascavel, PR, Brazil

**Willian Cézar Nadaleti**
UNIOESTE - Western Paraná State University - Post graduation Program, Master Course of Energy in Agriculture. Street Universitária, 2069, CEP: 85.819-130 Faculdade, Cascavel, PR, Brazil

**Jhonatas Antonelli**
UNIOESTE - Western Paraná State University - Post graduation Program, Master Course of Energy in Agriculture. Street Universitária, 2069, CEP: 85.819-130 Faculdade, Cascavel, PR, Brazil

**Fabiola Tomassoni**
UNIOESTE - Western Paraná State University - Post graduation Program, Master Course of Energy in Agriculture. Street Universitária, 2069, CEP: 85.819-130 Faculdade, Cascavel, PR, Brazil

**Piotr Salachna**
Laboratory of Ornamental Plants at the Department of Horticulture, West Pomeranian University of Technology in Szczecin, Papieża Pawła VI 3A, 71-459 Szczecin, Poland

**Agnieszka Zawadzińska**
Laboratory of Ornamental Plants at the Department of Horticulture, West Pomeranian University of Technology in Szczecin, Papieża Pawła VI 3A, 71-459 Szczecin, Poland

www.ingramcontent.com/pod-product-compliance
Lightning Source LLC
Chambersburg PA
CBHW080634200326
41458CB00013B/4621
* 9 7 8 1 6 8 2 8 6 1 6 0 8 *